PRIMATE
PARADIGMS

PRIMATE PARADIGMS

Sex Roles and Social Bonds

with a new Introduction

Linda Marie Fedigan

Illustrations by Linda Straw Coelho

The University of Chicago Press
Chicago and London

The University of Chicago Press, Chicago 60637
The University of Chicago Press, Ltd., London

© 1982, 1992 by Linda Marie Fedigan
All rights reserved. Originally published 1982 by
Eden Press, Montreal.
University of Chicago Press edition 1992
Printed in the United States of America

99 98 97 96 95 94   6 5 4 3 2

Library of Congress Cataloging-in-Publication Data

Fedigan, Linda Marie.
    Primate paradigms : sex roles and social bonds /
Linda Marie Fedigan : illustrations by Linda Straw
Coelho.
        p.   cm.
    Originally published: Montreal : Eden Press,
© 1982. With new introd.
    Includes bibliographical references and indexes.
    1. Primates—Behavior.   2. Sexual behavior in
animals.   3. Sex role.   4. Social behavior in
animals.   I. Title.
QL737.P9F35   1992
599.8' 0451—dc20   91-34490
    ISBN 0-226-23948-9 (pbk.)   CIP

Drawings © 1980 by Linda Straw Coelho
Photographs by permission of
Larry and Linda Fedigan

To my parents, John and Jolanda Wood,
with gratitude for their love and support

TABLE OF CONTENTS

INTRODUCTION TO THE 1992 EDITION

I began my 1982 Introduction to *Primate Paradigms* by stating that one of my aims was to redress a previous imbalance. While male primates were fairly extensively covered in the research and publications of the time, there was little record of how females played significant parts in the social life of their groups, and had played such parts in the evolution of their species. In retrospect, it is clear that my effort to remedy this lack was only one part of a larger movement in the 1980's to bring female primates onto center stage.

This Introduction will first outline in a very general manner what I see to be five major areas of change in the discipline in the past decade. Even though an exhaustive review of all that has happened in this time period is not possible, the five trends to be outlined are sufficiently encompassing to provide a sense of recent directions in primatology. After the presentation of these five general areas of change, I will turn to specific topics discussed in the book, and present in more concrete detail a selection of the new ideas and studies that can be used to update the material covered in the subsequent chapters.

The basic coverage of primate sex roles and social bonds that *Primate Paradigms* first provided ten years ago is still sound; little of it has been refuted or rejected by subsequent research. Nonetheless, much of the material in the book may be better appreciated by today's reader in light of the research trends that have occurred in primate behavior since 1982.

RECENT TRENDS IN PRIMATE RESEARCH

The first trend, as noted above, is the growing research focus on females of the species. For example, in the past decade several books have been published that document the lives and roles of female animals, especially those of female primates. The reader is referred specifically to J. Altmann's *Baboon Mothers and Infants* (1980), S. Wasser's *The Social Behavior of Female Vertebrates* (1983), S. Hrdy's *The Woman Who Never Evolved* (1981), N. Tanner's *On Becoming Human* (1981), M. Small's *Female Primates* (1984), E. Shaw and J. Darling's *Strategies of Being Female: Animal Patterns, Human Choices* (1984), I. Elia's *The Female Animal* (1985), and B. Kevles's *Females of the Species* (1986).

In fact, the coverage of female as well as male behavior, and the conceptualization of research questions and primate societies from a female perspective, have now been highly developed and incorporated into the structure of the discipline. This is true to such an extent that primatology is sometimes singled out for praise as an "equal opportunity" science, and/or described as a discipline in which a female perspective has made a major impact (e.g., Rowell 1984, Morgen 1989, Bleier 1984; cf. Sperling, in press). Although a comparative analysis of primatology as a science is beyond the scope of this introduction, I can outline here some of the key questions about gender and primate studies that are currently under discussion. Is primatology distinct from its parental disciplines of psychology, anthropology, and animal behavior, in terms of the amount of attention that is devoted to the roles of female subjects? And if primatology *is* different, why might it be so? Suggested answers to the latter question have been that there are proportionately more women primatologists than there are women in other disciplines, and that (whatever the sex ratio) the women in primatology have been highly influential as researchers, thinkers, and role models. Another possible answer is that the subject matter itself—the social life of primates—and the prevalence of phenomena such as female-bonded groups in primates have influenced the nature of the discipline. Finally, it has been suggested that the perception of women as differentially influential in the field of primatology may be an illusion created by all the media attention devoted to our field, especially to the women responsible for the study of the great apes (e.g., Montgomery 1991, Mowat 1987). There are no easy answers to these questions, and only a small handful of people have begun to address them (seldom in toto).

It is, however, safe to say that in the past decade the place of females in primate societies has been given due coverage. The prior emphasis on male primates in the early days of our discipline has been largely redressed. And it is surely a measure of the maturity of our discipline that historians and philosophers of science, as well as primatologists, have begun to take an interest in the science of primatology as a subject of study itself (e.g., Asquith 1986, Haraway 1989, Hrdy 1986, Hrdy and Williams 1983, Fedigan and Fedigan 1989, Schubert and Masters 1991, Strum and Latour 1987).

A second trend is that the discipline has grown in the theoretical sophistication of the research questions that can be asked, and has advanced far beyond its original natural history foundation. As is common to many disciplines, primatology in the earlier stages had to pass through a "basic data collection" phase, when the primary focus was on accumulating descriptive behavioral data on as many different species as possible. Natural history has played an essential part in our science, and it always will be important to continue collecting such baseline descriptive data. But the social systems of many species are now sufficiently described that primatologists have turned to more problem-oriented studies in which hypotheses are clearly tested. We need no longer ask only very broad questions such as "what is the characteristic group composition or mating system of this species?" Now we can ask more theory-driven and narrowly focused questions; for example, questions about the impact of specific social or ecological factors on particular group composition or mating patterns.

Some individual groups and populations of primates have been studied for sufficient periods of time that highly refined questions can be asked about particular animals and cohorts. For particular groups of macaques and baboons, we could, if we wished, compare the reproductive patterns of a cohort of known females to those of their great-grandmothers and determine what social/demographic and ecological variables might have influenced the different generations. Another aspect of the growing research sophistication of primatology is reflected in the increasing number of successful experimental studies in the field: experiments that are carried out in conjunction with purely observational work.

A third trend of the 1980's is a continuation of one that had already begun in the 1970's, and that is a move away from a reliance on the ideas and concepts developed in the social sciences, and a more firm grounding of primatology within the theories of modern evolutionary biology. The constructs of the social sciences, such as role and socialization theories and structural/functional models of society, are used less and less by primatologists. Much research is conceptualized in terms of modern evolutionary thinking; for example, testing the predictions of sexual selection, kin selection, parental investment, and reciprocal altruism theories, which focuses on the adaptive consequences of competition and cooperation among primates (for an overview of sociobiological research in primatology, see Gray 1985). While this evolutionary perspective has been highly productive and has led to many important insights about primate behavior, there still is an important part of primatology that focuses on proximate explanations; that is, upon the immediate experiences, motivation, and social interactions of the animals. For this latter approach, ideas have been, and still are being, borrowed from the fields of human psychology, anthropology, sociology, and political science.

A fourth characteristic of primatology in the 1980's is the flourishing of what had been somewhat underdeveloped themes within the discipline. Ecological and demographic principles are better understood and incorporated into our discipline now, and we have a great deal more empirical data on a variety of species from which to generalize and theorize about foraging adaptations and population dynamics. Our coverage of the different taxonomic categories of primates is better distributed, the last decade having seen major strides forward in our knowledge of prosimians, neotropical monkeys, and the arboreal Old World species. The broadening of our data base across more species, along with the increasing depth of our knowledge about variation between groups and change over time, has led to a greater interest in life history studies and the evolution of social systems. The accumulation of many years of field data on well-known social groups has also contributed to a growing interest in the social uses of intelligence. There are also a few topics which, although not entirely "new" to the 1980's, came much more to our attention in this decade, such as the investigation of deception, reconciliation, and social strategies.

The final trend that I will briefly describe is also not confined to the most recent decade, but has come much more to the forefront with every passing year. And that is recognition of the absolute necessity that we take action to ensure the continuation of the primates in their natural habitats. In the past few years whenever I teach an introductory primate behavior course, there is always at least one enthusiastic student, a potential recruit to the discipline, who asks me if there will be any primates left for them to study outside of the laboratories and the zoos. I answer that there is still hope, but that we all have to do everything within our power right now to halt and reverse the destruction of primate habitats around the world. Primatologists of all types have come to realize that conservation is the number one issue, lest our very raison d'être should cease to exist in the natural world.

THE ORDER PRIMATES: ORIGINS AND TAXONOMIC RELATIONS

In this book, which is focused almost entirely on social behavior, I have largely avoided discussion of taxonomic controversies, and in particular have not made reference to the continual splitting, aggregating, and renaming of primate species. However, one taxonomic issue is sufficiently noteworthy that it should be brought to the reader's attention, and that is the place of the family Tarsiidae in the Primate Order. Until recently tarsiers were usually categorized with the lemurs and lorises, as prosimians, in the suborder Prosimii. The other suborder is known as Anthropoidea and includes the monkeys, apes, and hominid forms. Some features of the tarsiers, in particular some external physical characteristics, are similar to those of prosimians; however, other features, especially those at the molecular/genetic level, are more reminiscent of monkeys. It is important to remember that all taxonomic categories, except at the level of species, are products of the human mind rather than of nature, and thus it is possible for some types of organisms to exhibit characteristics of more than one of our categories and fail to fit neatly within one cell of our system. This is the case with the tarsier. Recently many taxonomists and introductory primate texts have separated the tarsiers from the lorises and lemurs, and placed the tarsiers together with the monkeys, apes, and humans in a suborder called "Haplorhini" (the name refers to a simple nose with dry skin). The lorises and lemurs are still classified together in one suborder, which is now called "Strepsirhini" (named in reference to "twisted" nostrils surrounded by wet or damp skin of the muzzle). The reader who is interested in knowing more about primate classification can pursue the topic in any or all of these excellent references, which provide more detail: Wolfheim 1983, Kavanagh 1983, Napier and Napier 1985, Fleagle 1988, Martin 1990a.

Another significant change in our understanding of the primate order is part of an ongoing controversy about primate origins and the basic primate pattern. Recently, some of the fossil forms that were thought to be our earliest primate ancestors from the Paleocene have been rediagnosed as being most closely similar to "flying lemurs," which are neither lemurs nor fliers, but a type of tree-gliding mammal better known as dermopterans. As noted by several researchers, this reclassification of the Paleocene fossils "throws our interpretations of primate origins up in the air again" (Shipman 1990, Martin 1990b, Beard 1990, Kay et al. 1990). Years ago LeGros Clark (1962) developed a list of primate trends that characterize the order and suggested that these features evolved as adaptations to an arboreal existence. This idea is described in *Primate Paradigms*, along with Cartmill's (1975) theory that the original primate traits evolved to facilitate nocturnal visual predation on insects in the trees: the "visual predation theory." Some alternatives to the visual predation theory have been suggested, such as Sussman's (1990) argument that modern primates, bats, and plant-feeding birds all arose around the time of the Paleocene-Eocene boundary in conjunction with appearance of the angiosperms (flowering plants). Sussman sees primates as having co-evolved with the angiosperm rainforests, by exploiting the resources of the flowering plants (fruits, nectar, seeds, and the insects that utilize flowers), as well as acting as major seed dispersers for them.

Now, some of the fossils (known collectively as plesiadapiformes) on which earlier theories about primates were based are said to have been *not* primates after all, but gliding mammals. It may be that dermopterans and primates (and bats and tree shrews) are all related through common ancestors, or that the fossil forms are not related to either primates or dermopterans, or to any extant mammals at all. Whatever the final decision on phylogenetic affinities, it is interesting that the same features of primates that have been described as adaptations to arboreal, visual predation, or to feeding on the fine terminal branches of flowering plants—grasping hands, forward-facing eyes, hand-eye coordination—may also be those useful for gliding between trees. While it has not yet been seriously suggested that we evolved as tree gliders who later became visual predators, or angiosperm mutualists (see Shipman 1990), the recent radical reclassification of our supposed earliest primate ancestors demonstrates that we still have a long way to go toward understanding the origins of the features that characterize us as primates.

PRIMATE BEHAVIOR AND THE COMPARATIVE METHOD

In Chapter 1 of *Primate Paradigms*, I noted that one of the main drawbacks to the use of the comparative method in our discipline is that it is not possible for all two hundred or so primate species to be equally studied, which leaves the comparative process open to an involuntarily or voluntarily biased selection of examples. And I suggested that the open-country, ground-dwelling species, such as the macaques, baboons, vervets, and chimpanzees, have been unduly influential in our generalized understanding of the Primate Order. Prosimians, New World monkeys, and highly arboreal Old World species were singled out as badly underrepresented in the primate literature. I am happy to say that in the years since I made that statement, a concerted effort to rectify that bias has been underway. Indeed, scientists have made notable progress toward understanding some of the difficult-to-study species that were so little known to us a decade ago.

In the area of prosimian studies, the last years have seen many important advances, only two of which will be outlined here. After a number of years during which foreign researchers were restricted from working in Madagascar, non-Malagasy scientists are once again admitted to the island. Great strides have been made in creating and maintaining reserves to protect the native flora and fauna, in setting up field sites with known and habituated individuals, and even in discovering new species of lemurs, or rediscovering species thought to have been extinct (Meier et al. 1987, Simons 1988, Wright 1988). Several long-term projects, such as those at Berenty and Beza-Mahafaly, are being conducted by Alison Richard, Robert Sussman, and Alison Jolly (who were some of the original researchers on Madagascar in the 1960's and 1970's) as well as by their students. Such longitudinal collection of genealogical, demographic, and social information on individually recognized animals and groups is beginning to result in a much better understanding of lemuriformes and how they are similar to and different from the other types of primates.

Another major event in the world of prosimian studies, resulting in part from the increased access to free-ranging lemuriformes, is a controversy that has been brewing over the possible causes of and the universality of female dominance in lemurs. It has been known for some time that the well-studied ringtailed lemur, *Lemur catta*, lives in female-dominated societies. And it has been suggested more than once that the same may be true of all lemuriformes (see discussion in Richard 1987 and also Kaufman 1991). In particular, the adult females of a ringtailed lemur group were noted to have priority of access to food ahead of the males. This is a rarity among primates, and a special theory was developed for its occurrence in lemurs. Hrdy (1981) and Jolly (1984) argued that female dominance may occur in a situation including two factors: males who suspend intrasexual competition outside of short mating seasons and females who experience strong seasonal reproductive stresses. Thus, they suggested, male lemurs may be deferring to females who are experiencing heavy energy demands from fast-growing offspring in an extremely seasonal environment. More recently, Kappeler (1990) proposed that female dominance in *Lemur catta* is more than just female feeding priority due to male deference. Instead, he argued, the females of this species exhibit agonistic superiority (the ability to evoke submissive signals from males) across annual cycles and in all social contexts. Also, as the data begin to come in from diverse types of lemurs, it is being suggested that not all lemuroids are female-dominant (e.g., Pereira et al. 1990, Young et al. 1990). It does occur in two families of lemurs, one of which is also characterized by monogamy, but clearly more research on additional species will be necessary before researchers can fully determine the phylogenetic distribution of its occurrence, much less its ecological correlates. Female dominance is a rarity for mammals in general, which is one reason it is so noteworthy in lemurs. Since exceptions usually help us to understand the nature of the rule, the study of the relations between and within the sexes in lemur species is proving to be a highly significant area in primatology.

The study of New World or neotropical species has also made major strides in the past ten years, but only a few highlights can be mentioned here. The muriqui or woolly spider monkey (*Brachyteles arachnoides*), which is a severely endangered species and which is the object of a major international conservation campaign, has also been the subject of long-term study in Brazil (Strier 1987a, b, 1990; and see Milton 1984, 1985). The muriqui has been found to be similar to spider monkeys in respect to female dispersal and male intrasexual competition. Many people in Brazil have awakened to the decimation of their forests, and thus Brazilian as well as foreign scientists are very active in both primate conservation and primate research in Brazil. One of the more pioneering endeavors has been the successful re-introduction of captive-bred golden tamarins (*Leontopithecus*) to their former habitats in the Atlantic forests of Brazil.

Further north, in Venezuela, the long-term field site of Hato Masaguaral, a cattle ranch and wildlife refuge, has yielded excellent data and many publications on the ecology, life histories, and social dynamics of wedge-capped capuchins, *Cebus olivaceus* (e.g., Fragaszy 1986, O'Brien 1991, Robinson 1984, 1988a, b, deRuiter 1986). From these studies we know that capuchins are not only superb extractive foragers, but that they live in complex, hierarchical societies (see also Fragaszy et al. 1990).

Probably the best-known recent project on neotropical primate studies has been the collaborative research conducted in Manu Park in Peru (e.g., Goldizen 1987, Janson 1985, 1986, McFarland Symington 1986, 1987, van Schaik and van Noordwijk 1989, Terborgh 1983). There are thirteen species of primates living at the lush and isolated site of Coche Cashu. It was here that a new type of primate social system was recognized—polyandry. It had been thought that all callitrichids are monogamous, as described earlier in *Primate Paradigms*. However, a study of the saddle-backed tamarin, *Saguinus fuscicollis*, by Terborgh and Goldizen (1985) revealed variable group compositions, with the most common form being several adult males residing in a breeding group with one adult female. It is likely that polyandry also characterizes the related *Saguinus imperator* species, and perhaps other members of the family Callitrichidae. Terborgh and Goldizen argued that polyandry may be an extension of the "helper at the nest" pattern that characterizes callitrichids. Tamarin females give birth to relatively large twins and produce rich milk, which results in significant energetic stress on the mother. Older offspring and adult males help the mother by carrying the infants, and the presence of a second adult male in the group may help to ensure the survival of young. Even if a male is not the father of every young that he helps to carry, it may be to his lifetime benefit to cooperatively rear young with other males and their female mate.

Terborgh's 1983 book, *Five New World Primates*, also helped to alleviate past deficiencies in our knowledge of neotropical species. In this book, Terborgh compared five sympatric species that are rather similar in their general omnivorous diet but nonetheless differentiated by their size, social systems, and foraging strategies. His finely detailed analyses of dietary overlap, feeding specializations, and polyspecific associations of these monkey species has become a classic in primate community ecology.

Finally, the past decade has seen more studies of the difficult-to-observe arboreal Old World monkeys, including species in both the family Colobidae and the genus *Cercopithecus*. Most of these species live in polygynous societies, although they exhibit a wide array of dietary adaptations and social relations. The interested reader is referred to the following: Cords (1987), Gautier-Hion et al. (1988), Roonwal and Mohnot (1977), Struhsaker and Leland (1987).

EVOLUTION OF GROUP LIVING AND SOCIAL SYSTEMS

Hand in hand with the increasing natural history information on additional species and social systems, and our growing understanding of how intergroup differences are related to ecological and demographic factors, has come increasing interest in the "ultimate causes" of group living and the evolution of different forms of social systems. (For good introductions to primate socioecology, a field that is not covered in *Primate Paradigms*, see Dunbar 1988, Jolly 1985, Richard 1985, Rodman and Cant 1984, Smuts et al. 1987, Standen and Foley 1989.) The past decade has seen quite a few papers presenting theoretical explanations of why primates live socially rather than solitarily, and why they live in groups of a particular size and structure. Taking the inevitable risk of oversimplification by providing a summary explanation of a complex set of publications that have come out in recent years, it can be said that group living has been theoretically related to three major factors: protection from predation, defense and

xvii

exploitation of resources, and cooperative rearing of young. All of these explanations recognize that there are competitive costs for individuals to live in social groups, but argue that either the enhanced predator defense system, facilitation of foraging, and/or improved reproductive success brought about through cooperative parenting and allomothering offset these costs. All three explanations also lead to specific predictions as to optimal group size and structure under different ecological conditions.

Some researchers have stressed the significance of predator protection as the ultimate cause for social living rather than group defense of resources (e.g., Terborgh 1983, Moore 1984, Terborgh and Janson 1986, van Schaik and van Hooff 1983, van Schaik 1989) whereas other researchers have focused on resource defense as the single most important factor (e.g., Wrangham 1983, 1987), two differing foci which have resulted in an interesting set of publications. When read in chronological order, they almost argue back and forth as in a dialogue.

The paper which, in some respects, sparked this ongoing debate on the evolution of primate social living is Wrangham's 1980 publication: "An Ecological Model of Female-bonded Primate Groups." In this article, Wrangham argued, among other things, that female behavior is adapted more directly to ecological pressures than that of males because food availability is a major limiting factor on female fitness. Male behavior, on the other hand, is most directly adapted to finding mates and achieving mating success, as predicted from sexual selection theory. This means we should not assume that ecological pressures have equal significance for females and males, and that, in some sense, female adaptations to the environment may be seen as "prior" to those of males. This argument may be conceptualized as follows: The environment presents certain ecological challenges to each species, and females distribute themselves to take best advantage of the ecological resources necessary for survival and reproduction. Thus, the pattern of female distribution is like the first overlay on a map of the resource distribution. Males then distribute themselves to take best advantage of the distribution of female mates. What emerges from these three correlated and overlaid patterns of distribution is what we recognize as social structure. In the case of most primate species, females receive benefits from exploiting food resources cooperatively with their female kin, and thus form female-bonded groups to which males attach themselves either singly (as in polygynous systems) or multiply (as in multi-male societies). In some cases, however, resource distribution is such that females forage singly with their dependent offspring. Males then can overlap several adult female ranges singly, as in galagos and orangutans; or they can overlap several female ranges cooperatively, as in chimpanzees and spider monkeys; or they can permanently attach themselves to one female and cooperatively defend a range with her, as in monogamous species.

Not everyone agrees with this formulation of how primate societies have come to assume their myriad forms. Terborgh (1983), for example, has developed a model postulating that social structure is constrained by group size. And group size is set by an optimal trade-off between increased predator protection and decreased foraging opportunities, exacerbated by greater travel costs. Although Wrangham was not the sole originator of the idea that social structure results from the consecutive interactions of resource distribution with female bonding patterns and male mating strategies, his 1980 paper has been pivotal in revitalizing and stimulating thinking about the evolution of primate social systems. It certainly should be noted as one of the key publications that brought female adaptation to the foreground of evolutionary explanation in primate studies.

COMPETITION, AGGRESSION, AND DOMINANCE

The past decade has also seen significant advances in our thinking about competition and cooperation. In 1982, when *Primate Paradigms* was first published, the concept of dominance had just weathered a storm of criticism, and in 1981, a key paper by Bernstein, entitled "Dominance: The Baby and the Bathwater," had reviewed the several contentious aspects of the issue. It stressed, among other things, that the practical difficulties which primatologists have experienced in attempting to operationalize the concept of hierarchical ranking should not lead us to abandon the study of dominance, because competition, conflict resolution, and asymmetrical power relationships are very real aspects of primate social lives. His paper was followed by commentary from many of the participants in the ongoing debate. Several of these referred to how fatigued and impatient everyone was becoming of the seemingly endless argument.

This 1981 comprehensive treatise of opinions on dominance can be seen as somewhat of a "capping exercise," or a turning point; it both reviewed the previous approaches and heralded the changes that were to come. The interrelated topics of dominance, aggression, and competition were to be revitalized in the 1980's by two linked reconceptualizations of competition in primates: the analysis of social strategies and the study of reconciliation.

For several reasons, it was to be expected that early studies of dominance in primates would focus on the easily discernible, physical characteristics of the participants in a conflict or competition. For one, the rule of parsimony suggests that we should always start out with the simplest explanation for natural phenomena, and the relative size and strength of combatants is the most straightforward, obvious, and easy-to-measure variable in any competition. What is more, the bigger, stronger animal has often been argued to win competitions in the nonprimate world. For at least some bird, amphibian, and ungulate species, the bigger ones can be reliably predicted to dominate the smaller ones. However, far before 1980, it had been recognized that this rule did not hold true in many conflicts among primates. Thus, other possibly causal variables had been introduced and examined, such as age and tenure in the group, presence of allies, history of previous interactions, and even assertive temperament. Finally, in a series of studies that began in the 1970's but were mainly published in the 1980's researchers turned to the possibility that primates actively construct social strategies for winning competitions, and that the best tactics rather than the largest size is what leads to success. This requires that nonhuman primates be able to perceive the nature and the multiple dimensions of relationships between the members of their groups, that they be able to predict the combined effects of different types of interactions between more than two members of the group, and that they can plan and manipulate such interactions to their own ends. It is now commonly accepted that they are capable of such skills, and the study of dominance and other aspects of competition has thus advanced from a rather mechanistic and reductionist model of "brute" force to an

xix

approach that sees many parallels to the study of human politics, with all its complexities of motivations, objectives, and manipulations. (See Schubert and Masters 1991 for a political scientist's view of the similarities and differences between human and non-human primate "politics.")

As examples of this new approach to dominance, Altmann (1980), Strum (1982, 1987), and Smuts (1985), among others, have argued that baboon males rely more on strategy than on strength to achieve competitive and mating success. Strum, who has followed her study groups of baboons for twenty years now, documented in 1982, and in her book, *Almost Human* (1987), how the elaborate social skills exhibited by males alleviate the need for them to use force and aggression to succeed. Female baboons, Strum suggested, can rely on their relatives to support them during conflicts. But males must "finesse" their way into the group and into the favor of females by offering them friendly gestures and aid in times of conflict, and also by relying on their abilities to observe and predict the reactions of other baboons.

Smuts's work too, especially her book on *Sex and Friendship in Baboons* (1985; see also 1987a, b), is a convincing demonstration that baboon males are sentient beings. Smuts has argued that they pursue cross-sex friendships as a form of reciprocal social exchange, which sometimes but not always results in enhanced mating success. Her work is unusual in that she applies a quantitative and selectionist approach that succeeds in avoiding reductionism of social complexities. Both Smuts and Strum have supplied data and arguments to contradict the assumption that dominant males achieve greater mating success as a result of superior force.

Studies of the importance of coalitions and alliances to power relationships in primates also have developed considerably in recent years. De Waal, in a book entitled *Chimpanzee Politics* (1982), described the history of power relationships in one zoo colony of chimpanzees, and Goodall (1986, 1990) described similar relationships in the free-ranging Gombe chimpanzees, in such a manner that the reader could not help but draw parallels to stories of intrigues and political machinations in human arenas— expedient coalitions, conditional reassurances, deceptions, false friends, attempted coups, and fatal gang beatings of adversaries are all presented. In a later article, de Waal (1984) noted a sex difference in the pattern of coalition formation of chimpanzees. He suggested that male chimpanzees tend to form flexible coalitions for immediate status enhancement, whereas female chimpanzees form more stable coalitions to protect their relatives and friends. Other studies and reviews of primate alliances include those by Datta (1983 a,b,c), Dunbar (1984), Fairbanks (1980), Harcourt (1989), Moore (1982), Seyfarth and Cheney (1984), and Silk (1982). In addition to these largely observational studies, Chapais (see 1991 summary) conducted an elegant series of experiments on alliances. These document that alliances between related female Japanese macaques (and the potential for, or actual occurrence of, agonistic interventions based on such alliances) play a key role in dominance rank acquisition and maintenance.

Related to this more "strategic" approach to primate competition has been the idea that we should devote at least equal attention to the mechanisms by which primates cooperate and thus manage to maintain and perpetuate the relationships and social systems in which we find them living (see Hinde 1983). De Waal in particular (1986, 1987, 1988) has argued that the strong focus on ultimate or selective explanations for aggression and competition has led us to almost neglect the highly obvious fact that primates do successfully live in social groups, and that they likely have mechanisms—processes, principles, and skills—for doing so. De Waal noted that most observers have focused on the conflicts and failed to examine the resolution of these conflicts. He followed the simple but innovative practice of turning on his stopwatch after a fight had ended and watching until the former adversaries came together again in some interaction that could be classified as reconciliation. Much of his research in recent years, and that of some others, has been devoted to documenting the ways in which primates manage to get along with each other, and how they reconcile after conflicts—what he calls "peacemaking" (de Waal 1989, de Waal and Ren 1988, de Waal and van Roosmalen 1979, de Waal and Yoshihara 1983; and see Aureli et al. 1989, Cords 1988, Judge 1983, York and Rowell 1988).

COMMUNICATION, COGNITION, AND SOCIAL INTELLIGENCE

The area that has blossomed more than any other in the past ten years is the study of primate communication, cognitive abilities, and social intelligence. The advances in our knowledge and understanding of these phenomena are not unrelated to the growing awareness of primate social strategies just discussed. As in the study of primate conflict and conflict resolution, so in the study of communication and cognition, scientists have now been able to document just how complex are the interaction patterns and how sophisticated are the social skills of primates.

It has been generally agreed for some time that primates are very clever because they do so well on laboratory tests of intelligence. However, field studies are now beginning to show us that these animals also regularly rely upon their intelligence in social situations in the wild. Traditionally, there have been two major explanations for the evolution of intelligence in primates: the social skills theory and the practical skills (foraging) theory. The former holds quite simply that social pressures selected for intelligence in primates (described in Byrne and Whiten 1988, Essock-Vitale and Seyfarth 1987, Cheney and Seyfarth 1990), and the latter argues that primates evolved their cognitive skills in order to better exploit food resources that were patchy and ephemeral but also predictable (Milton 1988, and see Parker and Gibson 1990).

Although the nonhuman primates are, with the exception of the chimpanzee and perhaps the capuchin, not particularly technologically skilled, Kummer (1982) pointed out that they are adept at using each other as "social tools." Kummer's early ideas on social intelligence (1967) and those of Jolly (1966, and see 1985, 1988) and Humphrey (1976) helped to stimulate research on what monkeys really know about the structure of their societies, about the nature of their own and others' relationships, and about the mental states of their conspecifics.

At the same time that primatologists were being inspired to document the use of social intelligence in the field, work on language parallels in animals was advancing the study of primate communication (e.g., Snowden et al. 1982 and Harre and Reynolds 1984). The research by Cheney and Seyfarth on vervet monkey communication is probably the best known (summarized in their 1990 book), and see also Gouzoules et al. 1984, 1986, on rhesus macaque communication. By recording the vocalizations of individual monkeys and playing them back to other group members from hidden speakers in the absence of the original stimulus, these researchers were able to document that monkey calls are not just expressions of arousal, but contain and convey specific information about the environment and about the social contingencies of an interaction. For example, Cheney and Seyfarth demonstrated that vervets give different alarm calls for different types of predators and respond differentially to such alarm calls, and that they not only recognize individuals, but can recognize relationships between other members of the group. The Gouzoules's research demonstrated that the screams of macaques during conflicts convey information about the rank and physical contact of the participants, and that listeners act upon this information. Through a rigorous experimental protocol and the use of sophisticated technology for recording, analyzing, and playing back vocalizations, this type of research has now legitimized the impression, held by researchers for many years, that monkeys communicate more than simply immediate emotional arousal. Rather, they remember past interactions, recognize relationships along kinship and dominance lines, and predict the responses of others on the basis of this knowledge. While this would seem to be commonly agreed upon by most primatologists today, a yet unsettled issue is the extent to which the nonhuman primates can be said to practice deception, that is to deliberately mislead their conspecifics. For discussions of deception, see Mitchell and Thompson (1986), Byrne and Whiten (1988, 1990), Cheney and Seyfarth (1990), Snowden (1990); and for wider discussions of cognition in a variety of animal species, see Griffin (1982, 1984) and Roitblat et al. (1984).

Another branch of primate research that has contributed to our growing awareness of the cognitive capacities of our subjects is represented by the attempts to teach language and language analogues to the great apes, especially the "chimpanzee language research." This is a large field that has been active for many years, and is far beyond the scope of this introduction. Nonetheless, the study of the language capacities of primates, which has witnessed significant events in the past decade, is an important aspect of our discipline. For an overview of these, the following recent references are suggested: Savage-Rumbaugh 1986, Gardner et al. 1989, Parker and Gibson 1990, Snowden 1990.

SEXUAL SELECTION THEORY AND MATE CHOICE

As part of the trend to increasingly conceptualize primate behavior within the framework of evolutionary biology, some work has been done in the past ten years toward furthering our understanding of how sexual selection might operate in primates. Bradbury and Andersson (1987) noted that sexual selection is one of the most rapidly developing topics in evolutionary biology, and that it also continues to be one of the most controversial issues in this field, especially the concept of female choice. In a very important, "state-of-the-art" compendium on the topic, Bradbury and Andersson gathered together some of the better-known researchers and theorists in the study of sexual selection, and attempted to formalize and clarify alternative hypotheses and how these might be tested. In the introduction, the editors outlined four problematic issues that are treated in their book and need to be addressed in further research:

(1) Mating preferences and attractive traits—do females choose mates preferentially, using cues that will lead to more adapted, viable offspring? Or do females choose traits that result in "runaway" selection on males for extravagant and "bizarre" characteristics (random evolutionary trajectories)?

(2) Conflicts within and between the sexes—is parental investment ("who gets stuck with the baby") the only significant conflict that arises between the sexes? What other forms of competition within and between the sexes are related to mating systems and determine the nature and strength of sexual selection?

(3) The measurement of sexual selection—how might we measure not only variance in mating success, but the value of the traits that are being sexually selected?

(4) Constraints on sexual selection—what are the physiological, energetic, and life history constraints that operate on the expressions of female choice and male competition?

Some of these questions are beginning to come under investigation in the field of primatology, although the theoretical and mathematical models are still far in advance of empirical research. And, as was the case ten years ago, there are still far more studies of intrasexual competition than there are of mate choice. Indeed, Smuts (1987a) suggested that the relationship between male dominance and male mating activity has received more attention than any other aspect of primate social behavior. This topic is covered in Chapters 7 and 17 of *Primate Paradigms,* under the rubric of the priority-of-access-to-estrous-females ("PAEF") model, and has been reviewed by Berenstain and Wade (1983), Dewsbury (1982), Fedigan (1983), Robinson (1982), Shively (1985), and Smuts (1987a). Also, the diverse chapters in Clutton-Brock (1988) provide excellent coverage of life history studies of reproductive success in male and female vertebrates.

Small (1989) recently summarized the research on female choice in nonhuman primates. She distinguished between "preferences" for certain male partners and "choice," the latter being defined as an action on the part of the female that affects her fitness and that of her partner. Having made such a distinction, she concluded that whereas quite a few studies of different primate species have documented female preferences, no study cited in the review was specifically designed to support or refute the significance of female choice. Further, Small argued that female sexual assertiveness, which is also well documented for primates, is not the same thing as evidence for female choice, and that female primates are probably more sexually assertive and promiscuous than they are "choosy" (see also her 1988 paper on female sexual behavior, and commentary from other primatologists). Small's review of female choice is organized around a set of questions: What do females want from males—resources, parental care, protection from aggression, familiarity, novelty, status, superior genes? How might they evaluate males for these traits? And how might they exert choice? Smuts (1987a) considered these same questions, and reviewed the evidence that female primates exert choice both directly, through their actions in mating contexts (they initiate, accept, and/or refuse mating opportunities), and indirectly, through their influence on male membership in social groups. Empirical studies with the clear objective of documenting female preferences or female choice in primates are few and far between, but the interested reader is referred to research by Huffman (1987, 1991), Janson (1984), Keddy (1986), and Manson (in press).

PRIMATE MODELS OF HUMAN EVOLUTION

A primate book almost inevitably ends with reference to the implications for human behavior, and true to form, *Primate Paradigms* concludes with a chapter on models of human social life derived from our knowledge of particular primates. In the ten years since that chapter was originally penned, the chimpanzee model has continued to be the one of preference, especially as our knowledge and understanding of the common chimpanzees and the related bonobos continue to grow and convince us of the similarities between these animals and ourselves (see, for example, Heltne and Marquardt 1989, Goodall 1986, Nishida 1990). The "woman the gatherer" model, which is largely based on the chimpanzee analogy, also has been developed beyond the description originally provided by Chapter 19, in works such as those by Dahlberg (1981), Tanner (1981), and Zihlman (1983); reviewed in Fedigan (1986). Other important theoretical advances and suggestions as to how we can best use the nonhuman primate data to understand the early social life of hominids are summarized in a series of papers edited by Kinzey (1987).

THE NEXT TEN YEARS AND BEYOND

We have learned much about our primate relatives in the past decade, and yet we are humbled by the knowledge of how much more there is to learn. Most primate books also must end with a heartfelt plea for the conservation of these animals and their habitats. It is perhaps ironic that our discipline should be growing in practitioners, data base, and theoretical sophistication in almost inverse proportion to the number of free-ranging primates left in the world. For at least 3½ million years humans have lived sympatrically with other primate species; we have eaten them, worshiped them, caged them, been entertained by them, and used them extensively as substitutes for humans in research. But only in the past sixty years have we gone out among them to observe these special animals in their native habitats, in order to better understand their place in nature, and thus our own. I hope that this book will help to convince readers of the significance of this endeavor, and persuade them to join us in the attempt to comprehend and conserve these animals.

LITERATURE CITED

Altmann, J. 1980, *Baboon Mothers and Infants*. Cambridge: Harvard Univ. Press.
Asquith, P. J. 1986, Anthropomorphism and the Japanese and Western traditions in primatology. IN: *Primate Ontogeny, Cognition and Social Behavior* (J. G. Else and P. C. Lee, eds.) Cambridge: Cambridge Univ. Press, pp. 61–71.
Aureli, F., van Schaik, C. P. and van Hooff, J. A. R. A. M. 1989, Functional aspects

of reconciliation among captive long-tailed macaques (*Macaca fascicularis*), *Am. J. Primatol.* 19:39–51.

Beard, K. C. 1990, Gliding behaviour and palaeoecology of the alleged primate family Paromomyidae (Mammalia, Dermoptera), *Nature* 345 (24 May 1990): 340–341.

Berenstain, L. and Wade, T. D. 1983, Intrasexual selection and male mating strategies in baboons and macaques, *Int. J. Primatol.* 4:201–235.

Bernstein, I. S. 1981, Dominance: The baby and the bathwater, *Behav. Brain Sci.* 4:419–457.

Bleier, R. 1984, *Science and Gender: A Critique of Biology and Its Theories on Women.* New York: Pergamon Press.

Bradbury, J. W. and Andersson, M. B. (eds.) 1987, *Sexual Selection: Testing the Alternatives.* Chichester: John Wiley and Sons.

Byrne, R. W. and Whiten, A. (eds.) 1988, *Machiavellian Intelligence: Social Expertise and the Evolution of Intellect in Monkeys, Apes, and Humans.* Oxford: Clarendon Press.

Byrne, R. W. and Whiten, A. 1990, Tactical deception in primates: The 1990 database, *Primate Report* 27 (May 1990).

Cartmill, M. 1975, *Primate Origins.* Minneapolis: Burgess Publ. Co.

Chapais, B. 1991, Matrilineal dominance in Japanese macaques: The contribution of an experimental approach. IN: *The Monkeys of Arashiyama: Thirty-Five Years of Research in Japan and the West* (L. M. Fedigan and P. J. Asquith, eds.) New York: SUNY Press, pp. 251–273.

Cheney, D. L. and Seyfarth, R. M. 1990, *How Monkeys See the World: Inside the Mind of Another Species.* Chicago: Univ. of Chicago Press.

Clutton-Brock, T. H. (ed.) 1988, *Reproductive Success: Studies of Individual Variation in Contrasting Breeding Systems.* Chicago: Univ. of Chicago Press.

Cords, M. 1987, Forest guenons and patas monkeys: Male-male competition in one-male groups. IN: *Primate Societies* (B. B. Smuts, D. L. Cheney, R. M. Seyfarth, R. W. Wrangham, and T. T. Struhsaker, eds.) Chicago: Univ. of Chicago Press, pp. 98–111.

Cords, M. 1988, Resolution of aggressive conflicts by immature long-tailed macaques (*Macaca fascicularis*), *Anim. Behav.* 36:1124–1135.

Dahlberg, F. (ed.) 1981, *Woman the Gatherer.* New Haven: Yale Univ. Press.

Datta, S. B. 1983a, Relative power and the acquisition of rank. IN: *Primate Social Relationships: An Integrated Approach* (R. A. Hinde, ed.) Oxford: Blackwell Scientific Publ., pp. 93–103.

Datta, S. B. 1983b, Relative power and the maintenance of rank. IN: *Primate Social Relationships: An Integrated Approach* (R. A. Hinde, ed.) Oxford: Blackwell Scientific Publ., pp. 103–112.

Datta, S. B. 1983c, Patterns of agonistic interference. IN: *Primate Social Relationships: An Integrated Approach* (R. A. Hinde, ed.) Oxford: Blackwell Scientific Publ., pp. 289–297.

Dewsbury, D. A. 1982, Dominance rank, copulatory behavior, and differential reproduction, *Quart. Rev. Biol.* 57:135–159.

Dunbar, R. I. M. 1984, *Reproductive Decisions: An Economic Analysis of Gelada Baboon Social Strategies*. Princeton: Princeton Univ. Press.

Dunbar, R. I. M. 1988, *Primate Social Systems*. Ithaca: Comstock Publ. Assoc.

Elia, I. 1985, *The Female Animal*. Oxford: Oxford Univ. Press.

Essock-Vitale, S. and Seyfarth, R. M. 1987, Intelligence and social cognition. IN: *Primate Societies* (B. B. Smuts, D. L. Cheney, R. M. Seyfarth, R. W. Wrangham, and T. T. Struhsaker, eds.) Chicago: Univ. of Chicago Press, pp. 452–461.

Fairbanks, L. A. 1980, Relationships among adult females in captive vervet monkeys: Testing a model of rank-related attractiveness, *Anim. Behav.* 28:853–859.

Fedigan, L. M. 1983, Dominance and reproductive success in primates, *Yrbk. Phys. Anthropol.* 26:91–129.

Fedigan, L. M. 1986, The changing role of women in models of human evolution, *Ann. Rev. Anthropol.* 15:25–66.

Fedigan, L. M. and Fedigan, L. 1989, Gender and the study of primates. IN: *Gender and Anthropology: Critical Reviews for Research and Teaching* (S. Morgen, ed.) Washington, D.C.: American Anthropological Assoc., pp. 41–64.

Fleagle, J. G. 1988, *Primate Adaptation and Evolution*. San Diego: Academic Press.

Fragaszy, D. M. 1986, Time budgets and foraging behavior in wedge-capped capuchins (*Cebus olivaceus*): Age and sex differences. IN: *Current Perspectives in Primate Social Dynamics* (D. M. Taub and F. A. King, eds.) New York: Van Nostrand Reinhold Co., pp. 159–174.

Fragaszy, D. M., Robinson, J. G. and Visalberghi, E. (eds.) 1990, Adaptation and adaptability of capuchin monkeys, *Folia Primatol.* 54.

Gardner, R. A., Gardner, B. T. and van Cantfort, T. E. 1989, *Teaching Sign Language to a Chimpanzee*. Albany: SUNY Press.

Gautier-Hion, A., Bourliere, F., Gautier, J.-P. and Kingdon, J. (eds.) 1988, *A Primate Radiation: Evolutionary Biology of the African Guenons*. Cambridge: Cambridge Univ. Press.

Goldizen, A. W. 1987, Facultative polyandry and the role of infant-carrying in wild saddle-back tamarins (*Saguinus fuscicollis*), *Behav. Ecol. Sociobiol.* 20:99–109.

Goodall, J. 1986, *The Chimpanzees of Gombe*. Cambridge: Belknap Press.

Goodall, J. 1990, *Through a Window. My Thirty Years with the Chimpanzees of Gombe*. Boston: Houghton Mifflin.

Gouzoules, H., Gouzoules, S. and Marler, P. 1986, Vocal communication: A vehicle for the study of social relationships. IN: *The Cayo Santiago Macaques: History, Behavior, and Biology* (R. G. Rawlins and M. J. Kessler, eds.) Albany: SUNY Press, pp. 111–129.

Gouzoules, S., Gouzoules, H. and Marler, P. 1984, Rhesus monkey (*Macaca mulatta*) screams: Representational signalling in the recruitment of agonistic aid, *Anim. Behav.* 32:182–193.

Gray, J. P. 1985, *Primate Sociobiology*. New Haven: HRAF Press.

Griffin, D. R. (ed.) 1982, *Animal Mind–Human Mind*. New York: Springer-Verlag.

Griffin, D. R. 1984, *Animal Thinking*. Boston: Harvard Univ. Press.

Haraway, D. 1989, *Primate Visions: Gender, Race, and Nature in the World of Modern Science*. New York: Routledge, Chapman, & Hall.

Harcourt, A. H. 1989, Social influences on competitive ability: Alliances and their consequences. IN: *Comparative Socioecology: The Behavioural Ecology of Humans and Other Mammals* (V. Standen and R. A. Foley, eds.) Oxford: Blackwell Scientific Publ., pp. 223–242.

Harre, R. and Reynolds, V. (eds.) 1984, *The Meaning of Primate Signals.* Cambridge: Cambridge Univ. Press.

Heltne, P. G. and Marquardt, L. A. (eds.) 1989, *Understanding Chimpanzees.* Cambridge: Harvard Univ. Press.

Hinde, R. A. (ed.) 1983, *Primate Social Relationships: An Integrated Approach.* Oxford: Blackwell Scientific Publ.

Hrdy, S. B. 1981, *The Woman That Never Evolved.* Cambridge: Harvard Univ. Press.

Hrdy, S. B. 1986, Empathy, polyandry, and the myth of the coy female. IN: *Feminist Approaches to Science* (R. Bleier, ed.) New York: Pergamon Press, pp. 119–146.

Hrdy, S. B. and Williams, G. C. 1983, Behavioral biology and the double standard. IN: *Social Behavior of Female Vertebrates* (S. K. Wasser, ed.) New York: Academic Press, pp. 3–17.

Huffman, M. A. 1987, Consort intrusion and female mate choice in the Japanese monkey, *Ethol.* 75: 221–234.

Huffman, M. A. 1991, Mate selection and partner preferences in female Japanese macaques. IN: *The Monkeys of Arashiyama: Thirty-Five Years of Research in Japan and the West* (L. M. Fedigan and P. J. Asquith, eds.) New York: SUNY Press, pp. 101–122.

Humphrey, N. K. 1976, The social function of intellect. IN: *Growing Points in Ethology* (P. P. G. Bateson and R. A. Hinde, eds.) Cambridge: Cambridge Univ. Press, pp. 303–319.

Janson, C. H. 1984, Female choice and mating system of the brown capuchin monkey *Cebus apella* (Primates: Cebidae), *Z. Tierpsychol.* 65: 177–200.

Janson, C. H. 1985, Aggressive competition and individual food intake in wild brown capuchin monkeys, *Behav. Ecol. Sociobiol.* 18: 125–138.

Janson, C. H. 1986, The mating system as a determinant of social evolution in capuchin monkeys (*Cebus*). IN: *Primate Ecology and Conservation* (J. G. Else and P. C. Lee, eds.) Cambridge: Cambridge Univ. Press, pp. 169–179.

Jolly, A. 1966, Lemur social behavior and primate intelligence, *Science* 153: 501–506.

Jolly, A. 1984, The puzzle of female feeding priority. IN: *Female Primates: Studies by Women Primatologists* (M. F. Small, ed.) New York: Alan R. Liss, pp. 197–215.

Jolly, A. 1985, *The Evolution of Primate Behavior,* 2nd ed. New York: MacMillan Publ. Co.

Jolly, A. 1988, The evolution of purpose. IN: *Machiavellian Intelligence: Social Expertise and the Evolution of Intelligence in Monkeys, Apes and Humans* (R. W. Byrne and A. Whiten, eds.) Oxford: Clarendon Press, pp. 363–378.

Judge, P. J. 1983, Reconciliation based on kinship in a captive group of pigtail macaques, *Am. J. Primatol.* 4: 346.

Kappeler, P. M. 1990, Female dominance in Lemur catta: More than just female feeding priority? *Folia Primatol.* 55: 92–103.

Kaufman, R. 1991, Female dominance in *Varecia variegata, Folia Primatol.* 57: 39–41.

Kavanagh, M. 1983, *A Complete Guide to Monkeys, Apes and Other Primates*. London: Jonathan Cape.

Kay, R. F., Thorington, R. W., Jr. and Houde, P. 1990, Eocene plesiadapiform shows affinities with flying lemurs not primates, *Nature* 345 (24 May 1990): 342–344.

Keddy, A. C. 1986, Female mate choice in vervet monkeys (*Cercopithecus aethiops sabaeus*), *Am. J. Primatol.* 10: 125–134.

Kevles, B. 1986, *Females of the Species: Sex and Survival in the Animal Kingdom*. Cambridge: Harvard Univ. Press.

Kinzey, W. G. (ed.) 1987, *The Evolution of Human Behavior: Primate Models*. Albany: SUNY Press.

Kummer, H. 1967, Tripartite relations in hamadryas baboons. IN: *Social Communication among Primates* (S. A. Altmann, ed.) Chicago: Univ. of Chicago Press, pp. 63–71.

Kummer, H. 1982, Social knowledge in free-ranging primates. IN: *Animal Mind-Human Mind* (D. R. Griffen, ed.) New York: Springer Verlag, pp. 113–130.

LeGros Clark, W. E. 1962, *The Antecedents of Man*, 2nd ed. Edinburgh: Edinburgh Press.

Manson, J. H. In press, Measuring female mate choice and its costs in Cayo Santiago rhesus macaques, *Anim Behav.*

Martin, R. D. 1990a, *Primate Origins and Evolution: A Phylogenetic Reconstruction*. Princeton: Princeton Univ. Press.

Martin, R. D. 1990b, Some relatives take a dive, *Nature* 345 (24 May 1990): 291–292.

McFarland Symington, M. J. 1986, Ecological determinants of fission-fusion sociality in *Ateles* and *Pan*. IN: *Primate Ecology and Conservation* (J. Else and P. C. Lee, eds.) Cambridge: Cambridge Univ. Press, pp. 181–190.

McFarland Symington, M. J. 1987, Sex ratio and maternal rank in wild spider monkeys: When daughters disperse, *Behav. Ecol. Sociobiol.* 20: 421–425.

Meier, B., Albignac, R., Peyrieras, A. and Rumpler, Y. 1987, A new species of *Hapalemur* (Primates) from South East Madagascar, *Folia Primatol.* 48: 211–215.

Milton, K. 1984, Habitat, diet and activity patterns of free-ranging woolly spider monkeys (*Brachyteles arachnoides* E. Geoffroy 1808), *Int. J. Primatol.* 5: 491–514.

Milton, K. 1985, Urine washing behavior in the woolly spider monkey (*Brachyteles arachnoides*), *Z. Tierpsychol.* 67: 154–160.

Milton, K. 1988, Foraging behaviour and the evolution of primate intelligence. IN: *Machiavellian Intelligence: Social Expertise and the Evolution of Intelligence in Monkeys, Apes and Humans* (R. W. Byrne and A. Whiten, eds.) Oxford: Clarendon Press, pp. 285–305.

Mitchell, R. W. and Thompson, N. S. 1986, *Deception: Perspectives on Human and Nonhuman Deceit*. Albany: SUNY Press.

Montgomery, S. 1991, *Walking with the Great Apes*. Boston: Houghton Mifflin Co.

Moore, J. 1982, Coalitions in langur all-male bands, *Int. J. Primatol.* 3: 314.

Moore, J. 1984, Female transfer in primates, *Int. J. Primatol.* 5: 537–589.

Morgen, S. 1989, Introductory essay. IN: *Gender and Anthropology: Critical Reviews for Research and Teaching* (S. Morgen, ed.) Washington D.C.: American Anthropological Assoc., pp. 1–20.

Mowat, F. 1987, *Woman in the Mists*. New York: Warner Books.

Napier, J. R. and Napier, P. H. 1985, *The Natural History of the Primates*. Cambridge: MIT Press.

Nishida, T. (ed.) 1990, *The Chimpanzees of the Mahale Mountains: Sexual and Life History Strategies*. Tokyo: Univ. of Tokyo Press.

O'Brien, T. G. 1991, Female-male social interactions in wedge-capped capuchin monkeys: Benefits and costs of group living, *Anim. Behav.* 41 : 555 – 567.

Parker, S. T. and Gibson, K. R. (eds.) 1990, *Language and Intelligence in Monkeys and Apes*. Cambridge: Cambridge Univ. Press.

Pereira, M. E., Kaufman, R., Kappeler, P. M. and Overdorff, D. J. 1990, Female dominance does not characterize all of the Lemuridae, *Folia Primatol.* 55 : 96 – 103.

Richard, A. 1985, *Primates in Nature*. New York: W.H. Freeman and Co.

Richard, A. F. 1987, Malagasy prosimians: Female dominance. IN: *Primate Societies* (B. B. Smuts, D. L. Cheney, R. M. Seyfarth, R. W. Wrangham, and T. T. Struhsaker, eds.) Chicago: Univ. of Chicago Press, pp. 25 – 33.

Robinson, J. G. 1982, Intrasexual competition and mate choice in primates, *Am. J. Primatol. Suppl.* 1 : 131 – 144.

Robinson, J. G. 1984, Diurnal variation in foraging and diet in the wedge-capped capuchin *Cebus olivaceus*, *Folia Primatol.* 43 : 216 – 228.

Robinson, J. G. 1988a, Demography and group structure in wedge-capped capuchin monkeys, *Cebus olivaceus*, *Behaviour* 104 : 202 – 232.

Robinson, J. G. 1988b, Group size in wedge-capped capuchin monkeys *Cebus olivaceus* and the reproductive success of males and females, *Behav. Ecol. Sociobiol.* 23 : 187 – 197.

Rodman, P. S. and Cant, J. G. H. (eds.) 1984, *Adaptations for Foraging in Nonhuman Primates: Contributions to an Organismal Biology of Prosimians, Monkeys and Apes*. New York: Columbia Univ. Press.

Roitblat, H. R., Bever, T. G. and Terrace, H. S. (eds.) 1984, *Animal Cognition*. Hillsdale: Lawrence Erlbaum Assoc.

Roonwal, M. L. and Mohnot, S. M. 1977, *Primates of South Asia: Ecology, Sociobiology, and Behavior*. Cambridge: Harvard Univ. Press.

Rowell, T. E. 1984, Introduction: Mothers, infants, and adolescents. IN: *Female Primates: Studies by Women Primatologists* (M. F. Small, ed.) New York: Alan R. Liss, Inc., pp. 13 – 16.

de Ruiter, J. R. 1986, The influence of group size on predator scanning and foraging behaviour of wedgecapped capuchin monkeys (*Cebus olivaceus*), *Behaviour* 77 : 240 – 258.

Savage-Rumbaugh, E. S. 1986, *Ape Language from Conditioned Response to Symbol*. New York: Columbia Univ. Press.

van Schaik, C. P. 1989, The ecology of social relationships amongst female primates. IN: *Comparative Socioecology: The Behavioural Ecology of Humans and Other Mammals* (V. Standen and R. A. Foley, eds.). Oxford: Blackwell Scientific Publications, pp. 195 – 218.

van Schaik, C. P. and Van Hooff, J. A. R. A. M. 1983, On the ultimate causes of primate social systems, *Behav.* 85 : 91 – 117.

van Schaik, C. P. and vanNoordwijk, M. A. 1989, The special role of male *Cebus* monkeys in predation avoidance and its effect on group composition, *Behav. Ecol. Sociobiol.* 24 : 265 – 276.

Schubert, G. and Masters, R. D. (eds.) 1991, *Primate Politics*. Carbondale: Southern Illinois Univ. Press.

Seyfarth, R. M. and Cheney, D. L. 1984, Grooming, alliances and reciprocal altruism in vervet monkeys, *Nature* 308 (5 April 1984) : 541 – 543.

Shaw, E. and Darling, J. 1984, *Strategies of Being Female: Animal Patterns, Human Choices*. Brighton: Harvester Press.

Shipman, P. 1990, Primate origins up in the air again, *New Scientist* 23 June 1990 : 57 – 60.

Shively, C. 1985, The evolution of dominance hierarchies in nonhuman primate society. IN: *Power, Dominance and Nonverbal Behaviour* (S. Ellyson and I. Dovido, eds.) Berlin: Springer Verlag, pp. 67 – 87.

Silk, J. B. 1982, Altruism among female *Macaca radiata:* Explanations and analysis of patterns of grooming and coalition formation, *Behav.* 79 : 162 – 187.

Simons, E. L. 1988, A new species of *Propithecus* (Primates) from northeast Madagascar. *Folia Primatol.* 50 : 143 – 151.

Small, M. F. (ed.) 1984, *Female Primates: Studies by Women Primatologists*. New York: Alan R. Liss, Inc.

Small, M. F. 1988, Female primate sexual behavior and conception: Are there really sperm to spare?, *Curr. Anthropol.* 29 : 81 – 100.

Small, M. F. 1989, Female choice in nonhuman primates, *Yrbk. Phys. Anthropol.* 32 : 103 – 127.

Smuts, B. B. 1985, *Sex and Friendship in Baboons*. New York: Aldine Publ. Co.

Smuts, B. B. 1987a, Sexual competition and mate choice. IN: *Primate Societies* (B. B. Smuts, D. L. Cheney, R. M. Seyfarth, R. W. Wrangham, and T. T. Struhsaker, eds.) Chicago: Univ. of Chicago Press, pp. 385 – 399.

Smuts, B. B. 1987b, Gender, aggression, and influence. IN: *Primate Societies* (B. B. Smuts, D. L. Cheney, R. M. Seyfarth, R. W. Wrangham, and T. T. Struhsaker, eds.) Chicago: Univ. of Chicago Press, pp. 400 – 412.

Smuts, B. B., Cheney, D. L., Seyfarth, R. M., Wrangham, R. W. and Struhsaker, T. T. (eds.) 1987, *Primate Societies*. Chicago: Univ. of Chicago Press.

Snowden, C. T. 1990, Language capacities of nonhuman animals, *Yrbk. Phys. Anthropol.* 33 : 215 – 243.

Snowden, C. T., Brown, C. H. and Petersen, M. R. (eds.) 1982, *Primate Communication*. Cambridge: Cambridge Univ. Press.

Sperling, S. In press, Baboons with briefcases: Feminism, functionalism, and sociobiology in the evolution of primate gender, *Signs* 17.

Standen, V. and Foley, R. A. (eds.) 1989, *Comparative Socioecology: The Behavioural Ecology of Human and Other Mammals*. Oxford: Blackwell Scientific Publ.

Strier, K. B. 1987a, Demographic patterns in one group of muriquis, *Primate Conservation* 8 : 73 – 74.

Strier, K. B. 1987b, Activity budgets of woolly spider monkeys, or muriquis (*Brachyteles arachnoides*), *Am. J. Primatol.* 13 : 385 – 395.

Strier, K. B. 1990, New World primates, new frontiers: Insights from the woolly spider monkey, or muriqui (*Brachyteles arachnoides*), *Int. J. Primatol.* 11 : 7–19.

Struhsaker, T. T. and Leland, L. 1987, Colobines: Infanticide by adult males. IN: *Primate Societies* (B. B. Smuts, D. L. Cheney, R. M. Seyfarth, R. W. Wrangham, and T. T. Struhsaker, eds.) Chicago: Univ. of Chicago Press, pp. 83–97.

Strum, S. C. 1982, Agonistic dominance in male baboons: An alternative view, *Int. J. Primatol.* 3 : 175–202.

Strum, S. C. 1987, *Almost Human: A Journey into the World of Baboons.* New York: Random House.

Strum, S. C. and Latour, B. 1987, Redefining the social link: From baboons to humans, *Social Science Information* 26 : 783–802.

Sussman, R. W. 1990, Primate origins and the evolution of angiosperms. *Am. J. Primatol.* 23 : 209–223.

Tanner, N. M. 1981, *On Becoming Human.* Cambridge: Cambridge Univ. Press.

Terborgh, J. 1983, *Five New World Primates: A Study in Comparative Ecology.* Princeton: Princeton Univ. Press.

Terborgh, J. and Goldizen, A. W. 1985, On the mating system of the cooperatively breeding saddle-backed tamarin (*Saquinus fuscicollis*), *Behav. Ecol. Sociobiol.* 16 : 293–299.

Terborgh, J. and Janson, C. H. 1986, The socioecology of primate groups, *Ann. Rev. Ecol. Syst.* 17 : 11–135.

de Waal, F. 1982, *Chimpanzee Politics: Power and Sex Among Apes.* London: Unwin Paperbacks.

de Waal, F. 1984, Sex differences in the formation of coalitions among chimpanzees, *Ethol. Sociobiol.* 5 : 239–255.

de Waal, F. 1986, The integration of dominance and social bonding in primates, *Quart. Rev. Biol.* 61 : 459–479.

de Waal, F. 1987, Dynamics of social relationships. IN: *Primate Societies* (B. B. Smuts, D. L. Cheney, R. M. Seyfarth, R. W. Wrangham, and T. T. Struhsaker, eds.) Chicago: Univ. of Chicago Press, pp. 421–429.

de Waal, F. 1988, The reconciled hierarchy. IN: *Social Fabrics of the Mind* (M. R. Chance, ed.) London: Lawrence Erlbaum Assoc. Publ., pp. 105–136.

de Waal, F. 1989, *Peacemaking Among Primates.* Cambridge: Harvard Univ. Press.

de Waal, F. and Ren, R. 1988, Comparison of the reconciliation behavior of the stumptail and rhesus macaques, *Ethol.* 78 : 129–142.

de Waal, F. and van Roosmalen, A. 1979, Reconciliation and consolation among chimpanzees, *Behav. Ecol. Sociobiol.* 5 : 55–66.

de Waal, F. and Yoshihara, D. 1983, Reconciliation in rhesus monkeys, *Behav.* 85 : 224–241.

Wasser, S. K. 1983, *Social Behavior of Female Vertebrates.* New York: Academic Press.

Wolfheim, J. H. 1983, *Primates of the World: Distribution, Abundance, and Conservation.* Seattle: Univ. of Washington Press.

Wrangham, R. W. 1980, An ecological model of female-bonded primate groups, *Behav.* 29 : 904–910.

Wrangham, R. W. 1983, Ultimate factors determining social structure. IN: *Primate*

Social Relationships (R. A. Hinde, ed.) Oxford: Blackwell Scientific Publ., pp. 255–262.

Wrangham, R. W. 1987, Evolution of social structure. IN: *Primate Societies* (B. B. Smuts, D. L. Cheney, R. M. Seyfarth, R. W. Wrangham and T. T. Struhsaker, eds.) Chicago: Univ. of Chicago Press, pp. 282–296.

Wright, P. C. 1988, Lemurs lost and found, *Natural History* July 1988 : 57–60.

York, A. and Rowell, T. 1988, Reconciliation following aggression in patas monkeys, *Erythrocebus patas, Anim. Behav.* 36 : 502–509.

Young, A. L., Richard, A. F. and Aiello, L. C. 1990, Female dominance and maternal investment in strepsirrhine primates, *The American Naturalist* 135 : 473–488.

Zihlman, A. L. 1983, A behavioral reconstruction of *Australopithecus.* IN: *Hominid Origins: Inquiries Past and Present* (K. J. Reichs, ed.) Washington, D.C.: Univ. Press of America, pp. 207–238.

PREFACE

This is a book about sex differences and similarities in the behavior of primates, principally of nonhuman primates. My objective has been to survey and synthesize current research findings in a number of areas of primatology which seem relevant to questions concerning male and female patterns of behavior. In order to avoid presenting an encyclopedia of undigested and possibly indigestible data, I have tried to approach the topics conceptually. Therefore I have often outlined the theoretical approaches of different research efforts, and, where appropriate, explained how they have come into conflict, as well as offering my own understanding of the data. My intention has been for the book to be accessible and interesting to non-specialists, yet acceptable and even helpful to my colleagues in primatology.

Thus I hope to attract readers of diverse backgrounds without being guilty of the oversimplification and sensationalism that I fear has marred some of the writings on both sex differences and ethology. Attempting to maintain such a balance often has left me feeling as I imagine a beginning tightrope-walker must; an excess of technical "jargon" will tilt me to one side, a burst of rhetoric to the other side, and a flurry of unwarranted generalizations will topple me right over. At the same time I have tried to keep the end of the wire in sight while remaining aware of my hoped-for audience of non-specialists of all kinds, of students, and of specialists in Primatology and Women's Studies.

Primatology is a fast-growing discipline, and popular descriptions and interpretations of the data on sex differences in the behavior of primates often have not kept pace with the research findings. In particular, a reorientation from short-term or cross-sectional studies, emphasizing only the immediately visible aspects of social life, to longitudinal efforts to understand the underlying mechanisms of cohesion and continuity in primate social groups, has led to a reassessment of the significance of female behavior patterns. It is the latter perspective, especially a reappraisal of female behavior, that I have tried to convey in this book.

There are many individuals, events, and even places, without which this book would not have been written. Much of the writing took place in 1979 on the Arashiyama West Primate Research Station at LaMoca, Texas. For this I must thank the Dryden Family of Laredo, and the University of Alberta which granted me a leave-of-absence from my position in the Department of Anthropology. Several colleagues read and commented upon drafts of the chapters, and I would like to thank Michael Asch, Ben Blount, Harold and Sarah Gouzoules, Joanne Prindiville, Emöke Szathmary, Adrienne Zihlman and anonymous reviewers for their thoughtful criticisms and suggestions. Many friends gave of their time to help prepare the manuscript, and for this I thank Joan Baxter, Inge Bolin, Kay Brink, Pam Dunk and Russanne Low. My friend, Linda Straw Coelho generously provided the drawings, interpreting with great skill my sometimes vague descriptions and impossible requests for desired illustrations.

Several colleagues sent me pre-publication manuscripts which were of great help and for that kindness I would like to thank Stuart A. Altmann, Montagu Demment, Harold and Sarah Gouzoules, Donald Lindburg, David Post and David Taub. Harold and Sarah Gouzoules and Claud and Sharon Bramblett lent unflagging support, advice and friendship throughout the two years of writing, as did the book editor, Sherri Clarkson, adding calm, patience and good sense. Finally and first, I would like to thank Larry Fedigan, whose encouragement made me believe the book was possible, whose continual support made the writing bearable, and whose listening, criticism and suggestions made it better. In a sense, there is also a debt to the monkeys, who have never let me rest complacent in my understanding of them, and to LaMoca, where for several years since 1972 I have found inspiration, instruction and, at times, a welcome "life exempt from public haunt."

Edmonton
April, 1980

INTRODUCTION

People today are greatly interested in the behavior of our primate relatives. These animals offer us not only a perspective upon ourselves but also upon the options and alternatives present in all the primate species, of which we are one. This book describes in particular the variable patterns of female and male behavior found in different primate species. Such a topic is not without a diversity of theory, interpretation and opinion, and I would like to begin by making my own approach and my own "biases" on the subject explicit.

A fair amount of nonsense has been written for the public in books attempting to popularize primate behavior and demonstrate relevance to humans, and this is especially true where sex-related behaviors are concerned. Until recently the discipline of primatology itself has suffered from (and some would claim, continues to suffer from) an over-emphasis on males and the significance of their behavior, and this bias has been magnified in popular reports which reach far beyond the discipline. Thus, for example, I have found that many people have read or heard that baboons and chimpanzees, like men, are aggressive and that primate societies are based on male bonds and physically-enforced male dominance hierarchies. But very few people know that some, and probably many, primate societies are structured around female kinship groups as well as male hierarchies, and that the mother-offspring bond (and not the sexual bond) is likely the foundation of mammalian social life.

Thus, at the risk of raising the impatient eyebrows of some of my (not only male) colleagues, one of my aims in writing this book has been to document how females, as well as males, play significant parts in the social life of their groups, and have played such parts in the evolution of their species. Many primatologists, both female and male, have contributed to the change in perspective which is occurring today. There does appear to be a general growing awareness of the ways in which cultural assumptions color our view of even empirical data, and also of the manner in which certain scientific findings quickly achieve public prominence and others do not. I do not believe this to be a conspiracy to suppress certain findings, it is rather the long-observed human disposition to see and hear preferentially that which fits our preconceptions.

I have heard it said that studies emphasizing sex roles, women, or females are "trendy" or just a passing fad. I think rather that questions of differences and similarities between the sexes have been a topic of intense interest as far back as can be traced, and likely always will be a significant concern. More importantly, I hope that the study of what constitutes one-half of the animal population of the world, the females, will be considered important enough to survive passing fads in interpretation. Finally, the findings of scientific endeavors will always be used to speak to popular or "trendy" topics. Rather than shrugging off responsibility, I think it is up to us as scientists to see that our findings are explained in a way that satisfies our own demand for rigor, and our students' and public's demand for clarity, even if we would rather be off studying our animals.

A second major characteristic of my approach to the material in this book, which will soon become apparent to the reader, is that I do not think the simplest interpretation or simplest explanation is always the best one, especially if it requires any important sacrifice of the complexity, variability or subtlety of our research findings. While parsimonious explanations are indeed elegant and seductive, sometimes we are too eager for quick and simple answers to our questions, especially in our quest for knowledge about ourselves. Simple answers often require over-simplified questions and data, and the resulting reductionism is not to be confused with parsimony. I think this is true of many problematic areas of primate behavioral research, and in this book I have tried to provide an accurate description of different theories and different points of view, as well as an indication of the variability in our data about primates. In doing this, I have not failed to offer my own interpretations and opinions, for while I have tried to measure my advocacy, I feel that to completely stifle it would result in a somewhat bland, boring or rudderless guide to the topics. Rather than leave the reader guessing as to which approach or interpretation I favor, I have attempted to make my opinions clear and comprehensible whenever it seemed appropriate.

Thirdly, while I have worked both in the laboratory and in the field, on short-term and long-term projects, my preference is for data and insights obtained from non-caged social groups of primates, and documented over the longest periods of study possible. The laboratory offers the researcher the possibility of much better control over the multiple variables of any behavioral study, and also the possibility of experimentation as well as observation of behavior. Nonetheless, the basic information about the behavior of a species, and the generation of important questions and hypotheses to be tested in the laboratory, are best obtained, whenever possible, from studies of free-ranging primates in the habitats where we find them living. Also, many of the things we wish to know about primates relate to the life-histories of individuals or even of social groups — for example, discovering the factors which influence reproductive success, or determining how groups originate, are maintained and fission into daughter groups, or finding out when and why males of many primate species leave their groups of origin. None of these questions could be answered in one field season, or even in one long-term study of a single group for that matter. We not only have to be patient in foregoing oversimplified answers to such questions, we also must have the forebearance to wait for the answers to develop gradually as we pursue our research questions over the years of individual primate lives and social group histories.

Finally, since most of the discussion in this book concerns the behavior of "non-human primates," I would like to explain briefly my use of the term. It is a term which I do not like, but the alternatives, infrahuman ("beneath humans"), and prosimians-monkeys-and-apes, are in the former case equally problematic, and in the latter case too cumbersome. So, when I refer to nonhuman primates, it indicates I am making a generalization from which I wish to exclude humans. For example, I might say that nonhuman primates do not talk (although they communicate very adequately), or that most female nonhuman primates experience discrete and periodic phases of sexual motivation known as "estrous periods." While many of the concepts and theories discussed in this book are of great interest to humans (especially when we apply them to ourselves), I have taken as my goal in writing to furnish comparative information from other primate species. This is not so much with the aim of suggesting congruences or even homologues with human societies, but more in the hope that it will be of interest in itself, and that it may lead to a broadening of our vision of the possibilities nascent in all primate species including our own.

The objective of the first part of the book is to introduce in succinct form basic data and some fundamental issues which are necessary to understand a discussion of sex differences in primates. Species are generally referred to by their common names, however, the first time they are mentioned in the book, their Latin name is given as well. Thus Chapter 1 introduces the reader to the various tupes of primates living in the world today, and to the use of the comparative method. Chapter 2 deals with the difficult subject of the use of language, especially terms or "labels" as applied to the behaviors of animals and the sources, definitions, and implications of such labelling. Chapter 3 explains how ethologists currently approach the issue of inherited and environmental causes of behavior, and how that approach has changed, while Chapter 4 is an introduction to the study and the classification of primate societies. Finally, Chapter 5 surveys the data on morphological differences between females and males (sexual dimorphism) and the theories which have been proposed to explain these differences.

The second part of the book introduces several major concepts in the study of primate behavior, and discusses their implications for understanding sex differences in behavior. Reviewed and discussed in this part are the relevant findings and the interpretations of concepts such as aggression, agonism, dominance, alliance, roles, kinship, and sociosexual behavior in primates.

If it is found that adult females and males of a species differ in certain behavioral patterns, and we wish to know how these differences arise in individuals, some of the very best clues comes from studies of the development or ontogeny of young primates in that species. Thus Part III of the book surveys three approaches to the study of developing sex differences in young primates: experimental and physiological studies, learning and socialization studies, and deprivation research. Part IV follows with detailed descriptions of nine primate life-ways, with particular attention to the various parts that females and males may play in these different species and social systems.

Finally, Part V surveys the perspectives on sex differences offered by several aspects of evolutionary theory. The major area of evolutionary thought concerned with sex differences has been sexual selection theory, first developed by Darwin nearly a century ago. A chapter discussing sexual selection is followed by one reviewing more recent evolutionarly theories, known as sociobiological theories, and discussing how such modern thinking has viewed and explained differences betweeen females and males. The last chapter of the book treats schemas of the evolution of human social life, and describes and criticizes in some detail three models for early human societies which use three different nonhuman species as the "primate mirror." The significance and limitations of studying nonhuman primates as mirrors for understanding our own behavior is well stated by Haraway:

People like to look at animals, even to learn from them about human beings and human societies . . . We polish an animal mirror to look for ourselves. The biological sciences' focus on monkeys and apes has sought to make visible both the form and the history of our personal and social bodies . . . The science of nonhuman primates, primatology, may be a source of insights or a source of illusions. The issue rests on our skills in the construction of mirrors (1978: 37).

PART I

CONSTRUCTING THE PARADIGM

Chapter ONE

An Introduction to Primate Studies

Squirrel monkey
Saimiri sciureus
These small, arboreal monkeys are common in parts of Central and South-America. The female (right) is about 3/4 of the male weight.

PRIMATE BEHAVIOR AND THE COMPARATIVE METHOD

Many discussions of primate behavior begin with the claim that the study of pro-simians, monkeys and apes can reveal to us the "biological roots" or bases of human behavior. I shall not be making this claim as I do not think that it is appropriate within the framework of evolutionary theory or of any theories of behavior. Yet, given such assertions, it is small wonder that many people continue to think incorrectly of monkeys as not-quite humans, as primitive creatures retarded on the path to humanness. This of course is not what primatologists mean when they say we study monkeys and apes to reconstruct our ancestral or original conditions; rather they are referring to the use of animal models, and it is often thought that the more closely related to us the animal, the more plausible the model. Several primate models for early human society will be described in the last chapter of this book; however, an inherent problem in the applica-tion of all such models is that every species travels its own unique evolutionary route. No matter how closely chimpanzees (*Pan troglodytes*) may appear to resemble us, they are certainly not on their evolutionary way to becoming humans, and their present situation is not one of the way-stations along which our own species travelled. Nor does their behavior represent the biological core of our own behavior, stripped of its cultural orna-mentation. Simply it represents the behavior of an animal with which we share some relatively recent common ancestors.

Why then study the behavior of primates? There are so many possible answers to this question that I will make my answer a personal one. I study these animals firstly because they fascinate and delight me, and I think this may be true of all primatologists, and secondly because I am greatly interested in how and why creatures live in social groups, which may not be true of all primatologists. Had I been trained as a zoologist rather than an anthropologist I might have studied other socially-living animals such as elephants or herring gulls; however, the study of primates and not the study of proboscids is one of the subdisciplines of anthropology. For primates do constitute the order of animals to which humans also have been taxonomically consigned and the study of humankind's agreement with, and divergence from, other animals is one of the definitions of anthropology. Thus, it is reasonable to ask what we learn about *ourselves* from the study of these animals. While I have said that our primate relatives are not simplified versions of our own species, that sort of reductionism does not rule out other approaches to the relevance of nonhuman to human primates. So I would suggest that a third reason for studying the behavior of prosimians, monkeys and apes, is to enrich our perception and understanding of human behavioral patterns in the expanded perspective of the vast array of patterns existing in other primates, animals whose evolutionary paths separated from our own more recently than those of other mammals. Such a perspective is possible through the use of the comparative method.

Primatology is the study of the biology, evolution, distribution, classification and behavior of primates. Primate behaviorists may be trained in the academic disciplines of psychology, zoology or anthropology; they may prefer to work with animals in captive or in field conditions; and they may approach the study of behavior experimentally or through observational techniques. However, the comparative method is one approach which the various behavioral and life sciences share in common. Using the comparative method, principles (in this case, of behavior), are developed through the extended study of several examples of a given behavioral phenomenon and the comparisons of these examples. Many assumptions about sex differences and social bonds depend on the comparative method, and many controversial questions could be partly clarified through comparisons of a variety of primate species. Consider the commonly-held belief that men hold various types of social power over women because they are physically larger and stronger than women. It happens that primate species present us with an array of size dimorphism (males may be larger, smaller or the same size as females of their species) and of social relations between the sexes, such that we can begin a comparative approach to the question of correlating sexual dimorphism with "dominance" relations between the sexes.

We can also take a comparative approach to the question of whether males are generally more "aggressive" than females, or whether females are generally more nurturant of young than males, and so forth. Of course how we interpret our answers will depend in good part on our larger theoretical approach to behavior. The universal or even general appearance of a behavioral trait is often interpreted in the life sciences as evidence for a common biological (i.e., genetic, physiological, morphological) determination of the trait. However social scientists may offer a different interpretation, as evident in this statement by Makurius on the widespread occurrence in animal societies of inbreeding avoidance: "The universality of a social practice signifies that it is co-existent with society, not that it is rooted in nature, outside of society" (1968: 443). Thus the comparative method may help us to determine the similarities and differences between certain phenomena in humans and other animal species, and to begin to understand in what ways we are like other primates and in what ways we differ. It will not, by itself, establish the *causes* of similarities and differences. Such interpretation depends on one's theoretical and disciplinary approach to the study of behavior.

There are at least two major problems with the sorts of comparative investigations suggested here and described throughout this book. The first is that the definition and application of concepts involved in these areas (for example aggression, dominance and nurturance) are very complex and problematic, so that to use such concepts in the kinds of questions I have just posed, and which are commonly asked, may be to oversimplify them. Because I consider them to be so important, Part II of this book is an extended examination of concepts such as aggression and dominance, and their use in the comparative study of sex differences in primates. The comparative method may be used by some social scientists for developing structural principles in social behavior as suggested by Makurius and also by Callan (1970), or it may be used by others for developing functional principles, in the manner of evolutionary scientists, who are mainly interested in the adaptive value of human and nonhuman primate behavior. However it is essential to remember that, in either case, to compare behavioral patterns is not to equate them.

The second problem with the comparative method is that not all possible examples can ever be described and compared, and thus the process is open to a voluntary or involuntary biased selection of examples. Primates are so varied that it is not difficult to find one or several examples of almost any point one wishes to make about or with them. Many disagreements in the interpretation of behavior between primatologists revolve around problems of the biased selection of species, or behaviors. Out of some 200 to 250 species of primates, the behavior of less than one-quarter (50) or so has been studied and thus our ability to use the comparative method to analyze primate patterns is still limited.

Given the vast array of cultural and behavioral patterns displayed in human societies, it is equally easy to find one supporting example from human behavior for any argument one might wish to put forward. The problem is that this one example is often introduced as if it represented a norm for all members of the human species, when in fact it may only occur in one society and not others. Comparative discussions of animal and human social behavior seem particularly prone to introducing generalizations derived from our own society with the assumption that they are true of all people (e.g., "humans are primarily competitive," or "females are dependent upon their male mates for sustenance"). Recently this biased selection has been somewhat inverted, with discussions of animal behavior often introducing one example of behavior from a group of "exotic" peoples described in the literature of cultural anthropology, in order to sustain an extrapolation from animal to human behavior. Isolating such an example from the context of the array of cultural patterns in human societies does not make for a particularly useful or valid comparison. Instead of the mutual suspicion that often appears to reign, greater communication between cultural anthropologists, who study human behavioral variability, and primatologists or ethologists who study animal behavior, would improve our ability to use the comparative method in primate studies.

Bias may also be present simply as a result of the lack of representation among the primate species that have been studied. The latter do not form a representative or even a random sample of all primate species. Primatologists have naturally concentrated on those animals which are easiest to locate and follow. This has tended to be the open-country ground-dwelling species, such as baboons (*Papio*) and macaques (*Macaca*). Arboreal (tree-dwelling) or nocturnal (night-active) or thick forest dwellers have been relatively understudied. This has led to neglect in particular of the New World monkeys which are all arboreal, and of the prosimians which are often arboreal and nocturnal. In the past decade, primatologists have made a concerted effort to rectify this bias, and many persistent animal-watchers are now tracking African prosimians through the night, craning their necks to observe monkeys silhouetted against the sky in tree-tops, or wading through swamps to catch a glimpse of elusive forest-dwelling primates.

As the data become available from these studies, we realize that many previous generalizations about primate behavior must be altered, and thus we become increasingly aware that primatology is a fast-changing field. The species described in this book are therefore not a representative sample of *all* primates. However, in so far as possible, I have tried to make them a representative sample of those species for which we have social behavioral data. I think I have unfortunately underrepresented the prosimians, for which many studies are just appearing, and I would refer the interested reader to the following examples of recently published books: Doyle & Martin 1979; Richard 1978; Charles-Dominique 1977.

A concentration on open-country, ground-dwelling species may also be one of the reasons why the importance of males in primate societies was initially overstressed. Males in macaque and baboon societies are quite a bit larger and apparently more noticeable than females, and early studies of short duration tended to emphasize the greater significance of male behavior than female behavior in primate societies. Again this bias is being rectified as studies of longer duration shift the focus from the transitory dominance relations among males, to the importance of stable, longitudinal bonds between relatives, mates and friends. This has led to a reassessment of the significance of females even in baboon and macaque societies, while the study of other species is beginning to demonstrate just how variable the social behaviors of females and males may be in different social systems.

Thus even with its inherent limitations, the comparison of different primate social lives can shed light on the nature of social life in general and also on the questions which, as human beings, concern us in particular. A perception of the very familiar may become more informed by closely observing the less familiar. For example, as a child I had a very limited perception of my own culture until I was forced to live in other cultures. I then learned not only about these other cultures, but through comparison, about my own. As an adult, my perception of my own species was considerably enriched through careful observation of other species. It is this, rather than the usually superficial analogies between the behavior of one monkey or ape species — doing justice to neither the monkey, the ape, nor the human — that I believe is the value of the comparative method to self-understanding.

The behavioral sciences have long considered it part of their task to offer formulations of aspects of "human nature," and people often turn to these sciences for the answers to some of life's important questions and for the solutions to their social problems. It is my opinion that humans are largely what they make of themselves; in other words "human nature" is not so much an empirical reality as a process of self-construction. This means that if people become what they think they are, *what* they think they are is exceedingly important.

The roles of women and men in many societies are changing, and people also ask scientists for explanations of past patterns and predictions of the future. Some scientists have offered by way of an answer, descriptions and interpretations of the behavior of female and male animals as part of a trend to look to animals as models for the understanding of our own behavior. As people think about what they can become, what roles they can play in society, they sometimes receive warnings from disciplinary specialists, widely publicized in the media, that, for example, female animals are "by nature" nurturant and male animals aggressive and competitive, and so any efforts to enter other spheres of behavior can only result in unhappiness for everyone. In contradistinction, all of the discussions which follow in this book are based on two guiding propositions: animal behavior until recently has been interpreted largely from the male's point of view (this has sometimes been true even when the scientist is a female); and, in a search for simple answers to complex problems, the processes of nature in general and the behavior of animals in particular, often have been greatly oversimplified.

Although I continue to believe that humans should look to their own best achievements for models of what they want to be and to become, clearly studies of animal behavior will continue to provide mirrors, models, and even myths for many of the puzzling aspects of human behavior. While I do not think ethological studies will deliver the final verdict about what humans can hope to achieve in any society, nonetheless I do think the study of similarities and dissimilarities can be one of the paths to self-understanding, and that a knowledge of the variability and the significance of different primate social systems can be a valuable aid in the process of self-construction for humans. In this respect, I echo Embler who said:

7

The use of analogy in all branches of human thought is indispensable; but the making of nice distinctions is equally important and has ever been a leading characteristic of the human mind. The knowledge of subtle differences . . . is a knowledge invaluable to mankind. What is more the knowledge of differences leads to an understanding of relationships (1951: 93)

AN INTRODUCTION TO THE ORDER PRIMATES

This is a book about the social lives of primates in particular relation to the behavioral patterns of females and males. Thus I will not cover in any detail certain topics in primatology of equal interest, such as primate ecology and evolution and zoogeography. These areas are well discussed in several other texts. However, in order to make this book intelligible to those unfamiliar with primates, in this section I will provide an introduction to the major features of the order Primates, including an outline of the taxonomic relations, distributions of those species most often mentioned in the book, and references to further sources. Those readers already familiar with this information, which is a necessary background to informed discussion of primate behavior, may wish to go on to Chapter 2.

The Primate Pattern

There is no one feature which distinguishes all primates from other mammals (excepting, possibly, the structure of the middle ear, see Cartmill 1975), and every general rule which we may care to describe for primates will find its exception in at least one species. LeGros Clark's (1962) solution to this difficulty was to describe a set of evolutionary trends or traits which taken together make up the "primate pattern." These trends have become the standard defining description for the primate order, and since most of these traits concern morphology, I shall group and describe them in relation to parts of the body.

Primates have *generalized skeletons*, especially the structure of the limbs, such that their bodies are capable of many movement and locomotor patterns. This simply means that in comparison to other orders of animals, primates have not tended to develop anatomical specializations which would sharply limit them to a very narrow lifeway. The capacity for flexible responses to environmental contingencies is one of the hallmarks of primates, whether we consider morphology or behavior. Different species do exhibit moderate degrees of anatomical adaptation for particular locomotor patterns, such as brachiation in which the body is suspended and propelled by the arms, or quadrupedalism which is locomotion on four limbs. However all primate species have the anatomical capacity for *upright posture* or truncal erectness, and this is a trait which Napier and Napier (1967) added to LeGros Clark's list.

Primates, as an order of animals, evolved in the tropical forests, probably as insect-eaters (Cartmill 1975), and a number of their features can be interpreted as part of the special primate pattern of visually directed predation in an arboreal habitat. The ability to sit or stand erect frees the upper limbs for exploration, manipulation or simply grasping, and primate hands are highly suited for these functions. The presence of five digits on each hand or foot is not an unusual trait in mammals. However, primates possess

8

additional features of the hand and foot which are special — *mobile prehensile digits*, usually with a divergent thumb and divergent great toe, and the replacement of sharp, compressed claws by flattened *nails with highly sensitive tactile pads* on the underside of the digits. These traits allow the primates to develop extensive skills in grasping and manipulating objects in the environment.

The upright sitting primate can also bring objects close to its face for visual inspection, and exhibits skilled hand-eye coordination. In primates, as compared to their ancestral forms, the snout or *muzzle* has become *reduced* in correspondence with the *decreasing use of the apparatus of smell* or olfaction, and the *reduction in the number of teeth*, usually resulting in a "flatter" face. Vision is a very important sense in all primates and the visual apparatus, especially of the brain, is elaborate. Reduction of the snout with corresponding movement of the eyes from the sides to the front of the head leads to varying degrees of *binocular vision* and depth perception. Most primates also have color vision.

Primates tend to rely extensively on social learning for survival. Young primates are carefully and usually individually nurtured over long periods of time during which they are highly dependent on the adults of the species. Thus primates exhibit adaptations associated with longer gestation and *better nourishment of the developing fetus*, relatively large and *complex brains*, especially the cerebral cortex, and *prolonged postnatal periods* of suckling and growth to adulthood. Again, these are statements of general trends held in common by the species of the order Primates, but not all traits are shown by all species.

Taxonomic Relationships of the Primate Species Discussed

That primate taxonomy is by no means free of controversy may be deduced from the fact that experts variously list from 176 to 250 living primate species. However the fundamental taxonomic relationships are agreed, and I will outline them here, using Figure 1 as a guide. Before explaining the taxonomic chart, it may be helpful to note that in non-specialist terms there are four types of primates: prosimians, monkeys, apes, and humans. The distinctions between these types (and some distinctions within the types) will become clear in the following description.

The order Primates is divided into two suborders, Prosimii or prosimians and Anthropoidea or anthropoids (the latter includes monkey, apes and humans). Prosimians, as their name implies (pro = before, simian = monkey or ape-like), appeared in the animal world before monkeys and apes, and although most of the currently living species have changed considerably from the ancestral species, prosimian lifeways may provide some clues and some models for what early primates were like. Today there are three major categories or infraorders of prosimians — the Lemuriformes (lemurs) found only in the Malagasy Republic (formerly the island of Madagascar), the Tarsiiformes (tarsiers) represented today by only one genus found in southeast Asia, and the Lorisiformes (lorises) found in Africa and Asia. Most behavioral studies have been conducted on the lemurs of Madagascar, where the animals are diurnal and sometimes live in large multi-male, multi-female societies. All of the prosimians which do not occur on Madagascar are nocturnal, elusive, and difficult to study. However, research such as that by Charles-Dominique (1971), on several species of lorisiformes (for example, galagos, or *Galago demidovii*) has shown that these animals are not completely asocial as once was thought, rather individual males and female-and-offspring units occupy overlapping ranges. It has been argued

9

Figure 1. TAXONOMIC RELATIONSHIPS OF THE PRIMATES DISCUSSED IN THIS BOOK

ORDER   SUBORDER   INFRAORDER   FAMILY   GENUS (common name)

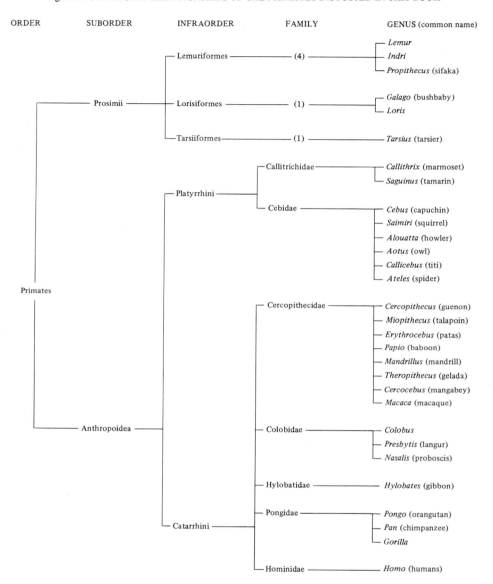

that this is the original primate social system, a suggestion which will be further discussed in Chapter 4.

From Figure 1 it can be seen that the suborder Anthropoidea is divided into two infraorders: Platyrrhini and Catarrhini. Platyrrhines are primates found in Central and South America, from southern Mexico to northern Argentina. These "New World Monkeys" are distinguished from other anthropoids in part by their characteristic nose shape (platy = flat, rhine = nose or nostril), by long tails, which are prehensile in some species, and by a dental pattern which includes three premolars. "Monkey" essentially means an anthropoid with a tail and it is confusing to many nonspecialists that all primates in the New World but only certain primates in Africa and Asia (the "Old World") are commonly referred to as "monkeys."

The infraorder Platrrhini, or the New World primates, is divided into two families, the Callitrichidae, or callitrichids, and the Cebidae or cebids. Callitrichids are the tiny marmosets and tamarins which are of unusual appearance ("marmosets" were small grotesque figures in medieval architecture) and some unusual traits, such as claws instead of nails on most digits, and the common birth of twins. Their social organzation is considered by most experts to be "monogamous" and will be described in Chapter 15. The other family of New World primates, the cebids, includes the more familiar species such as squirrel monkeys (*Saimiri*), spider monkeys (*Ateles*), and capuchins (*Cebus*, the "organ grinder's monkey"). All the New World primates are completely arboreal and have proven somewhat difficult to study in the field. However in the past decade our knowledge of the social lives of these animals has increased considerably, and today the study of New World monkeys is a very active area within primate behavior.

The other infraorder of anthropoids is the Catarrhini (cata = downward, rhini = nose) or Old World primates. While only monkey-like forms exist as native species in the New World today, the Old World primates are divided into two families of monkeys (Cercopithecidae, and Colobidae), two families of apes (Hylobatidae and Pongidae) and one family of humans (Hominidae).

Although humans are found in all parts of the world today, our family, the Hominidae, evolved in Africa. Nonhuman primates have in the past been found in all parts of the world where tropical forests ever flourished, with the exception of New Guinea, Australia and New Zealand. Today the nonhuman primates of the Old World are found only in Africa, on Madagascar, and in a band stretching across the southern part of Asia from parts of the Arabian peninsula to India and Japan.

The family Cercopithecidae consists of many familiar monkeys, such as macaques (*Macaca*) and baboons (*Papio*), and the forest-dwelling guenons (*Cercopithecus*). These Old World monkeys exhibit cheek pouches (sac-like openings on the inside of the cheek used to store extra food) and ischial callosities ("sitting pads," two thick, callused skin areas on the buttocks). Species of Cercopithecidae occur widely in Africa and Asia and exhibit a great range of lifeways, from arboreal to forest-floor dwelling, to terrestrial, open-country niches. This family of monkeys shows the most extensive adaptation to life on the ground other than the family Hominidae, and many species are relatively large-bodied monkeys, many live in multi-female, multi-male societies, and some exhibit extreme sexual dimorphism, with the female only half the size of the male.

The other family of Old World monkeys, the Colobidae, are often called the "leaf-eating" monkeys because leaves may form a substantial part of their diets, and they exhibit anatomical adaptations, such as sacculated stomachs to aid in the digestion of cellulose. Species of colobines occur both in Africa and Asia and they are not only arboreal, but extremely impressive leapers as they locomote through the trees. Partly because of their arboreality, colobines were relatively little studied until recently, when a number of studies documented their often polygynous (one male mating with several females) societies. Better known examples of colobines are langurs, such as the sacred Hanuman langurs (*Presbytis entellus*), and the beautiful black-and white colobus (*Colobus guereza*), often hunted for its splendid coat.

Of the two families of apes, the Hylobatidae and the Pongidae, the Hylobatidae are referred to as the "lesser apes" because of their relatively smaller size. The hylobatids, which have a lengthy separate evolutionary history from the pongids, consist of species of gibbons and siamangs found in Asia. These animals are all arboreal and are the most splendid aerial acrobats of the primate world. They are renowned as brachiators and also for their strong expression of territoriality (defense of space) and for their characteristically small monogamous groups. One species, the siamang (*Hylobates symphalangus*) is described at length in Chapter 16.

The family Pongidae consists of four living species of "great apes," the orangutan (*Pongo pygmaeus*), the gorilla (*Gorilla gorilla*) and two chimpanzees, the common chimpanzee (*Pan troglodytes*) and the pygmy chimpanzee (*Pan paniscus*). The orangutan is found today on Borneo and Sumatra, although the fossil record indicates it was once widely distributed across Asia. The gorilla and the chimpanzee are the African great apes. Although all the great apes show certain anatomical adaptations for brachiation, they actually use a variety of locomotor patterns, and all spend some time traveling and feeding on the ground. The orangutan does not live in cohesive social groups, but in dispersed overlapping female and male ranges somewhat reminiscent of nocturnal prosimians. Gorillas live in societies of one fully adult male with several females, and long-term studies have demonstrated these animals to be very unlike the mythology which surrounds them. Chimpanzees live in large, fluid "communities," which will be described in Chapter 14.

The family Hominidae today consists of one species, *Homo sapiens*, although other species of hominids are known from the fossil record. Anatomically we are a rather generalized species, however we are characterized by our habitual bipedal (two-footed) locomotion and various related adaptations, including weight-bearing rather than prehensile feet, and broadened pelvis and lengthened legs. We also have relatively large brains, flat faces and small anterior (front) teeth. Humans are comparatively large primates, but do not exhibit a striking degree of sexual dimorphism in size. The chimpanzees, our closest primate relatives by all available evidence (paleontological, biochemical) are also only moderately dimorphic in size, although somewhat more so than modern humans. The other great apes, the orangutans and the gorillas, are both extremely large in size (a male gorilla may weigh up to 200 kg) and both extremely sexually dimorphic. In Chapter 5 I shall turn to a discussion of the distribution and the possible reasons for and results of sexual dimorphism in primates. If the reader wishes to learn more than this very brief introduction to the primate pattern, distribution and classification, good sources are Schultz (1969); Rosen (1974); Napier and Napier (1967); Cartmill (1975); and Chiarelli (1972). Comprehensive descriptions, discussions and illustrations of primates and of primate behavior can be found in introductory textbooks by A. Jolly (1972);

Bramblett (1976); and Chalmers (1979). Many of the concepts used in these texts are discussed in later chapters. Because quite a few of the disagreements over concepts, such as aggression and dominance, reflect problems in our use and interpretation of language, language-use, especially in the study of behavior, is itself the subject of the next chapter.

Chapter TWO

Words

Black and White Colobus
Colobus guereza
Often hunted for their beautiful coats, these leaf eating arboreal monkeys are from Central Africa. The female (right) holds an infant still with its white natal coat.

. . . we are compelled to point out a universal problem in the development and use of words for scientific purposes. To describe observations, human beings, including scientists, employ various kinds of communication devices, including pictographs, diagrams, mechanical structures, and most typically, words. No matter which device is selected, certain implications generally follow from the properties inherent in the particular form of communication. When a word is chosen to represent an observation, it is nearly always a metaphor. Every metaphor carries its own implications.

Sarbin and Allen 1968

Language is thought to both reflect and structure our thinking. Scientists who are aware of its importance try to be careful in its usage in order to enhance their goal of greater objectivity, or at least the lesser goals of operationalisation and replicability. To some scientists, objectivity in the observation, understanding and ultimately, control of a wholly material world, remains the cornerstone of their belief. Others, seeing themselves as participants in a world whose perceptions are filtered and reformed by language, accept that complete objectivity is impossible. What they do hope, partially through careful use of language, is to build and share a more generalizable perception of the phenomena they observe. The objective extreme, for example the assertion that only empirically verifiable statements can be true, leads, as Scheffler noted, to the "relegation of metaphysics, poetry, religion and ethics to the limbo of the cognitively meaningless" (1967: 5). He went on to state that the subjective extreme "has the effect of isolating each scientist within an observed world consonant with his theoretical beliefs" (1967: 15). Indeed, at this extreme it becomes doubtful whether language allows any real communication between people. So, the question of whether there is an objective reality, independent of, and realizable by different observers, or whether each viewpoint and thus each scientist, to a greater or a lesser extent, creates its own reality, cannot easily be resolved. It does seem that most scientists work in the belief that some degree of objectivity is attainable and that therefore the excesses of subjective thinking are to be avoided. In the study of animal behavior such subjective thinking often takes the form of "anthropomorphism," or the attribution of human characteristics to animals or inanimate objects. Because so many of the problems of language usage in the study of animals derive from anthropomorphisms, a discussion of this topic here precedes a consideration of the manner in which communication devices such as metaphors and analogies are used to explain and create our perceptions of animals. That in turn is followed by a discussion of some examples of androcentric language in the study of sex roles and sex differences.

ANTHROPOMORPHISM

Many people consider one of the more important achievements of Euroamerican scientific and intellectual thought to be the realization that we are not the center of, nor the prototype for, all else in the universe. The earth is not the center of our planetary system, our species does not represent the necessary culmination of evolutionary forces toward which all species must inexorably gravitate, and our culture is not the final aim nor the final stage of all cultural development. Contrary assumptions are known as anthropocentric ("man-centered") and ethnocentric (belief in the inherent superiority of one's culture), and are closely related to, and perhaps the source of, anthropomorphisms, the attributing of human characteristcs to animals *because* we believe that all organisms must be in some ways like ourselves.

It has been argued in Chapter 1 that the comparative method in both cultural anthropology and ethology can offer us new forms of self-understanding as we come to see ourselves in the perspective of one pattern in a range of variants. It should follow that we can never understand animals or other cultures in their own right, nor apply that understanding to our behavior, if we begin with the assumption that others must be, or should be, just like us.

However, while anthropomorphism is to be avoided or minimized, it will not be eliminated, for we cannot stop thinking as human beings. We are limited by our human perceptions and senses, by our mental, physical and physiological abilities. We cannot smell what a galago smells when she sniffs a branch, we cannot feel what a spider monkey feels as she leaps a gap between trees. Instead, we transform these acts to our dimensions, and record perhaps an "olfactory inspect" and "arboreal locomotion"; terms which likely do not portray the essence of the act from the perspective of the animal. Ethologists often try to avoid value-laden or functional terminology in labeling and recording behavior and use instead motor descriptions (e.g., "open-mouth gape" instead of "angry face"). This is a necessary part of the scientific method which attempts to approach the study of material without prejudgements, but it can also result in what others regard as the cold and jargon-laden language of the behavioral sciences, as well as a false sense of objectivity.

It should also be noted that the ethological method is not an attempt to understand and describe animals as they would describe themselves if they could talk to us (what cultural anthropologists would call an emic study). It is rather an effort to study a variety of forms through the use of defined categories or common measures in the hope of achieving both a greater objectivity and a comparative perspective (what cultural anthropologists would call an etic study). On the other hand, some scientists have pointed out that there are also benefits in studying animals not as distant, different "objects," but as fellow creatures about which we can have highly important insights and intuitions. Indeed, it is often said that the creative insight or intuitive leap of the imagination, the hallmark of subjective thinking, also is the common characteristic of many of the greatest scientific discoveries. Thus, as respected a field worker in animal behavior as George Schaller has said that much of what we understand about primates we do through intelligent empathy (1964). Further, Frisch (1963) has noted that the Japanese concept of nature, different from our own Euroamerican tradition, has been judged anthropomorphic by many Western scientists, yet has given rise to methods of studying monkeys and to discoveries concerning precultural behavior and kinship which are gaining more and more respect throughout the world and which are very important to the discipline of primatology (see also Watanabe 1974).

Although anthropomorphic thinking has taken on virtually the status of a "taboo" in ethology, accusations of such thinking may be differentially directed at certain forms of anthropomorphism, while others go unrecognized. Particularly offensive to ethologists are assumptions concerning the emotions (e.g., the gorilla beats his chest because he is "angry"; the hamadryas male herds females away from other males because he is "jealous"), and sentimental views or interpretations of animals (e.g., "baby monkeys are sweet"). However, it is no less anthropomorphic to conceive and speak of an animal as "selfish" or "calculating" as many ethologists currently do, even if the argument is made that the animal is not consciously aware of being such but is so as a result of natural selection. Animals are not aware of being "sweet" either, both are our interpretations and our anthropomorphic labels. Furthermore, it is equally human-oriented and metaphorical, if very common, to think of an animal as a machine, or as behaving like a machine (computer, etc.). Humans, and not nature, invented machines, and it has been argued that machines are like humans because, in so far as they could, people designed machines in their own image (Embler 1951). Of course, that same argument is also used to defend

17

the use of mechanical models to explain behavior. It is clear that, in their efforts to be objective and to overcome past anthropomorphisms, scientists reject some terms and embrace others, which, while no less anthropomorphic, are selected from a more acceptable lexicon. Currently, labels which imply sentimentality, or even generally "good" qualities in our value system are rejected, while those which convey a hostile and combative imagery are embraced. This is evident in primatology; where much controversy, soul-searching and criticism surrounds the use of labels for behaviors which make animals seem "familiar and cozy" (Symons 1978), such as "play," "social roles," "leadership roles," "aunts," "cooperation," while very little criticism is leveled at equally anthropomorphic labels which make animal behavior seem undesirable in our human value system — for example, "murder," "rape," "slavery," "selfish."

One of the most common forms of unrecognized and rebounding anthropomorphism is what has been called the "aha! reaction" (Lorenz 1970; Callan 1970; Martin 1974). As Sarbin and Allen noted in the quotation given at the head of this chapter, scientists *choose* words to describe what they observe, and these words are often metaphors. Ethologists label the phenomena of animal behavior in order to study and communicate about them, and the labels are usually borrowed from the sphere of human experience. Thus a certain form of social pattern in ants is called "slavery," a particular interaction between two chickens is called "dominance," or a particular interaction between two gibbon groups is called "territoriality." These terms are chosen precisely because the ethologist perceives commonality between the familiar human phenomenon which forms the model, and the new observation of animal behavior which must be labeled. The first aha! occurs here — the animal is territorial just like us. Much ethological labeling is anthropomorphic, but ethologists attempt to give rigorous definition to the term as used in the discipline (a scientific or operational definition). When, as so often happens the supposed appearance in animals of such categories of behavior produces a thrill of recognition and extrapolation back to human behavior (aha! humans are territorial just like animals), or when the human example is seen as a special case of the general principle discovered in nature (see Allen et al 1976), then the logical fallacy known as circular reasoning is completed, and the rebounding anthropomorphism known as the aha! reaction has occurred.

Clearly the labeling and the defining of categories for the observation of behavior is both important and problematical. That confusion can result from different understandings even of the operationally defined categories of an ethogram for one species is well illustrated by Reynold's paper comparing four rhesus macaque (*Macaca mulatta*) studies (1976). In this paper he demonstrated that variability in choosing and defining categories, and in the application of similarly defined categories, leads to severe difficulties in comparing studies even within this one species. Indeed, in answer to his own question, "How can the many problems of obtaining accurate, objective records of behavior be overcome?" Reynolds saw no answer other than to counsel a more careful and judicious choice and use of description and categories (1976: 106).

DEFINITIONS, METAPHORS AND ANALOGIES

It is sometimes assumed that figures of speech are only to be found in the utterances of the poet and the creative writer, but this is not so. I could say that our everyday language is "full" of metaphors and I would have used one in the saying of it. The use of metaphors is so common a language habit that most of us are seldom aware of it, and perhaps it is a larger comment upon the phenomenon of anthropomorphism that semanticists believe we always attempt to describe and understand the unfamiliar through first likening it to the familiar. We are indeed obliged to, since unless we constantly invent new words for experiences, sensations, perceptions and events that are new to us, something which would render communication impossible, we must describe each new event with the words we already know. We often say of a new experience, "it reminds me of . . ." or "it is like . . ." in order to relate it to the things we know, and by so doing give it meaning to us. Thus the ethologist who classifies and describes formerly "unknown" animal behavior in terms of familiar human behavior in order to make sense of it, easily can be drawn into the anthropomorphic circle described as the aha! reaction.

To express one thing in terms of another is one of the purposes of metaphor and analogy, and because these always carry with them the associations, and even the conceptual structures of their source, we cannot be too careful in our choice of one to apply to perceived or hypothesized events. As semanticists constantly remind us, words not only have "denotations" or dictionary definitions, they also have "connotations," associated meanings derived from our previous encounters with and usage of, the words. Although scientists may be careful to define a word for their own use, they cannot easily shed the connotations of borrowed words, and they are occasionally cavalier, or as Callan (1970) has said, "complacent" about their initial labeling of phenomena. For example, in defense of his defiantly anthropomorphic description of gene selection theory (genes are described as ruthless, happy, loyal, gregarious, etc.), Dawkins (1976) has asserted that he can define words as he likes for his own purposes, provided he does so clearly and unambiguously. Although this is an extreme position Dawkins shares with Humpty Dumpty in *Alice Through the Looking Glass*, ethologists sometimes show a similar lack of awareness of the semantic principles that so strongly influence thinking and communication.

One of the better examples of this lack of awareness concerns the currently popular use in animal behavior studies of highly charged, emotive and value-laden words to describe observed behavioral patterns and interactions; words such as "murder," "rape," "passive," cheaters," "suckers," etc. The first two words in particular involve a subjective judgement of cognition and purposive behavior (i.e. murder = killing with malice "aforethought"; rape = sexual intercourse without "consent"). It is highly questionable if terms referring to volitional or premeditated actions and belonging to a class of socially unacceptable behaviours we call "crimes," should be applied to the behavior of animals.

Further these words are burdened with emotive connotations which make them difficult to apply even in our own society. Some courts of law are discarding the term "rape," for example, in an attempt to escape the social history of the word. Sylvia Ashton Warner (1963) discovered some years ago while attempting to teach young children to read, that some words have such a powerful emotional impact upon us that we do not learn, remember or understand them in the same purely cognitive manner as other words, but instead react to them almost as conditioned reflexes. Consider for example, your reactions to the statements: "animals kill each other," versus "animals murder each other"; or to the statement "she was assaulted," as opposed to "she was raped."

When we apply these emotive words to animals, they become powerful metaphors, and as such have a way of "coming true" in our minds, or as Embler has phrased it:

. . . those who find their attitudes implicit in the metaphor construe the metaphor to be a statement of identity, that is, a statement of fact. Figures of speech, when they are fitting and felicitous, and especially when they occur in print, give poetic sanction, as it were, to hitherto dimly felt, inarticulated beliefs. When a metaphor is new, and when the reader does not enjoy the perspective vouchsafed by time, the metaphor is taken literally, and its function is not that of rhetorical device, but of statement of fact, *prescribing* certain kinds of behavior. (1951: 85).

How metaphors which have become statements of fact may then become statements of judgement is shown by Stent (1977) in his criticism of the use of the words "selfish," and "altruistic" by sociobiologists. Some recent evolutionary theories, further discussed in Chapter 18, have proposed that the word "selfish" be used to describe behavior which "benefits the individual in terms of genetic fitness at the expense of the genetic fitness of other members of the same species" (Wilson 1975: 594). The word "altruism" is to be used to describe behavior which increases another entity's chances of survival at the expense of one's own survival (Dawkins 1976). However altruistic behavior such as this only *apparently* reduces one's fitness because it is differently directed to entities with similar genetic make-up, and thus in the long run perpetuates one's genes in relatives. All such altruistic behavior is in fact "selfish" in Dawkin's use of the word. What Stent criticizes is the capricious use of this word, giving it special meaning directly contrary to its conventional meaning; and also the attempt to use both of these words as though they were value-free or ethically-neutral. He brings up the problem of what is known as the "naturalistic fallacy" or the tendency to confuse what *is* in nature with what *ought* to be, the confounding of purported statements of fact with value judgements. Much talk is made in the biological and behavioral sciences of the naturalistic fallacy, as part of what I believe is a general attempt by scientists to distance themselves from ethical questions, in a sense, to claim their opinions are based on facts (what is) rather than values (what ought to be). However, scientists who insist on giving observations interpretative labels such as selfish, altruistic, etc., cannot claim to be avoiding the naturalistic fallacy. It often seems, as Allee (1978: 48) has observed, "we will profit by being less certain that the more unpalatable the interpretation, the closer the approach to truth." Moreover, as Stent points out, if we deliberately choose value-laden language then our ethics must be inherent in our statements of fact.

A final important point concerning the use of metaphors and analogies for understanding and communicating about animal behavior is that if *one* metaphor or analogy causes us to think about behavior in a certain manner because of its source, then the consistent choice of many such comparisons from one or similar sources has an even more powerful impact upon our perception. For example, biology is often described as the "core," the "root" or the "substrate" of behavior, causing many people to perceive biology as lying underneath, or anterior to, behavior. This "onion" model (Martin 1974) of the organism is popular with many science writers and is misleading in its projection of separate layers of behavior and biology, and of behavior and environment as superficial layers which can be peeled away.

As a second example, consider the different ways in which evolutionary theory has been described and explained. Several historians of science (Hofstadter 1959; Matson

1976; Eiseley 1961; Collingwood 1969) have noted that Darwin explained his theory through consistent metaphors, analogies, and images of struggle, conflict and warfare. Struggle, conflict, and victory in the battle for life was the leitmotif in his presentation of evolutionary theory, and not coincidentally, was a major theme in Victorian thought. In the first half of the twentieth century, important evolutionary theorists, such as J. Huxley, G. Simpson and T. Dobzhansky, spoke of the processes of evolution with very different images and figures of speech, borrowed largely from the philosophy of pragmatism, in which creatures were described as more "opportunistic" than "combative," and humans were to regard natural selection as a force to be harnessed and controlled rather than a destiny to be suffered (Matson 1976). Finally, in quite recent writings on evolutionary theory, we find that metaphors and analogies are drawn rather consistently from exchange theory and from what we might call "industrial economics." Many writers today describe animal behavior as if it were controlled by "cost/benefit analyses," "investment planning," and "efficiency models." For example, dependent offspring may be described at various times as "assets," "commodities," "liabilities" and so forth. A second modern source for analogies is game theory, as some scientists state that individuals (or individual genes) have "strategies," and can be labeled "cheaters," "losers," "winners," "suckers." Thus, calculated individualism and efficiency, or achieving the desired result of gene self-promotion with the least effort and waste, could be called the leitmotif of current evolutionary explanation. This leads easily to the perception and the belief that evolutionary and behavioral processes are just like our economic processes, or to the circular argument that our own particular Euroamerican economic system mirrors that of biological evolution (see Ruse 1979 for a defense of this form of argument).

The major point I have been attempting to make in this section is that while in one sense these descriptions and labels are only communication devices in the on-going attempt to understand behavior and evolutionary processes, in another sense "the metaphor is the message." Or, as Embler has said:

> A whole philosophy of life is often implicit in the metaphors of creative writers, the philosophy of an entire generation, indeed of an entire civilization . . . To future generations, an age may be known by the metaphors it chose to express its ideals. (1951: 85)

Thus the metaphors we use and the implicit or explicit assumptions about the systems they represent are very far from being value-free. If we accept, as does the current language of evolution, that natural systems, like the stock market, operate at or close to their limits, offering great benefits from marginal differences in a system whose constraints are always at their limits, then assessing the behavior of animals in terms of costs and benefits seems quite apt. If, on the other hand, our field observations and our experience lead us to doubt that natural systems are constantly so highly constrained or so uniformly competitive, then the current lexicon of metaphors and analogies may seem inappropriate and uncomfortable. This may be so especially for those researchers who feel that to a considerable extent the lexicon "imposes" its particular point of view.

LANGUAGE AND THE STUDY OF SEX ROLES

Much of this chapter has been devoted to the illustration of how language both mirrors and perpetuates our biases, and this reflection of biases, albeit unconscious, is no less evident in the scientific study of differences between females and males. It is generally accepted that many scientists, female and male, have taken, until quite recent times, an androcentric view of the world and have believed females to be biologically, behaviorally, and intellectually inferior to males. Therefore, it is not surprising to find that they have often used androcentric language, or language which implies the inferiority of females, in their scientific descriptions of their subject matter. Since many people are not continuously or consciously aware of implicit meanings in their choice of words, or may reject the importance of connotations, and since terminology in science quickly becomes traditional, it is possible for terms and descriptions suggesting female inferiority to persist in language even after "equality of the sexes" has become an avowed social policy or scientific tenet.

Simone de Beauvoir has noted that:

> It is, in point of fact, a difficult matter for man to realize the extreme importance of social discriminations which seem outwardly insignificant but which produce in women moral and intellectual effects so profound that they appear to spring from her original nature. (1974: XXX)

Perhaps this is one reason why few aspects of the feminist movement in North America and Europe have received as much ridicule as have efforts to reduce androcentrism in language (Miller and Swift 1976). Yet in spite of the ridicule and charges of overreaction when some women express opposition to, for example, the use of the word "man" to mean all humans, female or male, androcentric language is an important force in the study of sex roles and sex differences at all levels, conceptual, empirical and speculative. For this reason, I propose here to give some examples of how language use in the study of sex roles and sex differences in primates may influence our conclusions and thinking.

It is fitting to begin with an explanation of why I refer to behavioral and physical differences between female and male primates as "sex" rather than as "gender" differences. The latter term is usually used to denote societal or culturally-derived differences; one's gender related behavior may be characterized as "feminine" for example, and cognition concerning the differences between females and males is implied in the term. "Sex" however, is a term which usually refers to "aspects of maleness and femaleness of an organic kind" (Reynolds 1976: 119). Since in many cases in this book the derivation of differences between females and males is itself in question, or is considered to be both organic and societal, the dichotomy represented by sex versus gender is false, and I have chosen the term "sex differences" simply because it has fewer anthromorphic connotations of cognition and assumes less about the motivation or level of self-awareness of animals.

I have suggested that much controversy has surrounded efforts to encourage scientists, educators and all writers to use the words "people" or "humans" rather than "man" when referring to our species, efforts usually countered with the oft-repeated defense that "when I say man, I embrace woman." Although it would take us far astray to present the pros and cons of the latter argument, I would like to briefly describe two studies, relevant to primatology, which attempted to test the argument that in learning our language as children we all come to understand that "man" means humans and includes both women and men. In a set of experiments conducted with junior high school students studying human evolution (Harrison 1975), subjects were broken into three groups and asked to draw pictures of people engaged in seven different activities, and to assign first names to their created people. The first group were asked to draw "early man" engaged in making tools, gathering food, etc.; the second group were asked to draw "early humans" doing the same; and the third group drew "early men and women." The figures in the resulting drawings were assessed as male or female, depending primarily on the masculine or feminine first name assigned by the subject; names that might have applied to either sex were not counted. All three groups drew significantly more male figures than female figures, but the "early man" group drew the largest proportion of male figures with masculine names, even sketching in more males when the activity represented was "childcare."

Perhaps this is less surprising in view of a survey of beginning texts on prehistoric people, all having the word "man" in the title, which found that illustrations of males outnumbered illustrations of females by eight to one (Nilsen 1973). The same author conducted a picture-selection study with 100 children ranging from the nursery school level to seventh grade, and found that a majority of children of both sexes interpreted the word "man" in sentences like "man must work in order to eat," to mean male people and not to mean female people. This point will be taken up again in the last chapter, when I consider the fact that most theories of early human social life have considered women to be unimportant in the evolution of the unique human capacities which gave rise to our cultural achievements. That many people may not perceive women as taking part in the evolution of "early man" is not irrelevant, and is in part another example of how language and meaning interact.

In several chapters of this book, terminology or behavioral labels for sex differences which are value-laden and androcentric are discussed in detail. A few other instances are summarized here to show how they conform to and reinforce our cultural stereotypes of females as objects to be acted upon, and femaleness as a passive state.

In a description of the biological processes of sexual differentiation in fetal animals, Hutt said: "it is clear that the basic mammalian pattern template is female. It is male differentiation which is actively organized by androgenic hormones; female differentiation occurs by default, as it were" (1972: 17, 18); And, "in other words, masculine differentiation is an active process, female differentiation a passive one" (1972: 62). A discussion of the processes of sexual differentiation to which she is referring is taken up in Chapter 11. However, it can be noted here that rather than the terms "default" and "passive" for females and "active" for males, the process could be equally well described, as it has been by Money, as showing that "nature's first preference is to differentiate a female" (1980: 17). This description represents the same process with a different bias created by the words "first preference" as opposed to "default" and "passive." None of these words are operational terms defining biochemical or physiological processes; they are evaluative metaphors intended to dramatize a particular interpretation of the process of sexual differentiation. However, they are often used as if they were operational terms, giving the unwary reader the impression that "passive" and "active" are empirical categories instead of metaphors of personal and social stereotype.

When we find, as we do in primate species, females and males living together in stable social groups, the males are sometimes spoken of as "owning" the females (R. Fox 1972), and perhaps as "providing for" or "maintaining" the females and their young (Crook 1972). Further, the social organization of the entire group is often described as based on male behavior and male-male relations. More extreme versions of this lexicon and reflected conceptualization are found in the popular use of the word "harem" to describe polygynous animal societies, and in Fox's assertion that the basic difference between nonhuman and human kinship systems is the change from females as objects of sexual use to females as objects of sexual exchange (see Chapter 19). As demonstrated in Chapters 8, 9, 14, 15, 16, and indeed throughout this book, female nonhuman primates are not dependent upon males for sustenance or social cohesion, and there is no empirical evidence they are in any sense "owned" by the males of the group.

Finally, I would like to present briefly two examples of how differences between the sexes may be labeled or described in such a way that human females are perceived as passive or inferior. Neither of these originated with Hutt (1972), but they are both reiterated and supported in her book. The first concerns the finding that human males are more stimulated by visual and narrative erotic material, while human females are more stimulated by tactile interaction. Money's (1963) description of this states: "males are more readily *responsive* than are women to visual and narrative material, women being *dependent* upon tactile stimulation" (cited in Hutt 1972: 83, emphases mine). As a result of this description males will be thought of as readily active and females as passive. Now, if we were to switch the adjectives in the quotation, males would become dependent upon visual material and females responsive to tactile stimulation. The connotations would be reversed while the statement would remain perfectly congruent with the experimental findings. The best, unbiased expression of the results would have both females and males as either responsive or dependent.

The second example concerns the finding that in Euroamerican populations males tend to exhibit a wider range or greater variability of scores on many psychological, social and physical measures, such as I.Q. tests, height, academic examinations, social deviance, and so on. Heim (1970) has said that:

> men rather than women are found at the extremes. There are more male geniuses, more male criminals, more male mental defectives, suicides and stutterers, more color-blind males than females. This list is a long one, with relatively few exceptions. (cited in Hutt 1972: 90)

Whilst this of itself might be considered an intriguing finding, amenable to both biological and sociological explanations, what is curious is to find that Heim has labeled the tendency of males to exhibit a wider range of scores and females a narrower range of scores on such measures as the "Mediocrity of Women." Why not the "Inconsistency (or one of its synonyms: inconstancy, fickleness, instability, capriciousness, irregularity, inharmoniousness) of Men?" Further, the one conclusion Hutt draws from this study, perhaps resulting from Heim's label, is a good example of an aha! reaction which has nothing to do with the data alone:

The fact that males predominate in the intellectual and creative echelons seems to have a basis other than masculine privilege. (1972: 90)

What this chapter, and what these examples have attempted to show is that the choice of words or metaphors or analogies to describe behavior is far from unimportant in reflecting and determining our perceptions and our beliefs. That this is not generally accepted, even now, can be seen in Ruse's recent statement that objections to the andro-centric language of sociobiology "point to nothing that some routine copy-editing could not rapidly rectify" (1979: 93). We are not yet in an era in which the underlying biases of our thinking revealed in our language can be corrected by copy-editing, like a simple re-programming of the mind.

That we cannot re-invent our language is evident. We only can hope that through an awareness of the power of language, and care in its use, we can avoid both the propaga-tion of bias and the confusion of semantic with substantive differences described by Lehrman (1970). Unless that care extends from the simple motor categories of an etho-gram to the conceptual categories of a theoretical model, even workers in the same discipline will not be sure of sharing the same meanings, and as Scheffler comments: "without the sharing of meanings there can be no genuine exchange of understandings" (1967: 46).

Chapter *THREE*

Instinct, Innate
and Learned Behavior

Horsfield's tarsier
Tarsius bancanus
This little studied, nocturnal prosimian is found on Sumatra and Kalimantan.
They live in dispersed polygynous groups and show little or no dimorphism.

The emancipated woman of whatever nationality is the product of seventy million years of evolution within the primate channel in which status, territory and society are invariably masculine instincts, in which care of children has been a female preserve, and in which social complexity has demanded of the female the role of sexual specialist. Yet she must somehow struggle along in a human society which idealized in her behavior every masculine expression for which she possesses no instinctual equipment . . . she is the unhappiest female that the primate world has ever seen, and the most treasured objective in her heart of hearts is the psychological castration of husband and sons.

Ardrey 1961: 165

Do males dominate females because males are instinctively more aggressive than females? I have been asked this question in a variety of forms, including the rhetorical, and still find that it is essentially unanswerable because the question involves one misconception and two oversimplifications. Aggression and dominance are concepts which have been grossly oversimplified in most popular and in some scholarly writing, whilst that of instinct could more properly be called a misconception. Aggression and dominance will be discussed in the following unit, but here I shall try to show how instinct has been misconceived. I usually attempt to answer what I consider to be the intent of the opening question, that is: what is the relative importance of inherited and experiential factors to sex differences in behavior? But, in order to do this I need to begin with an explanation of the terms, "instinct," "innate" and "learned."

Instinct is, for many people, an example of a word with powerful, often pejorative, connotations because of its implications of behavior which is inevitable and inalterable. For many, instincts are inescapable, to be suffered, to be driven by; at best to be controlled by constant vigilance, at worst to be embraced or flaunted as a Faustian destiny. Instincts have been variously portrayed as raging forces within us we must constantly strive to subdue, or as deterministic forces which can only be deflected by the intervention of a powerful outside agency. In the most hopeful construction, instinctual forces may be capitalized upon, as the boxer with the "killer" instinct or the "born" artist. So it is not surprising that in the first half of this century the study of instinct became inextricably tied up with the nature/nurture controversy, which itself became linked to political views concerning solutions to social problems:

> Even within the academy, social biologists have typically preferred conservative solutions of social problems, while environmentalism has been the traditional ally of the liberals. Pastor's study in 1949 of twenty-four disputants in the "nature-nurture" controversy revealed that of the "hereditarians" only Lewis B. Terman qualified as a liberal, while John B. Watson was the sole environmentalist of conservative orthodoxy. (Bressler 1968: 180)

More recently, popular ethology has introduced many people to the idea of "aggressive instincts" and "territorial instincts," in what has been interpreted by many as an effort to assuage our feelings of worry about violence in urban societies, and as an effort to reinforce the dominant position of males in these societies. The message, to many critics, seems to be that violent pastimes, war and crime are the natural and necessary result of man's instincts (male man's that is, since females have no active role in this view of life). Thus:

> The Lorenz-Ardrey type of interpretation . . . is being used at present to justify our feelings of despair, attitudes of helplessness, and pessimism. If man is a creature with ineradicable traits that make him murderous, carnivorous, continually prone to war, and in the extreme, cannibalistic, then there is no need to do anything about him at all. We are stuck with him. We are going to continue to have a world that is unmanageable, whose inhabitants may destroy it. (Mead 1971: 373)

28

At the other extreme, there have been writers who claimed that inherited factors count for nothing compared to the effects of environment and learning in humans. Modern reformulations of the "tabula rasa" concept, the blank slate model of the human organism made famous by Jean Jacques Rousseau, still are to be found, especially in the literature of education and behaviorism. The opposition of instinct to learning, of genes to environment, or the "nature/nurture controversy," as it has been called, has a long and complex history but is best considered a false dichotomy and even something of a red herring. There are now, and have been for some time, many biological and social scientists who argue that behavior is strongly influenced by the *interaction* of environmental and biological factors, As many textbooks note, no scientist today argues that behavior is entirely biologically determined or entirely environmentally determined. Scientists have retreated from these extreme positions, if they ever held them at all, and have come to one which states in effect that all behavior is the product of environmental and heritable factors. Or, in Hinde's succinct statement "No behavior depends only on genes or only on the environment" (1974: 41). Indeed, environment and biology interact at so many levels in the production of behavior that it is often not possible, and some researchers (e.g. Piaget 1974) would argue, not conceptually useful, to attempt to separate them.

But does an end to a falsely conceived nature/nurture controversy mean that scientists now know exactly how different elements interact to produce behavior? No, it does not, and a reading of different authors on causes of behavior will quickly show that they ascribe varying degrees of importance to different factors in the development and production of behavior, and that they have widely differing views on just how flexible or labile behavior is in different species. To some extent we are still like the proverbial blind men describing an elephant; those researchers most concerned with, or most closely associated with, biological aspects will tend to emphasize and extrapolate from biology, while those with interests in learning and experience will tend to give emphasis to environmental factors. There are scientists working very hard toward a holistic view of behavior, toward a comprehensive description of the whole elephant; but it is a massive undertaking, for behavior is a large and complex phenomenon, indeed.

One of the ideas fallen by the wayside in the search for a more sophisticated understanding of behavior is the concept of instinct. Some have gone so far as to describe it as a "disreputable" concept (Hinde 1974), others have simply dropped it quietly from their vocabulary. However, as examples of a contrary point of view, see Brown (1975) and Cassidy (1979). The term, instinct, usually referred to a biologically determined drive to perform a specific behavior or specific type of behavior. It was used a great deal in early ethology and psychology to explain what occurs inside the behaving animal, its motivation, how it feels. Instinct was often associated with an energy model of motivation, which postulated that there are different energies in the body, specific to certain actions (sexual energy, aggressive energy, etc.), and that these energies build up and must be discharged (e.g. Lorenz 1950). Hence the animal is *driven* to do certain things.

This idea has become outmoded by a more thorough understanding of physiology and a more sophisticated approach to behavior. There is no evidence for action-specific energies driving and determining particular behaviors. The energy model of behavior is actually older than the study of animal behavior or instinct. Perhaps it is an intuitively satisfying model, as it seems to live on in our notions about behavior. For example, a runner told me that he ran to use up his aggressive energies, and we often talk about "letting off steam."

29

The word "instinct" has been largely replaced by the term "innate behavior," but this is not merely a matter of vocabulary change. Innate refers only to inherited potential or biological possibilities. This involves an important shift in emphasis from biological determinism to biological potential because innate is a relative not an absolute term, and indicates the degree of variability we can expect in a given behavior pattern in different environments (for a short but informative discussion of this difference see Gould 1977). But if all behaviors are the result of inherited and environmental factors, how can we determine the relative effects of each of these on each specific behavior? This is essentially the question with which we began this chapter.

There is no direct evidence which would allow us to define a behavior as being, for example, 80% gene-determined and 20% environment-determined, nor is there any generally applicable method for making that determination. As Hebb perceived more than 25 years ago:

> This is exactly like asking how much the area of a field is due to its length, and how much due to its width. The only reasonable answer is that the two proportions are one-hundred per-cent environment, one-hundred per-cent heredity. (1953: 44)

The best one can do is develop an inferential or circumstantial case for relative causation. That is, behaviors can be conceptualized as being on a continuum from very stable (even stereotyped) to very labile or flexible patterns. So that, if a behavior occurs in all members of a species (or all members of one sex, or age group, or order of animals, etc.) in the same or similar form, in spite of different environments, it is considered to be "stable" and one can infer that it is closely, innately circumscribed. If however a behavior is varibale (labile or flexible) in different individuals and/or environments, it is considered more environmentally influenced, that is, a greater range of expression is allowed by the genetic structure. For humans, it can be imagined as a spectrum, from behaviors such as simple reflexes like the eye blink and the knee jerk (which are thought to be very, but not totally, innately circumscribed), to complex cognitive tasks, such as comprehending the interactions of biology and environment in which we presently are engaged. The drawback to this inferential method is that similar behaviors are not always the result of the same genetic causes, and the same genetic instructions can have different results in different environments. Thus the stable-to-labile continuum is better viewed as a relative or probabilistic indicator than as an absolute scale.

Alison Jolly has stated the case similarly: "The relative innateness of a trait thus depends on the range of environmental conditions sufficient and necessary for the behavior to appear" (1972: 138). This means that rather than determining its inevitable occurrence, innate factors influence the potential for a particular behavior to appear. The more consistent in their influence the inherited factors are, and the narrower the range of phenotypic possibilities suggested by the genotype, the more likely that the behavior will appear whatever the environment. For example, nesting animals such as some birds and rodents may continue to attempt to nest in cages even in the absence of any appropriate nesting material, indicating that in this case the behaviors involved in nesting are quite stable, inflexible and closely innately circumscribed. This brings us to an important point about "the environment."

There is much more to environmental influence than just "learning" which is often defined for animals as a change in behavior resulting from experience. The social and the physical environment, in all its multifactorial aspects of appropriate companions, adequate nutrition, suitable material surroundings, and so on, is the necessary context for the appearance of behavior. The right mixture of environmental factors is necessary for the ontogeny of behavior in the individual. Ontogeny refers to the development of individual organisms and is distinguished from phylogeny, which refers to the descent, development or evolution of a species. However, closely innately circumscribed behaviors are not as demanding and specific in their environmental prerequisites and so they appear more stable. Other behaviors require more specific environmental conditions in order to develop, and since environments vary, so do these behaviors.

Viewed from this perspective we can see that it may not be a useful exercise to attempt to assign percentage weights to inherited versus environmental determinants of behavior. However, researchers have devised various experimental techniques which do attempt in some respects to tease out the intertwining strands of environment and inheritance, as they interact in the development of behavior. Deprivation and learning studies will be discussed in Chapters 12 and 13; here I will limit myself to describing what many scientists consider to be the experimental technique offering the best demonstration of genetic and environmental causes, by careful control of one or the other.

To control for genetic influences, researchers take individuals of identical genetic composition and rear them from birth in different environments. If differences in behavior appear, they may be ascribed to environmental effects. If similarities in behavior are maintained, they may be ascribed to genetic effects. Conversely, one can take individuals with different genotypes, raise them in identical environments, and assign any ensuing differences to genetic causes. By holding either the environment or the genotype constant, and systematically varying the other, one can attempt to trace the source of variability (Marler & Hamilton 1966). This method of experimenting with behavior and this perspective on behavior, has led scientists to the conclusion that it is not individual behavioral characteristics which can be called inherited or learned, but the *differences* between characteristics which may be referred to as inherited or learned. These factors are seen as sources of variability in behavior rather than as determinants of specific behaviors. An analogy by Kummer explains this difficult concept very well:

It takes a drum and a drummer to produce a sound. Nobody would try to differentiate between sounds produced by the drummer and sounds produced by the drum. But we can very well discuss whether two recorded performances sound different because of a new drummer or a new instrument. (1971: 12)

In theory this technique of holding either the environment or the genotype constant, is an ingenious way to sort out different causes; in practice its application is very nearly limited to Drosophilia (fruit flies) or other equally fast-breeding laboratory animals. Completely controlling genetic or environmental variation in primates, as will be seen in the description of the deprivation studies, is virtually impossible. The closest efforts to this technique in primatological studies may be those of Kaufmann and Rosenblum (1966, 1969a, b; Rosenblum 1971a) and Bernstein (1971b), in which different species of primates were raised in identical laboratory colonies and caged situations and their behaviors carefully compared. Differences in behavior do appear; in the comparative study by Kaufmann and Rosenblum bonnet macaques *(Macaca radiata)* huddled in body contact with all group members in one area of their cage, while pigtailed macaques *(Macaca nemestrina)* spaced themselves out around their cage and interacted preferentially with kin. One may conclude that the differences are to some extent due to inherited differences between the species. However, as these researchers know, some caveats are necessary — local social traditions which are not inherited, such as huddling, may become established in some groups, and no two laboratory colonies of primates are ever truly identical, either in genes or in social environment.

Earlier in the chapter I noted that scientists still do not know a great deal about the details, or have not agreed upon the details, of exactly how the biological properties of the organism interact with environmental elements to produce behavior. It is agreed that the process of interaction is very complex and that it often differs from behavior to behavior and species to species. But none of the conceptual schemes or experiments described thus far gives, I suspect, any real idea of how the processes which result in behavior are currently visualized. Thus I present the following generalized description, which while greatly simplified, still does justice, I hope, to the present understanding of this complex topic.

A given behavior may be said to be directly influenced and circumscribed by the present environmental context, by past experiences and by the physical aspects of the phenotype (sense organs, morphology, hormonal status, central nervous system, etc). Obviously, an animal only performs actions within the capabilities and predispositions of its nervous system, as well as other aspects of its physique and physiology. Behaviors are often triggered, in an immediate sense, by an external stimulus (for example, see the discussion of aggression in Chapter 6), but individuals are predisposed, to varying degrees, to respond to certain stimuli with certain behaviors. What is sometimes referred to as the "neural substrate" is particulary important. As noted in Chapter 2, neural substrate is a useful metaphor in the sense of a "framework" for behavior, but not a very accurate one in its implications of "lying underneath." This neural framework for behavior, and other aspects of the body which influence behavior, have in turn been developed through a multiple-feedback system involving environmental factors (e.g., nutrition, disease, microorganisms, other aspects of material and social surroundings), and biological properties of the organism, especially genetic directives. These are often called the "genetic substrate" of behavior, although genes do not really lie at the bottom or core either, because they in turn have been influenced by the environment through the process of natural selection.

Thus we have what might be visualized as multiple feedback loops: environmental pressures acting through natural selection, select for certain genetic material, and the genotype sets the directions and the limits for the development of the body. The most appropriate metaphor for the genotype is not that of a "blueprint," but rather something like an "information-generating device" (Klopfer 1973a,b; see also Archer & Lloyd 1974). In turn, the physical and physiological capacities of an animal set the direction and the limits of the expression of behavior. Further, the expression of behavior may itself influence the selection of the related genetic material for the next generation. It is important to emphasize that the route between genes and behavior is neither simple, direct, nor linear. In addition, while there is evidence of genetic bases for behaviors, there is no evidence of specific genes for specific behaviors, because behaviors are processes and not material structures. Or, as Klopfer has said:

> We tend to treat behavioral phenomena as if they were palpable, forgetting that "behavior" is not best regarded as a noun . . . the hope of finding an instinct in a chromosome is in any event illusory. (1973b: 59, 60)

There is a complex interplay betwen the genes themselves, the developing biological structures, and aspects of the environment. One result of this interplay is behavior. In this sense, behavior itself does not evolve and is not inherited; the material biological structures which influence, direct and circumscribe behavior are evolved and inherited. The expression, "evolution of behavior" is, in essence, a metaphor taken from the process by which proteins are produced (Klopfer 1973a,b; Lewontin 1974). Of course this does not mean that biology or evolutionary processes are unimportant to behavior, but it does mean that the relationship of genes to behavior is rather more complex and fluid or more "loosely coupled" (Klopfer 1973b) than is sometimes portrayed. And it means that when scientists refer to the genetics of behavior or to the inheritance of behavior, they are using a verbal and possibly mental shortcut of which we should be aware. The actual, and the correct reference, is to the genetics and inheritance of the chromosomes which influence the development of the corporeal structures, which in turn influence behavior. Or, as Count describes it, "Behavior is what is observed to occur, evolution is what happens to a potential for behavior" (1973: 9).

Several good examples of how biological and environmental factors interact in the behavior of primates can be found in studies of primate communication. For example, R.E. Miller et al (1967) carried out some very interesting laboratory experiments on monkey communication skills. They linked up monkeys two at a time through closed-circuit television such that monkey #2 could see the face of monkey #1. Monkey #1 in turn, was able to see an approaching stimulus, either food or an unpleasant electrical shock for both monkeys, but he had no control over the impending stimulus. Monkey #2 could not see which of the two stimuli was coming but he could see the face of #1 and he could press levers ensuring they both received food or both avoided shock. Socially-raised monkeys were able to communicate effectively to avoid shock and receive food, but isolate-reared monkeys neither sent nor responded to the appropriate facial expressions. The results of this experiment are even more striking when one considers that fear is normally a readily communicable emotion in animals. Indeed, perception of and response to fear in others is a basic survival skill for social animals.

Much research has established that primates reared in social isolation from their own species exhibit severe social deficits, particularly communication deficits, even though they may be capable of the basic species communication signals (see Chapter 13). Why is this so? My first insight into this problem came when I noticed that one of the geladas (*Theropithecus gelada*) I was studying on a zoo monkey island was the object of frequent attacks and chases by other monkeys. His behavior struck not only me, but even casual zoo visitors, as "eccentric." I learned that he had been hand-reared by zoo keepers before being introduced to gelada social life. He was capable of the very characteristic lip-roll or lip-flip gesture which is only seen in this species, as well as some general Old World monkey signals, such as "Brauenziehern face" (elevated eyebrows, covered teeth, usually interpreted as a threat), to which he added a few unique gestures of his own, such as crouching on his hind legs while biting the end of his tail. In spite of his basic capacity for some physically correct signals, he did not appear to use them appropriately in a social context. He often approached younger animals or occasionally even the highest-ranking male on the island with exaggerated fear expressions followed by exaggerated threat expressions followed by screams, (and sometimes his own unique tail bite) and then a hasty retreat. As a result, he was the object of frequent attacks or often was left completely alone. Although geladas live in uni-male, multi-female groups or sometimes all-male groups, he was always a lone male and was never seen to interact affinitively with other geladas.

A great number of deprivation experiments indicate that at least some monkey communication signals are very stable and closely innately circumscribed, because they appear even in isolate-reared animals. However, the ability to use these signals appropriately to communicate, especially the understanding of proper context, seems to be acquired through social learning. Just how sophisticated the knowledge and how subtle the use of proper social context can be in non-human primates became apparent to me when I observed a young Japanese monkey (Macaca fuscata) suffering from what may have been a case of spastic cerebral palsy (for a complete description see Fedigan and Fedigan 1977). In spite of the fact that his vision and his motor-development were severely impaired, this individual survived for thirteen months, without human interference, in a large, free-ranging social group, and finally disappeared quite suddenly, probably left behind during a foraging trip. Like the gelada described above, this young handicapped monkey was capable of some basic monkey communication signals, but he was largely incapable of the integration and appropriate use of these signals necessary for social interaction. Other than interactions with his mother, this monkey generally failed to perceive communications or to respond to them correctly. Nor did his abilities improve over time, as did those of his peers. Using a model of human cerebral palsy, we might say that his physical handicaps resulted in a form of learning disability. He was an infant and received little or no aggression from troop members, but it seems unlikey he could have continued to live in his group without developing better social skills. His failures, such as never appearing to learn the complex web of relationships of rank, family and roles, taught me a number of things about normal monkey development. By the age of six months to one year, most young macaques are deft social manipulators in that they have become skilled at integrating signals to give finely discriminated messages in the appropriate context. Also, they are equally adept at perceiving complex communications and multidimensional situations, and reacting to them suitably, a skill we might call "social perception."

While the examples given above tend to highlight the importance of social learning to successful primate communication, the following example illustrates the significant consequences of genetic and morphological limitations for primate communication. Until two psychologists, Alan and Beatrix Gardner, achieved their spectacular success in teaching a chimpanzee to communicate with them via American Sign Language (1971), several researchers had made long and strenuous efforts to teach young chimpanzees to speak, even going to the extent of rearing chimpanzees in their homes, sometimes with their own children. All these attempts ended in failure. In one study a chimpanzee, named Vickie, was able to produce three or four poorly-articulated words, with great difficulty, after six years of training. Meanwhile, her scores on non-verbal tests of intelligence were very high (Hayes & Nissen 1971). Several reasons were offered for the failure of chimpanzees to learn to speak: for example, that their vocal apparatus was inadequate (their tongues are said by some to be incapable of the rapid changes in shape needed for the quickly alternating sounds of speech); or that they were incapable of the cognitive processes involved in symbolization and abstract thinking.

Today, some research indicates that what chimpanzees are incapable of is not abstract thinking nor the production of speech-sounds per se, but *volitional* control of the sounds they produce (Myers 1978). It appears that vocalizations in nonhuman primates are controlled by the limbic system of the brain, a phylogenetically old or "primitive" part of the brain, and are inflexible and directly related to emotions. As Washburn and Strum have expressed it: "Attempting to teach a monkey to make more sounds is like trying to teach it to have more emotions" (1972: 475). However, not all researchers in animal communications would agree with even this interpretation (Gouzoules, personal communication). The capacity for speech is dependent upon the ability to alter the sounds emitted as a result of exposure to auditory models. Only humans and birds are known to be capable of this, although the ability in the two cases is related to very different neural mechanisms. In humans, speech production is dependent on specialized neural structures in the neo-cortex of the brain. Humans still do produce emotion-related sounds, such as cries of pleasure or of pain, but their primary sound production method is neocortical and volitional.

Why, then, did the Gardners succeed in teaching their chimpanzee to communicate with them? Probably because they chose a communication system which, since it was developed for deaf people, relies on gestural-output, visual-input. Chimpanzees and monkeys have excellent volitional control over the use of their hands (Myers 1978), and they are skilled at perceiving complex visual messages. All of the various language experiments with chimpanzees which have been successful (Premack 1971; Rumbaugh et al 1975; Fouts 1975) have relied on communication systems which require use of the hands, be it to place magnetized symbols on a board, to press computer buttons, or to form hand signals. Because chimpanzees can deliberately control the use of their hands, they can learn new hand uses. It appears that hand use is a flexible, labile behavior pattern, while vocal production is a more stable, inflexible pattern. What language experiments exemplify beautifully is that behavior, and here we are interested particularly in learning or the acquisition of new behaviors, occurs within the species-specific limitations set by the genome.

Primates learn more than any other order of animals, and humans learn more than any other species, but even the general reliance on learning is dependent on the genetic endowment of the species. A primate, such as a macaque or a chimpanzee or a human cannot rely on learning for survival, unless it has the biological system to support and allow for modification of behavior in relation to previous experience. Thus the study of learning is inseparable from the study of biology, and the study of primate biology can help us to understand flexibility in primate behavior, as well as stability.

The above examples have been given in some detail in order to illustrate why the study of behavioral causation is a fascinating, as well as a thorny issue, and why it requires all the reasoning of a logician, the inventiveness of an experimenter and the delicate skill of a surgeon to attempt to tease apart the interlaced forces which influence behavior. As well as fascinating scientists, the study of behavior and its causes continues to be of intense interest to the general public and to generate a great deal of controversy. Partly, this is due to the desire on the part of the public, and on the part of some scientists, to find simple, definitive answers to questions which are complex in the scientific sense, and complex in their sociopolitical implications as well. Partly also, it is due to the fact that the questions are posed by people, scientists and lay people alike, who are themselves social animals, whose social learning leads them to frame the questions they ask about behavior in the light of individual and social preferences.

To the regret of many scientists, no area of science is independent of these social and political overtones, and especially not one which relates to the mutability of behavior at a time when some people would like to see behavioral changes on a large scale in our society. The essential point of this chapter is to illustrate that many factors interact at all levels to influence behavior. It is inappropriate to think of "biology" as a totally inflexible and structured central core, or to think of "environment" as a superficial veneer. Not all comments are as blatantly androcentric as that of Ardrey, quoted at the beginning of this chapter, but unfortunately the behavioral literature is replete with overt statements and covert implications that humans, and in the case of this quotation, human females and males, behave as they do today as a result of the "biological roots of behavior." Woe unto anyone who would attempt to pull up these "biological roots" (see, for example, Crook 1972; R. Fox 1972; Hutt 1972; Wilson 1975). I would suggest that if we are to use this metaphor, then behavior should be seen as a complex and hybrid plant, from its roots to its fruits; and neither biological influences nor environmental influences are prior, simple, or immutable.

I have suggested in Chapter 1 that sex differences in behavior, especially in humans, are as yet poorly understood or documented, and that we are not in a position to describe a general "female nature," much less a "human nature." In the nineteenth century, many scientists warned of dire consequences to the family, and to women, if females were allowed into the workforce or institutions of higher learning; arguing that females had neither the mental capacity nor the physical stamina (i.e. the "biology") for such masculine endeavors. Similar spurious warnings from scientists in this century predict that a continued press for equality of opportunity for the sexes will produce a strain on and a disruption of female nature, which, it is said, is basically "nurturant motherhood," and which could not survive in a competitive male world. Consider the following statements:

> The biological roots of human sexual behavior are different in the two sexes and likely to be fundamental determinants in the differentiation of mature personality . . . Clearly the role of women must undergo major changes in modern society . . . Nevertheless the remodeling of society according to unwise policies could impose excessive strain on many individuals of both sexes. There may well be a limit beyond which the trends affecting women in our currently male-oriented educational systems could produce highly disruptive results. Those concerned with Women's Liberation would be wise to ponder the biological and psychological complementarity of the two sexes and their deep emotional needs for partnership as a counter to the notion of a poorly defined "equality." (Crook 1972: 274, 275)

It would be a pity indeed if women sought to make this less a man's world by repudiating their femininity and by striving for masculine goals. It seems that few women can or wish to compete in the competitive, assertive spheres of the male . . . Nurturance, whether in the wider educational sense or in the narrower domestic sense, is and will remain women's forte. (Hutt 1972: 138-39)

In hunter-gatherer societies, men hunt and women stay at home. This strong bias persists in most agricultural and industrial societies and, on that ground alone, appears to have a genetic origin. No solid evidence exists as to when the division of labor appeared in man's ancestors or how resistant to change it might be during the continuing revolution for women's rights. My own guess is that the genetic bias is intense enough to cause a substantial division of labor even in the most free and most egalitarian of future societies . . . Thus even with identical education and equal access to all professions, men are likely to continue to play a disproportionate role in political life, business and science. But that is only a guess and, even if correct, could not be used to argue for anything less than sex-blind admission and free personal choice. (Wilson 1975b: 49-50).

Such warnings, even when presented as just "a guess," by established scientists, writing as scientists, with their confounding of causes and consequences, sound like little more than apologias for social inequalities. They are no more objective, so should be given no more credibility, than those of the last century. It is informative to note how Crook's view of sex roles appears to have changed over the years, apparently not as a result of new scientific data (compare Crook 1972 to Crook 1977). Nor will the forced acceptance of supposedly "complementary but equal" roles be any more beneficial to women than the policy of "separate but equal" institutions has been to ethnic minority groups. We do not know anything yet about the outer boundaries or possibilities of "female nature." Given radical changes in biological and social phenomena, such as the possibility of more women spending less of their lives bearing and raising children, who can predict what behaviors may become characteristic of women?

That aspect of evolutionary theory concerned with the adaptiveness of behavior tends to operate far better as an "a posteriori" explanatory device than as a predictor of future events. That is, it offers explanations of why existing behavioral and morphological structures may be adaptive in their present environments, or were so in past contexts, rather than predicting what structures will exist and be adaptive in the future in different environments. As Martin has said:

. . . most hypotheses about the biological bases of human behavior are selected according to their *plausibility* rather than their *testability* . . . it is apparent that we have not progressed far beyond the "satisfying explanation syndrome." (1974: 230)

If an hypothesis is developed explaining why a given behavior exists and is adaptive, even if we accept the hypothesis as plausible, it does not automatically follow that alternative behaviors are not possible. Nor does it follow that the given behavior will always be adaptive. To argue that an individual is potentially nurturant or potentially aggressive given the appropriate environment, is not the same as arguing that the individual is inevitably nurturant or inevitably aggressive given any environment. However, many discussions of biological influences on female and male behaviors appear to assume the two arguments are the same. Inherited factors may provide the potential and even the predisposition for behavior, but they do not determine the inevitable occurrence of behavior. Humans have (through their intelligence, technology and other aspects of culture) often stepped beyond the boundaries of apparent biological limits. No scientist would argue that humans will commit an offense against "nature" with resultant unhappiness, if they fly in the air or skindive under the water because humans did not evolve over hundreds of centuries with wings and fins. However, this is precisely the sort of reasoning that is used by some scientists with respect to changes in female and male roles.

Margaret Mead is reported to have remarked that one of the wonders of biology is that it can be used to justify any point of view. I find this an encouraging, rather than a depressing thought. Biological arguments have certainly been used to justify the niche women occupy in modern Western society, and, by extension, why they should remain in this niche; but biological information need not always be conservative in implication.

Consider the implications of the human brain and the primate neural capacity for learning. These biological phenomena are evidence of the tremendous range of possible behaviors in non-human and human primates. Consider the complexity and the pliability of the route between genetic material and behavior in animals. And we have not yet considered the interaction of human culture with the human phenotype. People who wish to implement significant behavioral changes and societal changes have nothing to fear from the data of biology. It is the distortion of biological potential into biological inevitability that is conservative, not the study of biology. As Scott so succinctly phrased it: "to explain human behavior by saying 'I do it because my ancestors did it' is at best an untested hypothesis, and at worst a poor excuse" (1974: 430).

Males do "dominate" females in our society because their ancestors did. Like their ancestors, they maintain the political system or the structure of power which gives them this control. But are they enabled to do so because they are "innately more aggressive" than females? The question, as we have seen, is unanswerable in that general form, and in fact is something of a red herring. Behavior is the result of interactions in which social learning can be as important as the physical and genetic endowment of the species. And since aggression, as we shall see, is a constellation of behaviors with multiple definitions and multiple causes, it is unlikely that there is any general trait we can label "aggressiveness," and so no meaningful way in which we can make such comparative statements. That the possibility of greater innate male aggressiveness resulting in greater male dominance, continues to be raised, may reflect only the discomfort of some at the realization that, in fact, men contine to "dominate" women for no necessary "biological" reason. As Kate Millet aptly notes, and Gould reiterates in his essay on biological determinism: "Patriarchy has a tenacious or powerful hold through its successful habit of passing itself off as nature" (Millet 1970: 58).

Chapter FOUR

Primate Societies

Gibbon
Hylobates lar
Agile and arboreal, these lesser apes are fairly widespread in Asia. They show little or no dimorphism.

Primates are notably gregarious animals. With few exceptions, primates are found in social groups which endure through the seasons and over the years. Even the most "solitary" of primates, the nocturnal prosimian species, are now recognized to live in dispersed polygynous groups. In this type of social system the adult male may not normally travel with adult females, but his range overlaps several female-and -offspring ranges, and the adults communicate and monitor each other's activities through vocal and olfactory signals (Charles-Dominique 1971, 1977). Most primate.species however do live in cohesive groups within which individuals travel, eat, sleep and interact in close visual proximity and coordination. For the average primate, all activities take place in a social context, and as Hall (1968b) emphasized years ago, all learning is in essence social learning. There are of course many ways to be social, and many types of social systems. Describing, classifying, and understanding the mechanisms which underlie the myriad social organizations found in prosimians, monkeys and apes has been one of the major goals of primatologists. This chapter will introduce definitions of some essential terms, describe some of the attempts at classifying nonhuman primate social systems, and discuss several of the major problems and controversies that have challenged, stimulated, and frustrated scientists engaged in the study of how and why primates live in groups.

TERMS

For most species of primates, social groups are not too difficult to recognize or to define. In these species, a social group shows lasting physical cohesion, coordination of activities between group members and spatial separation from outsiders, with most interactions, especially those related to reproductive processes, taking place between group members. Primate societies are often described as "semi-closed" reproductive units. While some individuals, especially adolescent males, may occasionally leave or join groups, most individuals are permanent group members and show hostility toward outsiders or strangers attempting to enter the group. Recruitment of new group members still does occur through births and through the usually difficult process of integrating sexually mature individuals who have transferred from other groups. Thus a primate social group does endure over time and over generations, or as Mason (1976) has said, in spite of a turnover in personnel.

The variability that characterizes the order Primates seems to provide at least one exception to every rule we might care to construct, and therefore a few species exhibit social systems in which the "social group" is not easy to recognize or define, and for which the social group may not always be a useful concept, in spite of the fact that the animals do live socially. Examples can be found in some *Cercopithecus* species in the African rain-forests, and some New World monkeys in the South American rain-forests which not infrequently travel in large mixed-species aggregations, although reproduction and other interactions occur almost entirely within the species. In this case is the "social group" the larger mixed species aggregation, or some smaller subunit consisting of only one species? Other problems of social group definition arise with chimpanzees and spider monkeys, who travel and interact in small fluctuating, temporary associations, but are closely familiar with a larger "community" of individuals. Then, too, geladas and hamadryas baboons *(Papio hamadryas)* of Ethiopia live in societies with three levels of social groups which are constantly fusing and fissioning.

A number of these social systems will be discussed in some detail in the chapters which follow, but the terms which have been developed to describe them can be briefly mentioned here. In geladas, the smallest, and the "core" social group consists of one adult male, several adult females, and their offspring. This type of social group was originally referred to as a "harem." This term largely has been dropped because of its anthropomorphic (and androcentric!) connotations, and such a social group is now called a polygynous group (see below) or a uni-male, multi-female reproductive unit. Several of these small polygynous groups in the geladas may associate together in what has been called a "clan," possibly linked through female kinship ties. In turn, many clans may sometimes join, to form the third and largest social grouping, called a "herd" (Dunbar & Dunbar 1975). Hamadryas baboons also exhibit three levels, which are known as uni-male groups, troops, and bands respectively. These social systems will be further discussed in Chapter 15.

For chimpanzees and spider monkeys, the small temporary associations of individuals who travel together are referred to as "parties"; while the larger and permanent but dispersed body of individuals who are familiar and non-hostile to each other is known as the "community." Another term should be mentioned here because it occurs commonly in the primate literature and that is "troop." This word is sometimes used to describe the large societies found in some ground-dwelling primate species, but is actually synonymous with "group." Indeed, although a few specialized terms have been described in this section, the term "social group" and the definitions given above will usually suffice in describing primate societies.

Throughout this book I will use the terms multi-female, multi-male or polygynous, or monogamous to describe various social groups. In animals, *polygamy* refers to mating with more than one individual of the opposite sex, or "multiple mating," and can be broken down into two possibilites: polygyny — one male mates with several females; or polyandry — one female mates with several males. The latter pattern in which one female mates, lives and/or raises offspring with several males does not occur in nonhuman primates, although there are rare examples of this system in other animal societies. Polygyny, in which only one male lives and mates with several females occurs in a number of primate species. Monogamy, in which one female and one male live and mate only with each other, also occurs in nonhuman primates and the couple are also referred to as a "mated-pair." Yet another type of primate mating system, which is believed to be fairly common in occurrence, has no satisfactory or agreed-upon name, but exists in multi-female, multi-male societies. In such groups, several females and several males live together and mate together. Such a system has sometimes been referred to as "promiscuous," but as Selander (1972) has noted, this word means mating indiscriminately or at random, and since individuals in such societies do show evidence of mate choice (see Chapters 10 and 17), the term promiscuous is inappropriate. In the absence of an appropriate descriptive term agreed upon by primatologists, I will simply refer to multi-female, multi-male or polygynous or monogamous social systems.

I should also explain here that these terms refer to different phenomena in animals than in human societies. In animal societies, polygyny, polyandry and monagamy, refer to different *mating* systems, whereas in the study of humans these terms refer to *marriage regulations*. Marriage is not merely a system of mating, but a social, political, and economic institution or contract between individuals, with implications reaching far beyond reproduction. It would be inappropriate to attempt to reduce the phenomenon of marriage, or the legal union of wives and husbands, simply to patterns of mating, such as are found in animals. Some social scientists (e.g. Leibowitz 1978) have argued the same for the term "family" which is also a political and economic institution in human societies, rather than merely a reproductive unit of mates and offspring.

Nonetheless it is acceptable to use the terms polygyny and monogamy to refer to mating systems in animals, since these words do have a long history of use in this context. So long as we remain aware that the words have diverse, special meanings for anthropologists, zoologists, and even botanists, they can be used to describe nonhuman primate mating systems.

TAXONOMIES OF PRIMATE SOCIAL SYSTEMS

Before the advent of any extensive primate field research, scientists largely believed that all prosimians, monkeys and apes lived in only one type of social group consisting of a single male, several females and their young. However in the early 1960's, as information from many field studies began to suggest the great diversity of primate social systems, primatologists began their first attempts to make sense and order of this variability, and to classify social organizations into a few types or grades of patterns.

One of the early and more influential attempts was contained in a paper by Crook and Gartlan (1966), in which they categorized the social systems of approximately thirty primate species into five adaptive grades. These grades were based on ecological and behavioral characteristics of the animals, such as mating system, sex role differentiation, habitat, diet and so forth. Alison Jolly (1972) later modified this taxonomy of primate social organizations into a six-grade system, and her grades are as follows:

A. nocturnal primates (e.g. galagos, usually dispersed polygynous groups)
B. diurnal, arboreal leaf-eaters (e.g. howlers and Nilgiri langurs, usually polygynous groups)
C. diurnal, arboreal omnivores (e.g. gibbons and titis, often monogamous groups)
D. semiterrestrial leaf-eaters (e.g. common langurs and gorillas, usually polygynous)
E. semiterrestrial omnivores (e.g. common baboons, macaques and vervets, usually large multi-female, multi-male "troops")
F. arid-country forms (e.g. patas, geladas, and hamadryas, all polygynous groups)

While these two taxonomies based on adaptive grades were a very useful first attempt to bring order to the description of primate social systems, they also have been criticized for over-simplification and ecological determinism. It is not really adequate to characterize an ecological variable such as habitat by the simple dichotomy of arboreal versus semi-terrestrial, or a behavioral variable such as sex role differentiation, by "slight" versus "marked." Nonetheless, these taxonomies are not so much wrong, as just early, in the history of primate studies. We have yet to develop another system of classification which does do justice to the complexity and variety of primate social organizations and ecological relations, although Clutton-Brock & Harvey (1977) have conducted an extensive correlational study of ecological and social variables in 100 species (see also Chapter 5).

Itani (1977) suggested a classification system founded on what he calls two types of basic social units. The first is the "pair-type" composed of an adult male, an adult female and their offspring; the second is the "troop-type" composed of matrilineal groups and associated males. All other varieties of social organization, he believes, were derived from these two founding types.

However, the attempt to infer the evolution or derivation of one social system from another is a problem which has confounded attempts to categorize primate social organizations. For example, multi-female, multi-male groups were often described as "breaking down," or turning into polygynous or uni-male groups under harsh ecological conditions. This was widely assumed to be the origin of uni-male groups in geladas, hamadryas and patas monkeys (*Erythrocebus patas*), species which occur in sparse, semi-arid habitats. It was believed to be an adaptation to such difficult conditions for multi-male groups to develop into uni-male groups. In an important theoretical paper, Eisenberg et al. (1972) argued quite the converse: uni-male groups are a far more common phenomenon in primates than we thought, and they have only occasionally given rise to multi-male groups, the latter being a specialized and rarer occurrence. In addition, Eisenberg et al developed a hypothetical scheme for the evolution of primate social systems, based on the increasing tolerance and propensity of adult males for permanent social life, first with adult females and later with other adult males. The tendency of adult males to live in permanent social groups is one of the outstanding characteristics of primates compared to other animals, yet Eisenberg et al. suggested that the supposedly widespread tolerance of adult males for other adult males, and thus the frequent occurrence of multi-male, multi-female groups has been exaggerated. In brief, their proposed evolutionary scheme is as follows: "solitary" species, such as nocturnal prosimians show overlapping ranges of adult females and males. This could give rise to mated pair groups if the male permanently attached himself to *one* of the females and her offspring. Alternatively, one adult male could periodically or permanently attach himself to *several* females and their offspring (matrifocal unit) whose ranges his own had previously overlapped. This would result in a polygynous or uni-male, multi-female group. Such a male might come to tolerate his own male offspring in the group, at least until their adolescence, and this would result in what Eisenberg et al call an "age-graded, male society," a new term they introduced into the literature. Finally, in a few cases, the adult male might come to tolerate the presence of other fully-adult but unrelated males in the group, and the multi-male, multi-female society would result. The concept of the age-graded, male society will be further discussed in Chapter 14.

Later survey or summary papers on primate social behavior, for example, that by Southwick and Siddiqi (1974) used the ideas proposed by Eisenberg to develop schemes of primate grouping patterns. Southwick and Siddiqi suggested the following six categories of social organization:

1. solitary, except for mother-infant pairs, and mating adults as temporary social groups;
2. monogamous adult pairs with offspring;
3. single-male groups with one adult male and bonded females plus offspring
4. aggregate single-male groups consisting of several single-male units grouped into a larger assemblage;

5. multiple-male groups, or age-graded male groups, consisting of several adult males hierarchically ranked, with associated females and offspring;
6. diffuse parties without stable or consistent group formation (1974: 308).

In addition, these authors suggested that certain behavioral tendencies, such as typical patterns of aggression (rare, overt, etc.) are associated with each of the grouping patterns. As is the case with the taxonomic schemes described earlier, this plan for classifying primate social systems requires a good deal of simplication of the behavioral and ecological data. Perhaps this is unavoidable in such taxonomies, which necessarily reduce the complexity of primate social life to a few comparable parameters. The question that remains is how useful these schemes will prove to be in helping us to understand the how and the why of primate social grouping.

"Social organization" tends to be an all-encompassing and rather vague concept, itself rendering classifications difficult. For example, social organizations may vary, and thus be categorized, with respect to at least the following factors:

1. the group size;
2. the group composition;
3. the mating system;
4. the social roles, especially for adult females and adult males, in relation to socialization, resource and predator defense, group movement leadership, etc.;
5. the major internal structuring principles, be they various types of dominance hierarchies, kinship networks, age/sex class associations, or alliance systems, etc.;
6. the permanence versus instability of group membership;
7. the tendency or ability of one level of social group to aggregate into larger social groupings;
8. the presence of only heterosexual reproductive units; or the additional presence of all-male groups; all-female groups; single individuals;
9. the pattern of interactions — who does what to whom, how often and under what circumstances?

Thus one could suggest primate societies be classified according to any one of these variables — according to group size (large troops, medium-sized groups, small units) or according to group composition (multi-female, multi-male groups; uni-male, multi-female groups; uni-female, uni-male groups) or according to mating system (multiple mating by both males and females, polygynous, monogamous) and so forth. Clearly to some degree the choice of a classificatory device is arbitrary. Therefore, simply because a choice has to be made, rather than out of a conviction that it is a better classificatory principle, I have decided or organize my description of selected primate societies in Chapters 14, 15, and 16, largely in reference to their mating systems. Group size and group composition tend to some extent to be correlated with mating system. Monogamous groups are small, normally containing only one fully adult female and one fully adult male and their offspring. Polygynous groups contain one adult male and several adult females and offspring and are moderate in size. A system of multiple matings by both females and males or multi-female, multi-male groups, are usually the largest of primate societies. However,

44

mating systems do not necessarily predict or correlate with other aspects of social organization, such as roles, internal structuring principles, and patterns of interaction.

SOME CAUSAL FACTORS IN SOCIAL VARIABILITY

Although differences between the social systems of different species were recognized early in the history of primate field studies, each species was still believed to exhibit a fixed, species-wide type of social organization. It is perhaps not unjust to suggest that the attitude was, "when you have seen one baboon group, you have seen them all." Such typological thinking, as it is sometimes called, is, after all, characteristic of most new disciplines. However, it was not very many years later that the second wave of field studies of the same species, whether common baboons, langurs, or vervets, revealed striking differences in the descriptions of social patterns from the first set of field studies. Although methodological differences between researchers is one highly plausible explanation for these variable descriptions, primatologists soon recognized the tremendous potential of the local environment to bring about change in the social behavior of any one particular primate group. In a sense, the study of primate ecology was born of this recognition, that even within one species, differential environmental pressures can bring about striking behavioral differences in the social groups. The classic cases of intraspecific variation in social behavior, believed to be caused by ecological differences, occur in the common baboon, the Hanuman langur, and the vervet monkey *(Cercopithecus aethiops).*

The society of common baboons living on the open savannah, was first described as tightly, hierarchically structured, male-oriented, and completely closed to outsiders. Later studies of the same species living in the forest suggested the opposite — loose structuring, no specialized male roles or male orientation and mobility of some individuals between groups (compare DeVore 1964, to Rowell 1966b; Paterson 1973, 1978; and Altmann & Altmann 1970 for details). The Hanuman langur of Northern India was described as living in peaceful and relaxed multi-female, multi-male groups. Later, the Hanuman langur in more densely populated study areas in Southern India was found to live in polygynous groups, with frequent outbreaks of violence in which infants may be killed (compare Jay 1965, to Sugiyama 1967; Yoshiba 1968; Hrdy 1977b). Finally, vervet monkey societies were found to differ, particularly in the expression of territoriality, depending, apparently, on the richness or poverty of the food resources in different areas (Gartlan & Bain 1968). Other examples of the impact of local environmental pressures on primate social patterns have been documented, and many animal behaviorists, such as the Altmanns (1970, 1974), and especially the British scholars Crook (1970a,b) and Clutton-Brock (1977) have helped to develop the study of such effects, known as the socio-ecology of primates.

Perhaps partly in reaction to the earlier idea that each primate species could only exhibit one type of social organization, the enthusiasm for finding current ecological causes for each aspect of social behavior sometimes became extreme. Inevitably a counter reaction set in, as some primatologists, for example Struhsaker (1969) and Kummer (1971a,b) stepped forward to reassert the highly important role that biological heritage, or the phylogeny of the species, can play in determining social patterns, or in circumscribing the expression of these patterns. Perhaps the best example of the form this argument can take concerns the question of why patas monkeys live in polygynous or uni-male, multi-female groups. Crook & Gartlan (1966) had suggested that since patas monkeys, like the polygynous hamadryas and geladas, live in semi-arid, treeless habitats, they too had developed polygynous societies as an adaptation to a sparse, harsh ground-dwelling niche. The similarities between these three species were suggested as convergent adaptations to similar ecological pressures. Then Struhsaker (1969) published a counterargument pointing out that, taxonomically, patas monkeys are closely related to *Cercopithecus* species, and that all but one of the many species in the genus *Cercopithecus* live in polygynous societies. Perhaps the pattern of patas monkey society is simply a phylogenetic trait which they maintained even while they moved out of the forests and into open, semi-arid habitats.

Is the social organization of a particular nonhuman primate group largely determined by the phylogenetic heritage of the species to which it belongs or by contemporaneous ecological pressures? This question, setting up an unnecessary dichotomy between biology and environment, is somewhat analogous to the nature or nurture question which was discussed in Chapter 3. The only answer that can be offered to such a question is that these two causal factors interact at every level to influence social behavior, along with other variables such as demographic or population structure differences, and unique events in the social history of each group.

NATURAL SOCIAL BEHAVIOR AND THE SOURCES OF STABILITY AND CHANGE

It does appear that different primate species exhibit different amounts of variability in social behavior, and thus possess differing potentials for varying their social patterns in response to external pressures. That this can affect the opinion of the researcher as to what degree primate social organization is species-specific and invariable, can be shown by these two contrasting quotations from Rowell and Kummer:

> It is important that we recognize at an early stage in the accumulation of comparative studies of social organization that there *may* be no such thing as a normal social structure for a given species, and that a description of social organization is only useful if accompanied by a description of the environment in which it occurs — and further, that we are still only guessing about which features of the environment will be essential in such a description (Rowell 1969c: 301)

> On the other hand, the evidence of hamadryas baboons suggests that the genetic potential of some nonhuman species may indeed be restricted to one type of society which environmental change would hardly alter (Kummer 1971: 145)

While Rowell's studies of social groups of common baboons in several different environmental settings, including captive and field conditions, found these animals to be quite variable in social behavior, Kummer's studies of social groups of hamadryas baboons, also in a variety of captive and field situations, found the latter to be highly inflexible in social patterns.

These contrasting findings lead us to a consideration of another controversial problem in primate behavior studies — how do we determine the "normal" behavior of the animal? Also, if some primates exhibit different behavior patterns in different environments, which environment can be considered natural? To be "normal" is to conform to a standard or to a common type, and "natural behavior" is generally considered to be that which is consonant with behavior seen in the natural habitat. The natural habitat in turn means the undisturbed environment in which the species under study evolved, and to which its behavior is generally adapted. By undisturbed environment, most people mean undisturbed by human activities, particularly the activities of industrial societies, or as the dictionary notes, nature "is the material world which exists independently of human activities."

Clearly, if one wishes to understand the significance of a particular behavior pattern, it is useful or sometimes necessary to see this behavior in the context in which it is thought to have evolved. Therefore, it is not surprising that most ethologists believe studies of the animal in its natural habitat are a prerequisite, or at least a necessary concomitant to, experimental or controlled laboratory studies of the species' behavior. However, only one very short paper by Bernstein (1966b) has discussed the problems inherent in actually finding, determining, and defining natural habitats. There is really no area of the world in which we find nonhuman primates, where humans have not altered "nature" in some manner with their technological prowess. Even in the most isolated tropical forests, aboriginal peoples may hunt animals with weapons, hardwoods may be logged, rivers may be dammed or polluted upstream, planes fly overhead, and pesticides may be carried in on high winds from distant agricultural fields. In this respect then, the completely undisturbed, natural habitat, apparantly sought by some ethologists, may prove forever elusive. However there are obviously differing degrees of human disturbance: the relatively minor and local effects of aboriginal hunting practices, compared to the massive defoliation of parts of Southeast Asia are an example. The small temporary clearing of forests by horticultural peoples in South America compared to the large scale clear-cutting of those forests by industry is another.

Other human interventions into the environment may be beneficial, at least temporarily, to nonhuman primates. Many field studies take place in national parks or game reserves, and these protected areas have no doubt saved many animals from extinction. However parks and reserves also result in altered population sizes, in changes in predator-prey relations, and in differences in competition levels between species, which must, in turn, affect the behavior of the animals. Rowell (1966b) has suggested, as one example, that common baboons in the national parks of Africa, where they are usually studied, behave very differently from baboon groups outside the parks.

Further, some nonhuman primates may have evolved alongside the human species for long periods, or it may now be difficult to reconstruct the environment in which the animal did evolve. Hanuman langurs and rhesus macaques have lived in, or adjacent to, urban centers in India for centuries. Has this then become their natural habitat? In some langur groups, usually living in densely populated areas, adult male langurs have been known to periodically kill infants. Primatologists are currently arguing whether this is a "natural behavior," that is, an adaptive reproductive strategy on the part of males, or an "unnatural behavior" or pathology resulting largely from the stresses of human disturbance of habitat and density of population (see Chapters 6, 18). In another example, the social system of common baboons has often been described as "savannah adapted," and indeed they are commonly found living in savannah regions today. But is this the habitat in which the species evolved? Rowell (1966c) has argued that common baboons evolved as a forest-dwelling species, and that the wide-spread savannahs in which we now find them are largely the result of relatively recent human burning practices. Humans have probably been hunting and eating this same common baboon for some three to five million years, to judge by the fossil remains from East Africa. Perhaps this is long enough to consider some human activities, such as hunting, to be a natural environmental pressure in the evolution of baboon social behavior patterns.

So, exactly where do we draw the line and suggest that human interference has resulted in unnatural behavior on the part of the animal? The positions taken in these examples show this is not easily resolved, and indeed may have no resolution. Rather than seeking to find and define that which is perfectly "natural," it is more feasible, if still not easy, to attempt to delineate which aspects of any given environment are influencing the behavior of the study animals, and in what manner. Perhaps, to be aware of, and if necessary to take account of, the contextual variables that affect behavior is a more realistic goal for primatologists, than the frequent and somewhat futile struggle to prove that one's own study animals are normal and natural in behavior, or that someone else's animals are abnormal and unnatural. It is in many cases understandable to be uncertain as to whether a behavior, or an individual, or a social group, should be considered representative of the norm for the species, but it is less understandable to be unaware of the variability which exists between behaviors, individuals and social groups, or to deliberately ignore consideration of the local contextual factors which might be influencing the variability.

Primatology has, in certain respects, suffered from a history of schism between the research efforts of field workers and those of laboratory workers. This, at least in part, is due to disputation over the interpretation of behaviors which occur differently in field and captive settings. The widely disparate approaches of some field workers and some laboratory workers to understanding primate behavior can be seen in these two quotations from what we might call two "founding fathers" of field and laboratory techniques respectively. C.R. Carpenter, the first primate field worker, stated that the following assumption is necessary to a discussion of monkey and ape social behavior:

> That for the valid investigation of some problems in comparative behavior it is obligatory not only to study "animals as wholes" but to observe whole animals in natural, organized undisturbed groups living in that environment which operated selectively on the species and to which the species is fittingly adapted (1942a: 177).

To this H.F. Harlow, famous for his laboratory studies of deprivation and isolation effects in monkeys, has postulated a quite contrary point of view:

> Many behavioral scientists, especially ethologists, believe or believed that the only normal environment was in the animal's natural habitat, and its indigenous feral home. The ethologists who have held this position have never lived in a feral environment and, indeed, if they had they would have been

sacrificed . . . Our own data show without question that the monkeys we raise in the laboratory are healthier, less disease-prone, more fearless, brighter, and presumably happier than any monkeys ever raised in a wild or feral state. These monkeys may not be in the "natural state" visualized by the ethologists, but they are in their element, in surroundings and situations suited to them . . . It is commendable to study subhuman animals in the feral environment, but it is better to do so in a laboratory where there are controls, as long as you do not destroy the capability for social interaction. The challenging problem is to produce laboratory environments in which the feral animal transcends its feral capabilities (Harlow & Mears 1979: 5,6).

What might be termed a middle-way was proposed by W. Mason, who has been viewed as a mediator between, or synthesizer of, field and laboratory research. This is his comment on how we should deal with the question of what is normal or typical, and how we should approach the research problem of variability in primate social patterns:

> The issue, then, is not whether grouping patterns are species typical . . . the issues are instead how a group manages to preserve some semblance of social order, to maintain a degree of cohesiveness and continuity, in spite of the vagaries of time and local circumstances. What are the sources of stability and change? How can the existence of modal grouping patterns be accounted for and at the same time the fact dealt with that deviations from these patterns are bound to occur? (1976: 433).

This solution, of course, does not resolve the discussion, which continues to oppose many field and laboratory workers.

The major problem at hand, then, is to determine by what mechanisms primate societies are maintained under various environmental conditions. Such mechanisms have been called the "infrastructure" of social life, and some of the suggested processes which make up the infrastructure will be the subjects of the chapters in the following unit. It is at least partly through the study of how primate groups experience both continuity and change over time, and how they exhibit both modal grouping patterns and their variations, that we may come to understand the commonalities, and the differences, between all primate societies.

Chapter FIVE

Sexual Dimorphism

Orang-utan
Pongo pygmaeus
This mainly arboreal great ape now is restricted to Borneo and Kalimantan (S.E. Asia). The female (right) averages just over half the male weight and usually travels alone or with her immediate offspring.

When people initially encounter an unfamiliar animal at the zoo, the first fact they usually wish to establish is its sex. "How do you tell the female from the male?" is one of the most frequent questions asked about animals. If the animal's genitalia are not immediately apparent, a number of other clues may be used to establish its sex, especially the relative size of the individuals in the exhibit. Males are somewhat larger and more robust, or more muscular, than females in our own species, and the assumption that this is true of all animals is a common one. In fact, it is a reasonably good rule of thumb for the adults of mammalian species at least, although there is variability from species to species, and in some, females may be the same size or even larger than the males. The sex of spider monkeys is usually incorrectly assessed by zoo visitors since adult female and male spider monkeys are more or less the same size and the females have a long, pink, pendulous clitoris, which resembles a penis. Then again, younger zoo visitors may use different criteria in order to determine sex. I once saw a large male mandrill pointed out as a female by a child, because of the prettier, brighter colors on his face, compared to those of the small, duller-colored female mandrill.

Sexual dimorphism is the term used when the females and the males of a species exhibit two distinct forms or patterns. The term should properly refer to differences in morphology, such as coloration, weight and body dimensions (Zihlman, personal communication) and I shall restrict its use to such phenomena. Behavioral differences between females and males I shall simply refer to as sex differences in behavior. In the primate order, sexual dimorphism is highly variable in its manifestation. Females and males in particular primate species may differ not only in aspects of body size, but also in coat color pattern, in seasonal physical features such as perianal sex skin swellings in females and brightening of the scrotal color in males, or in permanent physical features like shoulder capes, throat pouches, canine size, and shape of the skull (Crook 1972). Some species, like the tiny marmosets, the gibbons and the colobus monkeys may be very difficult to sex because the females and males look so much alike. Others, like the hamadryas baboons, are easy, because the male is nearly twice the size of the female and sports a long cape of hair around his shoulders, or the male stumptail macaque (*Macaca arctoides*), who is somewhat larger and has a balder, ruddier pate. Although all proboscis monkeys (*Nasalis larvatis*) have somewhat grotesque noses (hence their common name) the nose of the adult male in particular takes on Pinnochio-like proportions. Similarly both sexes of howler monkeys (*Alouatta*) look double-chinned, but the adult male has an especially large hyoid bone in the area of the throat, which aids in the characteristic vocalization for which howlers have received their common name.

For a number of reasons it is difficult to make generalizations about sexual dimorphism in primates. First of all, sexual dimorphism can take many forms, which are not necessarily related. A particular species may show sex differences in coloration but not in size (see Hershkovitz 1979) or sex differences in canine length but not in weight; or quite commonly, sex differences in weight but not in body length. For much of this chapter, the discussion will focus on sexual dimorphism in body weight, since this is the feature most commonly measured and described in the literature. As Gould (1975) has noted, body weight is a far more useful measure for analyzing differences *between* species than *within* species. However, the fact remains that it is for body weight that we have the most extensive data collected on sexual dimorphism in primate species.

Furthermore, until recently, few researchers have attempted to put together the data from different species and make any general comments about the nature of sexual dimorphism in primates. The first such attempts (Crook 1972; Clutton-Brock et al 1977; Leutenegger & Kelley 1977; Jorde & Spuhler 1974) show not only disagreements over the meaning and function of sexual dimorphism in primates, but even discrepancies in the baseline measurements of physical sex differences. These data were collected over many years by different primatologists, often using different techniques. Discrepancies in data can have major implications for our understanding of sex differences. For example, in both the common marmoset *(Callithrix jacchus)* and the black-handed spider monkey *(Ateles geoffroyi)* it has been claimed at various times that the sexes are the same size, that females are larger and that males are larger! (Clutton-Brock & Harvey 1977; Ralls 1976; Schultz 1969; Eisenberg 1976). In the case of the spider monkey, it is possible that in some of these studies subadult males were mistakenly measured as fully adult males (Eisenberg 1976). Although the size differences between females and males are no doubt slight in these two species, such differences can be given major and conflicting interpretations in various theories of primate sexual dimorphism.

Finally, it is difficult to draw general conclusions about primate sexual dimorphism because, as with so many other aspects of their lives, primate species are so variable, one from another, and we still have much to learn about most of them. Even taking that aspect of sexual dimorphism — body weight — which is probably the easiest to measure, or at least has been most often measured, we find that out of approximately 200 living primate species, the largest table available in the literature only contains 60 species for which body weight dimorphism has been calculated. All other such tables list only 20-30 species and different tables may be difficult to compare. In Figure 2, I have added my own short list to the literature by showing mean body weights of adult females and males for the species discussed in this book. Although this table suffers from the same multiplicity of sources and variety of methods in acquiring weight data as do other such tables, it is provided here simply to give the non-specialist some idea of the distribution of weight dimorphism in primate species, a necessary background to the discussion in this chapter. Sexual dimorphism in weight is calculated by expressing the female weight as a percentage of the male weight. (For example, the female common baboon weighs about 50% or half of the male's weight.) I find this percentage ratio the easiest to visualize, and suggest that glancing over this list will acquaint the reader with the variability which will be discussed for the rest of the chapter.

It can be noted from Figure 2 that in a very general sense, taxonomic affiliation partially predicts the amount of sexual dimorphism in weight; or, in other words, certain kinds of primates are more likely to be sexually dimorphic than others. Prosimians are by and large monomorphic, or similar in female and male weights, as are most New World monkeys, many Old World colobines (the family of leaf-eating monkeys), and all gibbons (hylobatids). In fact, sexual dimorphism is most characteristic of that large family of Old World monkeys known as Ceropithecidae, and also of great apes or pongids. However most primatologists believe that phylogeny or taxonomic and evolutionary relationship is a very inadequate predictor of whether a particular primate species will be sexually dimorphic. This is because very closely related species, such as the bonnet macaque *(Macaca radiata)* and the rhesus macaque *(Macaca mulatta)*, may show very different patterns of dimorphism. Or, one or two species in a family, such as the Red Colobus

(Colobus badius) and the proboscis monkey *(Nasalis larvatus)* in the Colobidae family, may be strongly dimorphic in a generally monomorphic family (Clutton-Brock and Harvey 1977). Or a superfamily may be quite variable; as in the Hominoidea, where chimpanzees and humans are not very dimorphic, while gorillas and orangutans are extremely so.

There must be more to the explanation of this variability in sexual dimorphism than just taxonomic affiliation, and different explanatory theories will be considered next.

Figure 2
Sexual Dimorphism by Body Weights of Nonhuman Primates Discussed in this Book

Common Names	Mean Weight ♀ (kg)	Mean Weight ♂ (kg)	♀:♂ Weight as percentage ♂ = 100%
PROSIMII			
Lemuriformes:			
Lemur catta[3] ringtailed lemur	2.10	2.10	100
Propithecus verreauxi[3] sifaka	4.5	4.5	100
Hapalemur griseus[1] gentle lemur	2.55	2.62	97
Lorisiformes:			
Galago crassicaudatus[2] bushbaby or galago	1.03	1.24	83
Nycticebus coucang[1] slow loris	1.24	1.34	92
Loris tardigradus[1] slender loris	.18	.22	82
Tarsiiformes:			
Tarsius bancanus[2] Horsfield's tarsier	.12	.13	92
PLATYRRHINI			
Callitrichidae			
Callithrix jacchus[1] common marmoset	.25	.27	93
Saguinus oedipus[1] cotton-top tamarin	.45	.40	112
Saguinus midas[1] red-handed tamarin	.33	.30	110
Cebidae:			
Cebus capuchinus[1,8] capuchin	1.72	2.24	77
Saimiri sciureus[3] squirrel monkey	.60	.80	75
Alouatta palliata[1] howler monkey	5.72	7.39	77
Aotus trivirgatus[1] owl or night monkey	1.00	.92	109
Callicebus moloch[6] dusky titi	.75	.75	100
Callicebus torquatus[6] yellow-handed titi	1.00	1.00	100
Ateles geoffroyi[1] spider monkey	5.82	6.17	94
CATARRHINI			
Ceropithecidae:			
Ceropithecus aethiops[6] vervet	5.6	7.0	80
Cercopithecus mitis[7] Sykes'	5.25	7.30	72
Cercopithecus ascanius[1] redtailed guenon	2.84	4.26	66
Cercocebus albigena[3] sooty mangabey	8.6	9.1	95
Miopithecus talapoin[4] talapoin	1.12	1.38	81
Erythrocebus patas[6] patas	5.6	10.0	56
Macaca niger[1] Celebes macaque	6.57	10.43	63

	Common Names	Mean Weight ♀ (kg)	Mean Weight ♂ (kg)	♀:♂ Weight as percentage ♂ = 100%
CATARRHINI (continued)				
Macaca fascicularis[1]	crab-eating macaque	4.10	5.89	69
Macaca fuscata[6]	Japanese macaque	12.30	14.57	84
Macaca mulatta[1]	rhesus macaque	7.51	8.22	91
Macaca nemestrina[1]	pigtailed macaque	7.77	10.37	75
Macaca radiata[1]	bonnet macaque	3.67	7.26	51
Papio cynocephalus[3]	common baboon	11.2	21.1	53
Papio hamadryas[3]	hamadryas baboon	10.4	18.0	58
Mandrillus leucophaeus[1,8]	mandrill	9.0	19.5	46
Theropithecus gelada[6]	gelada	13.6	20.5	66
Colobinae:				
Colobus badius[5]	red colobus	6.0	10.8	56
Colobus guereza[9]	black & white colobus	8.3	11.7	71
Presbytis entellus[1]	Hanuman langur	10.44	15.2	68
Presbytis johnii[6]	Nilgiri langur	11.1	11.2	99
Presbytis senex[1]	purple-faced langur	7.83	8.51	92
Nasalis larvatus[1]	proboscis monkey	9.98	17.65	57
Hylobatidae:				
Hylobates lar[1]	white-handed gibbon	5.51	6.11	90
Hylobates syndactylus[6]	siamang	11.0	11.0	100
Pongidae:				
Pongo pygmaeus[1]	orangutan	37.0	69.0	54
Pan troglodytes[6]	(common) chimpanzee	42.5	49.0	87
Gorilla gorilla[1]	gorilla	92.5	160.0	57
Hominidae:				
Homo sapiens[8]	(modern) humans	— —	— —	83-87

Principal Source of Data: [1]Napier and Napier (1967)
Other Sources: [2]Chiarelli (1972)   [3]Jorde and Spuhler (1974)
[4]Gautier-Hion (1971)   [5]Struhsaker (1975)
[6]*Primate Paradigms*, Chapters 14,15,16   [7]Coelho (1974)
[8]Schultz (1956)   [9]Oates (1977)

Although Clutton-Brock and Harvey (1977) have published the most extensive list of primate weights currently available, they ask that their data not be used as a reference source.

As can be seen from the data given in Figure 2, most primate species discussed in this book are slightly to moderately dimorphic. One of the arguments made in this chapter is that, while at the two extremes of the continuum — no dimorphism and extreme dimorphism — certain behavioral features, such as breeding systems, and male-female dominance relations may show some correlation with sexual dimorphism; in the middle (slight to moderate degrees of dimorphism) the associated behavioral factors are unpredictable. With this necessarily short introduction to the data on sexual dimorphism in primates, let us turn now to the various explanations of why it occurs in certain ways in certain species, and quite differently in others.

EXPLANATIONS OF SEXUAL DIMORPHISM

Sexual Selection

Most of the current suggestions regarding the nature and function of sexual dimorphism in primates relate to the theory of sexual selection. This was developed by Darwin over one hundred years ago especially to explain secondary sexual characteristics, or differences between females and males that are not directly necessary for reproduction. The theory of sexual selection will be considered at length in Chapter 17, and here I will just note that two major interrelated processes are considered to be involved: mate choice (usually described as female choice of male mates) and competition between members of the sex which is more numerous, for access to the members of the sex which is more limited (usually considered to be male-male competition for access to females). Darwin's explanation for greater male size and for male ornamention in many animal species, is that these features are chosen by females when they select males, and/or the features result from competiton between males for access to female mates. In practice most discussions of sexual dimorphism in primates, especially those concerning body size and canine size, have focused on male-male competition only, on the assumption that a bigger male is a more successful male because such an individual will win more fights. Only recently has the suggestion been raised (Ralls 1976) that in some situations a smaller male, through speed or agility for example, might have an advantage in male-male competition.

The usual explanation then for why orangutans, or baboons, or gorillas, are more sexually dimorphic than dusky titi monkeys (*Callicebus moloch*) or gibbons, is that male-male competition is stronger or more marked in the former species than the latter. Complications arise, however, when different researchers attempt to measure or to suggest how we might measure the degree of male-male competition. Crook and Gartlan (1966) suggested that male competition is stronger in large groups of open-country, ground dwelling primates which concentrate mating into a breeding season, and where everyone can see everyone else mating. Leutenegger (1977, 1978; Leutenegger & Kelly 1977) suggested that species which mate polygynously exhibit stronger male-male competition than monogamous species, and, indeed, he found a significant relationship between a polygynous type of breeding system and sexual dimorphism. Such a correlation between polygyny and sexual dimorphism has been found for animals other than primates also. Leutenegger however, did not distinguish between polygynous and multi-male, multi-female primate societies, but lumped the two together when in fact they are very distinct breeding systems and would be expected to have different links with sexual dimorphism.

Another approach to measuring the intensity of male-male competition and the degree of sexual dimorphism might be to compare socionomic sex ratios (the number of adult males in a group relative to the number of adult females) to the degree of sexual dimorphism. Gautier-Hion (1975) and Clutton-Brock et al (1977) did this and found correlations between socionomic sex ratios and degrees of size dimorphism. That is, the fewer the number of males relative to the number of females in the group, the *bigger* were the males relative to the females. However, a major complicating factor in this correlation is noted by Clutton-Brock et al (1977). It so happens that larger sized primates, indeed all larger-sized mammals, are more sexually dimorphic *and* are more prone to live in uni-male, multi-female groups. Smaller-bodied primates are less dimorphic and usually mo-nogamous. The apparent relationship between sex ratio and dimorphism may be partially an artifact of a confounding variable, overall body size.

In fact, a number of variables appear to be correlated, including at a minimum, ecological niche, breeding system, overall body size and sexual dimorphism – and Clut-ton-Brock and Leutenegger have argued back and forth in several papers as to the manner in which each one of these variables might lead to, or cause, another (Leutenegger and Kelly 1977; Clutton-Brock et al 1977; Leutenegger 1978; Harvey et al 1978).

At the risk of oversimplification, I will say that Clutton-Brock suggests that niche, especially diet and ground as opposed to tree-dwelling, leads to an optimum body size, which, along with the amount of male-male competition, as measured by socionomic sex ratio, leads to a characteristic degree of sexual dimorphism. Leutenegger argues rather that the intensity of male-male competition, as measured by the type of breeding system, determines the amount of sexual dimorphism, although female choice for smaller males may counteract the effects of male competition. Further, he argues the selection for male size (be it large or small) effects the overall body size of the species, which in turn can be counteracted by limited food supply. Both researchers of course recognize the confounding effects of numerous interactive, although secondary variables.

All of these arguments and suggested causal routes suffer from the inability to actually measure male-male competition directly. I would suggest that both breeding system and socionomic sex ratio are inadequate inferential measures of male-male com-petition. The underlying assumption is that in monogamous species every male is able to mate, while in polygynous species only one male in five or ten is able to mate. Lon-gitudinal research shows this is not necessarily so. Male-male competition in some mo-nogamous societies, such as marmosets and tamarins can be fierce, with one male suppress-ing breeding in others (see Chapter 16). And conversely, in many polygynous species, males join and leave groups as often as once every four years, so that every male who survives to adulthood may spend some years as a breeding male in a group and some years elsewhere. Finally, no one is really certain how many of the males in a multi-male, multi-female group are reproductively successful and to what degree. Until we have a means for accurately assessing biological paternity in many different kinds of primate societies, discussions of relative male reproductive success and the intensity of male-male competition must remain speculative and open to disputation.

Male Role

As one can see, it is easy to become trapped in complicated and circuitous lines of reasoning when several different major and nebulous variables are involved in apparent correlations, as between sexual dimorphism and breeding system, niche, body size, etc. This is no less true of the second major explanation for larger male size; that it is the result of males being selected for the role of protecting females and offspring of the group from predators. This has been suggested by several researchers, (DeVore 1963a; Crook 1972; Denham 1971; Hamburg 1972), most of whom were attempting to explain why terrestrial or ground-dwelling primate species tend to be more dimorphic than arboreal species (c.f. Bernstein's research on male roles described in Chapter 8). Much as socionomic sex ratio and breeding system have been used as inferential or "stand-in" measures of male-male competition, the "amount of time spent of the ground," or terrestriality, is often used as an indirect measure of predator pressure. A common assumption in primatology has been that predator pressure is greater in terrestrial environments than in arboreal ones. However we have no compelling evidence for this belief, and, being a terrestrial species ourselves, the assumption may merely reflect our own ground-dwelling biases, as well as reflecting our lack of knowledge about arboreal species.

A variation on the "male protector role" explanation for sexual dimorphism is Crook and Gartlan's (1966) suggestion that male-male competition *first* selects for large males, who are then *pre-adapted* for a protector role in a harsh environment. Jorde and Spuhler (1974) found a correlation between terrestriality and degree of sexual dimorphism for 29 primate species and Clutton-Brock et al (1977) did so also for their sample of primate species. However, neither a greater intensity of predation nor a greater protector role for males in terrestrial species has been proven. In fact, body size is again a confounding variable in this correlation, because ground-dwelling primate species tend to be larger in overall size than arboreal species, and as I noted, larger mammals tend to be more dimorphic. Thus terrestrial primates may be dimorphic simply because they tend to be larger animals than arboreal primates.

Bioenergetics and Sex Differentiated Niches

As overall species body size has come up more than once as an important variable in sexual dimorphism, it might be well to consider body size and its causes and consequences. First, it is important to point out that females do not just have different body sizes from males in many species, their bodies also perform different functions from those of males! While this statement seems so self-evident as to be inappropriate, it is, in fact, not often taken into account in discussions and theories of sexual dimorphism. The principles that apply to body size differences *between* species, such as between chimpanzees and gorillas, are usually applied "willy-nilly" to body size differences between females and males of the same species. Thus it is not at all uncommon to read that bigger bodies require more food or that males of dimorphic species need to eat more and use a larger share of the available resources than do females. It might be true that the larger animal eats more when different species are compared, but it is not necessarily true when females and males of the same species are compared. In fact, where data on food intake have been compiled for primate species, if sex differences do exist, females have been found to eat more and/or faster, and/or for longer periods than males. This is true for many species in which the male is bigger than the female, such as geladas, mangabeys, howlers and red colobus monkeys (see references in Clutton-Brock 1977). A nonhuman primate female spends most of her life pregnant or lactating, and the energetic costs of these reproductive functions are very high indeed. In small mammals, reproducing females may have feeding rates and metabolic rates that are up to 200% greater than nonreproducing females (Grodzinsky and Wunder, cited in Nagy & Milton 1979). Furthermore, how much an individual needs to consume also depends on its activity pattern. In the only primate study of which I am aware that attempts to take the extra reproductive and activity factors into account, Coelho (1974) found that female Sykes' monkeys, whose weight was about 72% of the male's weight, required more food and more energy than the male. Females not only had greater reproductive costs, but they tended to be more active. Thus there is more to energy requirements and resource utilization by females and males than body size alone suggests, and these other factors, especially female reproductive functions, may be implicated in sexual dimorphism. In fact, we might consider the possibility that we should be asking another question about sexual dimorphism. Instead of, or as well as, "Why have males become bigger than females in certain species?" we might ask, "What are the adaptive advantages of females having become smaller than males or being a different size from males in different primate species?" Several researchers (Ralls 1976; Post 1980; Demment 1978, 1979) are just beginning to ask this question, and some of the answers they offer, although tentative as yet, are very intriguing.

Ralls (1976) has considered the possibility that sexual selection might be acting upon females of a species instead of the males, especially in cases where mammalian females are larger than their male counterparts. It appears that in some kinds of birds such as phalaropes, sexual selection has operated upon females, resulting in species in which females are larger, more aggressive, and dominant over males. These species are polyandrous (one female mates with several males) and the parental role is left mainly to the males, a sort of reversal of what are believed to be the usual results of sexual selection (see Chapter 17). However in mammalian species, the presence of larger females is not correlated with any particular form of breeding system or any diminishing of female parental care, or with the presence of female dominance, aggressiveness, or greater development of weapons and ornaments. Thus Ralls (1976) does not think that sexual selection is the explanation when female mammals are larger than males of the same species. Rather she suggests that other selective pressures, such as intense competition for resources and advantages gained by larger mothers, may in combination, favor an increase in female size over that of males in some species.

Other researchers have focused upon different kinds of organisms, but have asked the same question — what is the advantage to females to be of a different size from males — and have suggested various, but not contradictory, answers. For birds, Downhower (1976) has suggested that females may be smaller than males in more fluctuating environments, because bioenergetic considerations result in smaller females being more successful in this situation. In a comparison of human populations, Hamilton (1975, cited in Ralls 1977) concluded that small size may be advantageous to females if resources are low during lactation. Post (1980) considered at length the relationship between overall, large body size and degree of sexual dimorphism in primates. He hypothesized that in larger species the energetic costs of pregnancy and lacation are differentially high for females, compared to the energetic costs in smaller-bodied species. Thus the optimal body size for females of large species is smaller than that of the males, large enough to bear the infants, but small enough to minimize costs of reproduction. In a paper on the bioenergetics of body size in common baboons, Demment (1978) argued among other points, that the smaller females (remember they have one half the body weight of males) must eat higher quality, non-fibrous foods that are rapidly processed. This is because of the bioenergetic relationship of their energy needs and their metabolic rates relative to their gut capacities. The relationship of feeding strategies, metabolic rates and body sizes (the

Jarman/Bell principle) in primates has also been discussed by Gaulin (1979) and Rae-mackers (1979). In a second paper, (1979) Demment suggested a reason for the correlation between body size and dimorphism. In order to divert energy to reproduction, females of already large species can be reduced in size, without much impact on their digestive capacities. However female size reduction in already small species would have a major impact on their digestive capacities because of altered relations between metabolic rates and abilities to process food.

In yet another consideration of female reproductive functions and body size, Leutenegger (1979) has argued that the primate characteristic of producing single-offspring births rather than litters of young depends on a minimum female body size. Restriction of female body size relative to neonatal weight sets a lower threshold for maternal size, below which the production and delivery of single-offspring would not be possible. The implication is that in larger primate species, females could undergo selection for reduced size, and sexual dimorphism would result. However, in the already small primate species, it is not possible for females to be any further reduced in size, and alternate reproductive strategies may evolve, such as twinning or multiple births, which occur in the "dwarfed" callitrichids (see also Ford 1980).

All three of these considerations of sexual dimorphism focus on overall species body size and female versus male bioenergetics, or what we might call the engineering problems and mechanisms of female and male bodies. Further, both Post and Demment consider the possibility that females and males in dimorphic species occupy somewhat different niches, especially that they eat different foods. The exploiting of different niches by females and males has been an alternative explanation of sexual dimorphism since Darwin (1871) recognized its possible existence in some sex differentiated species. Selander (1966, 1972) developed the "separate niches" hypothesis extensively for certain bird species. It has been occasionally mentioned in the primate literature (e.g. Rowell for common baboons, 1966b) and was considered but largely dismissed as a significant explanation for the pattern of primate dimorphism by Clutton-Brock et al (1977). Nevertheless, he described several cases where the sexes appear to occupy separate niches in his book (Clutton-Brock 1977). Research into the possibility of different eating patterns, or more generally, different niches, for female and male primates is in its infancy but several interesting reports on this subject have been published. McGrew has suggested that male chimpanzees forage differently from females in that they capture and eat more meat, while females gather and eat more insect matter (see Chapter 14). For orangutans, Galdikas (1979) has suggested "ecological separation between the sexes," in which males travel and forage much more on the ground and eat more of the foods found there, while females are more arboreal and eat more foods found in trees. Equally striking is Jouventin's (1975) description of male mandrills foraging along the forest floor, while the females and infants follow parallel paths in the trees above them. Yet another example is Gautier-Hion's (1980) finding that pregnant and lactating female Cercopithecus monkeys selected protein-richer foods such as young leaves and insects, while the males ate more fruit. As with so many of the other correlations, or apparent relations between sexual dimorphism and the various factors mentioned in this section, we are left with the chicken-and-egg question of whether niche separation leads to sexual dimorphism, or vice versa.

In fact, the best way to summarize this section might be to list the factors which have been suggested to be related to sexual dimorphism. Whether these factors cause, or result from, or simply coincide with, the presence of sexual dimorphism in body weight depends on the explanatory theory to which we subscribe. The factors are: taxonomic or phylogenetic affiliation; overall body size of the species; group size; group composition; breeding system or socionomic sex ratio; intensity of male-male competition; aspects of the species niche, such as diet and amount of time spent on the ground; specialized male predator-protection role; bioenergetic costs of pregnancy and lactation; bioenergetic principles and relationships in smaller and larger bodies; and finally, separation of feeding niches between the sexes. Western (1979) has argued that the main life history parameters are scaled to body size, and therefore size should be a central theme in ecological studies. In my opinion the relationship between large body size and the presence of sexual dimorphism is the most important correlation, and further, since the female primate bears high bioenergetic costs for pregnancy and lactation, it seems likely that it is the study of her body size and functions and the selection pressures upon her, in comparison to those acting on the male, that will best help us to understand the variable pattern of sexual dimorphism in primates.

DENTITION AND ORNAMENTS

As noted at the beginning of this chapter, primate females and males may differ not only in body size but also in specialized anatomical features such as canine size, and what Darwin referred to as "ornaments," such as throat pouches, capes of shoulder hair, head crests, and bright fur and genital coloration.

It is probably safe to say that Darwin regarded the larger size of male bodies and male teeth as "weapons" useful in male competitions, while aesthetically pleasing male ornaments were believed to be more the result of females choosing males who possessed such features. Recent interpretations of special ornamentation focus upon species recognition (Lack 1968) and especially the signalling functions of this aspect of dimorphism in a sexual context (see Mitchell 1979; and Chapter 17). In other words, the hamadryas shoulder cape, the "silverback" of the fully adult male gorilla, the swollen chest nodules of the gelada female, the brightened colors of the genitalia of one or both sexes, all have been described as functioning in the attraction and courtship and mating of females and males of these species. "Ornamentation" will be further discussed in Chapter 17.

Two major interpretations of sexual dimorphism in dentition, especially canine size, have been offered. Both should be familiar from the discussion in the previous section of this chapter. Males are believed to have larger canines than females of the same species as a result of male-male competition and/or as a result of their special defense role in the social group. First, of course, it should be noted that males in some species may have bigger teeth than females because of allometric principles, or simply because they have bigger bodies than do females. Leutenegger and Kelly (1977) measured the relationship between body size dimorphism and canine dimorphism for 18 species of monkeys and apes, and found quite a strong correlation ($r = .76$). However, there are definitely cases of primate species in which the males have extra large canines, larger than would be expected from body size dimorphism alone; spider monkeys and chimpanzees, for example. And conversely, there are species, such as the proboscis monkey and the orangutan, where the canine size of the male is less than we would expect from the body size dimorphism of the species. Finally, it is often true that the differences between female and male canine size is much greater than the female and male size differences in other teeth, such as molars.

All this leads one to suspect that there must be special sex-differentiated selection forces operating on canine size. Harvey, Kavanaugh, and Clutton-Brock (1978a) conducted a study of sexual dimorphism in primate teeth from which they concluded that male-male competition and predator pressure are the selective forces which bring about sex differences in teeth. Again they used socionomic, sex ratio and ground-dwelling niche as their "proxy" measures for male competition and predator pressure. In a second paper (1978b) they considered the possibility that selective pressures may be operating upon *female* canine size, and found that females living in polygynous societies have smaller canine size relative to body size than do females living in monogamous and multi-female, multi-male groups. Their interpretation of this finding is that females in monogamous societies participate more in territorial aggression, and that females in both monogamous and multi-female, multi-male groups engage in more aggression against males than do females in polygynous groups. Finally they argued that females in uni-male, multi-female groups are more closely related to each other than females in multi-female, multi-male groups and so the former grouping of females should be less competitive (see Chapter 18 on kin selection). There is no kinship data, only a hypothetical mathematical model, offered in support of this last statement on the kin relations between females, and I doubt that it is valid. Even the earlier statement on the amount of female aggressive behaviour to be expected in different societies is hypothetical. Finally, Smith (1980) reported the opposite findings on female canine size to Harvey et al, namely that females living in polygynous societies have *greater* maxillary canine development than do females living in multi-female, multi-male societies. The idea of looking for possible selective pressures on female canine size and the finding of greater-or-smaller-than-expected female canines in certain kinds of primate societies is important, but the suggested interpretations seem less than satisfactory.

Harvey et al (1978b) also mentioned the possibility that feeding differences between the sexes might result in differences in dentition. Post (1977) has suggested that sex differences in baboon teeth may be related to niche differentiation between the sexes in diet, and Butler (1975) has pointed out in a study of tooth wear, that females and males may use their teeth differently. As was the case with body size dimorphism, niche separation between the sexes is the third, and least investigated explantion for tooth size dimorphism.

One of the more intriguing mysteries of human evolution is why the first hominids of which we have any record, in the form of fossil remains, have very small canines in both female and male forms. Since our closest primate relatives, the great apes, and the closest fossil links to early humans, the Dryopithecines, all have large canines, and usually differentially large canines in males, human canines are considered to have been "reduced" through selection. But why? Suggestions have ranged from the idea that the invention of weapons replaced the need for large canines (Washburn and Ciochon 1974); to the theory that hominid teeth are adapted to rotary chewing action (Jolly 1970); or meat eating (Szalay 1975); to the recent hypothesis that early female hominids may have chosen as mates those males who had smaller canines (Tanner and Zihlman 1976). These and other theories of early human evolution will be discussed in more detail in the last chapter.

SEXUAL DIMORPHISM AND GROWTH PATTERNS

As well as the evolutionary mechanisms which may lead to sexual dimorphism, it is useful to consider for a moment the developmental or ontogenetic processes which bring about sex differences during the lifetime of individual primates. While it is true that human, rhesus macaque, and chimpanzee male infants are born somewhat heavier in weight and larger in size than female infants, the vast majority of adult sex differences come about because of different growth patterns in females and males after birth. In humans, girls mature faster at all stages of growth and development. Both sexes experience an adolescent spurt of growth, but it occurs at a later chronological age in males, and is more intense, resulting in their greater adult size (Malina 1975). In the only two non-human primate species whose growth patterns have been carefully monitored, the rhesus monkey and the chimpanzee, a similar pattern occurs, with later puberty and a more intense growth spurt in males than in females. In fact, Tanner (1977) has argued that there exists a characteristic primate growth curve, which is sexually dimorphic. While this may be true for those primate species in which adult males are bigger than adult females, the variability in adult sexual dimorphism between species should be reflected to some extent in different species' growth curves. Non-dimorphic or monomorphic species have almost never been measured during growth and maturation, but Hearn (1978; and see Chapter 16) has shown that commom marmoset *males* do mature somewhat *faster* than females, and as adults, are marginally smaller than females. Although the sexes are here reversed and would not lie correctly on Tanner's hypothesized primate growth curve, these data do support the general correlation between growth patterns and adult sexual dimorphism previously described for humans, rhesus and chimpanzee. Further, Rowell (1977) has argued that *different* Old World species, such as patas, talapoins, and rhesus macaques, may have quite different maturation patterns; for example all patas being extremely "fast" growers and early maturers, compared to rhesus. Indeed, Western (1979) concluded that life history parameters, such as growth rate and age-at-first-reproduction, are scaled to body size, so that we would expect different growth patterns in different sized species. Although these examples do not necessarily negate Tanner's general primate growth curve, they do represent first steps toward documenting some of the variability in primate growth patterns.

Schultz (1956), who did the original and the most extensive studies of growth in primates, has argued that sexual dimorphism in adult size is closely connected to the species-specific intensity of post-natal growth. The greater the changes during growth and development in a particular species, the more conspicuous will be the adult sex differences. In other words, the more they grow, the more different the sexes become. The more infant-like or neotenous a body feature remains, the less marked will be its sexual distinction in adults.

It seems fairly clear, therefore, that the immediate mechanism by which adult sex differences in physique arise is differential growth patterns. As well as asking why *adult* female and male primates are different sizes and shapes, we might well ponder why they grow differently in the first place. While growth is quite correctly thought of as a means to an end, perhaps it is the sex differentiated growth pattern, as well as the final adult size, which is adaptive. The bioenergetic theories of Post and Demment, described earlier, might suggest how this could be true, but Leibowitz appears to be one researcher who has tackled the question directly (see also Rowell & Richards 1979). Leibowitz suggested that sexual dimorphism in primate size results from selective pressures which favor early reproductive maturity and pregnancy in females, at a chronological age when males are still under selective pressure to grow:

> The hypothesis I am proposing therefore argues that in ecological settings which encourage males to forage more widely than females, reproductive advantages have fallen to those who are active enough to move around, large enough to do so safely and versatile enough to exploit alternative food resources and social stituations. At the same time reproductive advantages fall to those females who stop growing at pubescence, and are effective in using their limited food intakes for reproduction and nursing (Leibowitz 1975: 34).

To augment her hypothesis somewhat I suggest that we consider which growth and maturation patterns might result in the greatest reproductive success for females and males. Females can only produce a limited number of offspring in their lifetimes, and in some situations it might be adaptive for females to begin producing these infants as early as possible. In a sense they can be viewed as diverting the potential energy which might have gone into additional growth into infant-bearing and raising. Males, on the other hand, may become more reproductively successful by continuing to grow to a later chronological age and not attempting to attract female mates or to compete with other males, until they are in their prime. However, that these suggested female and male "reproductive strategies" are not in fact optimal for every species is demonstrated by the variable degrees of sexual dimorphism in different primate species. Perhaps, as was suggested earlier, in small species a female could not begin to bear single-offspring until she has reached a certain body size. Rather than increasing her reproductive success by starting early and growing less, the female could bear twins as in marmosets, or share the energy burdens of parental care with the male, as in many non-dimorphic species. In adaptive success there can be many means to the same end; forgetting this can be one of the pitfalls of evolutionary reasoning. In truth, I consider the preceding suggestion to be preliminary and not to be a completely satisfactory explanation for the diversity of sexual dimorphism patterns in primates. However, I do think that a consideration of sex-differentiated growth as an adaptive pattern in itself, may provide important new directions in the continuing mystery that is sexual dimorphism.

SEXUAL DIMORPHISM IN SIZE AND BEHAVIOR

It is stated or implied with some frequency in the literature that there is a relationship between sex differences in behavior and sexual dimorphism, with the assumption usually being that it is the physical factors which lead to, or cause the behavioral differences. Particularly common is the idea that in species where the male is larger than the female, he will also be more self-assertive than the female, and be dominant over her (Crook 1972; Goldberg 1973; Kummer 1971a; Mitchell 1979; Rodman 1973; Trivers

1972; Van den Berghe 1973). The lack of male dominance in non-dimorphic, monoga-mous societies such as gibbons and titis, and the presence of male "domineering behav-ior" in certain very dimorphic species such as the common baboon, would seem to lend credence to this idea. Several exceptions do exist, such as the patas monkey where the male is nearly twice the size of the female, and yet has been described as "hen-pecked" (see Chapter 15), and the spider monkey male who is nearly the same size as the female but is often belligerent toward her (Baxter 1979; Klein 1974). However, support for a hypothesized correlation between physical size and behavioral functions may come from some of the studies described in this chapter which have shown that there is a correlation between size dimorphism and the type of breeding system, the latter of course being a behavioral phenomenon. Specifically, very dimorphic species show some tendency to be polygynous and non-dimorphic species are often monogamous. The chapters which follow will stress the variability between primate species and the diverse meanings which terms, such as aggression, dominance, monogamy and multi-female, multi-male groups must take on in different species and different situations. Throughout the book there are examples of what initially look like similar or analogous patterns, breaking down under more careful and more in-depth consideration. As Oxnard (1979) has stressed, there is never a one-to-one relationship between physical structures and behavior in animals, and the former do not necessarily *cause* the latter.

Futher, in reading the literature on sexual dimorphism, I have received the impres-sion that extreme cases of the presence or absence of dimorphism are more likely to catch our attention, and perhaps even occasionally to distort our correlations. For example, in a "short list" of 20 primate species being tested for correlations between sexual dimor-phism, body size and ground-dwelling niche, the presence or absence of the orangutan (a very dimorphic, very large, very arboreal species) or the common marmoset (a very non-dimorphic, very small arboreal species) could have a tremendous impact on the results of statistical tests. As I noted at the beginning of this chapter, the majority of primate species are slightly to moderately dimorphic in body weight, and it is my belief, especially in these cases, that one cannot accurately predict female and male behavior patterns or the relationship between the sexes from the amount of sexual dimorphism. Ralls takes an even stronger stand for all mammals when she asserts: "At least at the present level of knowledge, it is not possible to predict the roles played by the sexes either from the social system or the degree of sexual dimorphism" (1977: 921).

Consider for a moment a case in point using primate examples. Talapoin monkeys and chimpanzees show similar degrees of sexual dimorphism in body weight, females weighing about 81% to 87% of male weight. It is a generalization of some researchers who study these species (e.g. Wolfheim 1978; Goodall 1975) that talapoins live in a "female-dominated" society in which females show a greater capacity for self-assertive behavior than males, while chimpanzees live in "male-dominated" society in which males show a greater capacity for self-assertive behavior than females. As I shall claim in several places in this book, the use of such generalizations is always questionable, however for those looking for relevance to the human species, we might note that humans exhibit about the same amount of sexual dimorphism in body weight as talapoins and chimpanzees.

Hamburg and McCown (1979), puzzling perhaps over the slight degree of body weight dimorphism in the chimpanzee, have argued that in this case it is inadequate to assess sexual dimorphism on the basis of body weight alone. It is surely true that it would be better if we could take more than one dimorphic feature into account when discussing the nature of sexual dimorphism in a particular species. However, Hamburg's further argument that chimpanzee male "fighting anatomy" is what should be considered in discussions of behavioral and physical dimorphism, is difficult to pursue. He lists canine size, skull features, musculature and strength as examples of highly important features with supposed behavioral repercussions. Chimpanzee male canines are unexpectedly large (Leutenegger and Kelly 1975) but then talapoin male canines are even larger (Gautier-Hion 1975), so what firm behavioral conclusions can we draw from this physical feature? Skull features such as sagittal crests tend to follow from size of dentition, due to weight pressures on the developing skull, and are not an independent measure of sexual dimorphism. Muscular robusticity will be to some extent taken into account in a body weight measure, as compared to a body size measure. Indeed, it is a common enough phenomenon in primate species for the male to be only slightly dimorphic in body length but more dimorphic in weight, reflecting his greater robusticity.

This discussion brings us back to a point made at the beginning of the chapter, that sexual dimorphism is very difficult to measure because there are so many features in which it can be manifested. In fact it seems very unlikely that there is an entity, known as "sexual dimorphism," for which we can say there exists selection in a given species. Rather, there may be selection for larger canines in males, for a faster maturation pattern in females, for a greater chest girth to body length ratio in males, for sex skin swellings in females, and so forth. These various aspects of sexual dimorphism need not be correlated or simultaneously selected in one particular species. I seriously doubt that a relation between size dimorphism and sex differences in self-assertion or dominance would hold for primates, even if we could find a way to meaningfully test this relation statistically. Furthermore, when we attempt to generalize about different species in such a variable order as Primates, the complexity of the topic and the futility of seeking for one explanatory mechanism for sexual dimorphism becomes even more apparent.

CONCLUSION

Sexual selection theory has been the primary explanation for sexual dimorphism. While, no doubt, the theory does offer a plausible explanation for some aspects of female and male differences, more direct measures of the intensity of male-male competition are needed before we can really determine what role this process plays in male features. And a greater consideration of the role of mate selection, especially female choice, would be germane to the issue. Also, as Ralls (1977) has noted, sexual selection does not account for all the variations in sexual dimorphism in different species. More recent considerations, although still in their initial stages, of the bioenergetic constraints on body size and the relationship of body size to sexual dimorphism, have offered some new insights. Finally, the overt recognition that female and male primate bodies have different reproductive functions to perform, and the realization of what may follow from this, including the fact that selective forces operate directly upon females as well as males in sexually dimorphic features, will surely lead to a more complete explanation of those features which make females and males different.

PART II

SOCIAL BEHAVIOR: SOME MAJOR CONCEPTS AND THEIR IMPLICATIONS FOR SEX DIFFERENCES

Chapter SIX

Aggression

Hanuman langur
Presbytis entellus
This leaf-eating monkey holds special status in Hindu beliefs and is protected throughout most of its range. The female, cradling the infant, weighs 68% of the average male weight.

In general, too little attention has been paid to the agonistic behavior of females, partially because it tends to be less violent and spectacular than that of males, and partially because male observers have been primarily interested in the behavior of their own sex.

J.P. Scott 1974: 422

Aggression is certainly one of the most, if not *the* most, researched and discussed topics in the study of animal behavior. Numerous books (such as Montagu 1968; Carthy and Ebling 1964; Holloway 1974; Johnson 1972; Megargee and Hokanson 1970) and even a journal, *Aggressive Behavior*, are devoted exclusively to the subject. The reason is obvious, for almost every paper on the topic begins by pointing out that humans today are highly concerned about our own aggressive behavior and that we would like to understand the nature of aggression in the animal world. "Instinctive aggression" in man, as the cause of many of our woes, has been the theme of two of the most widely-read popular ethology books, Ardrey's *African Genesis* and Lorenz's *On Aggression*. These books have been amply and justifiably criticized (for example, see Ashley Montagu's *Man and Aggression*) for presenting misleading oversimplifications of a large and complex problem.

In this chapter, I will examine definitions of aggression, and summarize what is known about its causes, expressions and functions, in an attempt to delineate, and hopefully clarify, our understanding of aggressive behavior in primates.

DEFINITIONS

Aggression is a good example of a word used a great deal on the assumption that everyone knows, recognizes, and agrees upon its meaning, but when the time comes to define it we find its meaning is both controversial and elusive. It is also a word with both a technical and a common meaning, and a term particularly subject to the aha! reaction (see Chapter 2), because we borrowed it originally from descriptions of human behaviors in order to apply it to a variety of animal behaviors, and now we are engaged in the reverse leap of analogy, from animal to human behavior.

Ethologists have generally agreed on a technical definition of aggressive behavior as being: "behavior directed towards causing physical injury to another individual," or sometimes "behavior where the goal is physical injury to another individual" (e.g. Southwick 1972; Carthy & Ebling 1964; Hinde 1974). The "directed towards" is included in the definition because sometimes aggressive behavior, for one reason or another, does not actually cause physical injury (the aggressor is not able to catch the intended victim, for example) and, conversely, sometimes non-aggressive behavior accidentally does cause physical injury. We should notice however that this inclusion means that "intent" or motivation is part of the definition. To say that an animal behaves aggressively is to speak not only of what it does but also of what it *intends* to do. These can be murky and unsafe waters for an ethologist, given our limited understanding of the subjective aspects of animal behavior. Not only is intention assumed, but given the emotional connotations of the word aggression, an aggressive animal also is assumed to *feel* angry.

Hinde suggests that while animal behaviorists generally restrict their use of the term aggression to its narrow, technical sense of physical injury, popular parlance uses an extended definition of aggression as any self-assertive or "go-getting" behavior. We should not assume this proves a common causal basis for violence and such assertive behavior. This delineation of two different understandings of aggression might be an interesting clue to what often appears to be an ambiguous attitude toward aggression in our own culture, because while we generally deplore physical injury (at least in within-group violence), we usually value self-assertive behavior in our competitive social system. Unfortunately for this nice distinction, a reading of the literature shows that in primatology the term is not restricted to behavior directed toward causing physical injury, but is frequently applied to all sorts of self-assertive behaviors — displays, supplantation (the taking of another animal's space), and territorial vocalizations, for example. These are often called aggressive behaviors, and yet as Barnett (1968) notes, the intent of these signals may not be to cause physical injury, but to induce withdrawal of another animal. In addition, Marler (1976) suggests that perhaps the most harmful forms of aggression in animals are the indirect result of competition for limited resources; that birds, for example, who fail to win territories may die of starvation or predation, and therefore have been physically harmed through an act, albeit indirect, of social aggression. If we stretch the "causing of physical injury" to this extent, then we may as well say that all self-assertive behavior inevitably hurts someone else, and the distinction between the technical and the non-technical definitions of aggression has disappeared. In humans, and perhaps in all animals which rely extensively upon learning, we also have the knotty problem of distinguishing psychological from physical harm.

Thus what appears to be a self-evident category of behavior is really not that clear, either in the minds of scientists, or of people in general. Because the concept of aggression is nonetheless intuitively satisfying to most people, and does appear to describe an important behavioral phenomenon (or phenomena), I do not think the term will ever disappear from our lexicon. But, certainly, it would further our communication and understanding in this area if we were clear in our own statements of what we mean by aggression, and if we demanded the same clarity from others. It is probably advisable to avoid the use of the word aggression when a better defined, more operational term such as agonistic behavior (see below) is suitable but this will not be entirely possible in a chapter devoted to discussing the research on aggression. Therefore, I shall use the word in its most expanded sense, as does most of the research under discussion, to mean any sort of self-assertive behavior.

EXPRESSION AND CONTEXT OF AGGRESSION

As we shall soon see, the contexts in which aggressive patterns occur in primates are varied, but Hamburg (1973: 193) has presented the following preliminary list, useful for its description of situations in which aggressive patterns are likely to occur:

1. In daily dominance transactions,
2. In redirection of aggression downwards in a dominance hierarchy,
3. In the protection of infants, most often by females,
4. When sought-after resources, such as food, or sexually receptive females, are in short supply,
5. When meeting an unfamiliar animal,

6. In defending against predators,
7. In the killing and eating of young animals of other species,
8. When terminating severe disputes among subordinate animals,
9. In the exploration of strange or dangerous areas,
10. When long-term changes in dominance status occur, especially among males,
11. When an animal has a painful injury,
12. When there is crowding of strangers in the presence of valued resources.

It appears that what these situations have in common is conflict, or disturbance, either between opposing individuals, or opposing impulses in an individual. In fact, focusing upon physical injury, and the intent to cause it, may take us away from, rather than toward, an understanding of a class of behaviors which are, in various ways, all related to *conflict situations*. A partial solution to this is the term, "agonistic" which Scott (Scott and Frederickson 1951; Scott 1974) has proposed to describe behaviors occuring in a conflict or competition situation between members of the same species, be they self-assertive or self-defensive behaviors. Scott argues that defensive patterns, such as escaping, freezing, and crouching, are alternative patterns in the same behavioral system as offensive patterns such as assault and threat. They are all part of the same repertoire or set, and represent alternative possibilities in situations involving conflict. Indeed, animals may even combine attacking and withdrawing elements. Thus, one or the other pattern (defensive or offensive) or even both may be the most adaptive in a given situation, but they are all part of the same complex of behaviors. This seems the best explanatory mechanism for disputes in Old World monkeys for example, where it often is not possible to label one monkey the aggressor and the other the defender because an individual may alternate or even integrate assertive and defensive behaviors as its confidence waxes and wanes (see Chapter 7 on dominance for further discussion).

When we develop a concept and a label for a behavior or a series of behaviors such as aggression, we are actually engaged in a type of cognitive taxonomy, that is, we are mentally categorizing behaviors which appear to be related. Thus, in proposing the term agonistic behavior, Scott is drawing an outline around a new grouping of behaviors he calls: "a system of behaviors which are adaptive in conflict situations between members of the same species" (1974: 417). This is in sharp distinction to our old grouping of "behaviours directed towards causing physical injury" or "self-assertive behaviors."

Because agonism is restricted to describing conflicts within the species (intraspecific), it is also somewhat less of a heterogeneous category than aggression, which is used to describe many different kinds of conflicts. It is actually very important to separate and distinguish between conflicts which occur:

1. between individuals of a social group (intragroup),
2. between social groups of the same species (intraspecific or intergroup),
3. between different species (interspecific).

The first type of conflict is usually understood through the explanatory mechanism of *dominance* hierarchies, which we will discuss in the next chapter. The second type is usually referred to as *territorial* behaviors, and the third may take the form of competition between species or *predator-prey* relationships.

It is particularly important to separate the third type of conflict from the other two, as Hinde makes clear:

> since in any one species predatory behavior and intraspecific fighting are usually elicited by different external stimuli, depend on different internal states, usually involve some different movement patterns, and may involve different neural patterns (1974: 250).

Studies of carnivores have shown us that there is no necessary relationship between predatory behavior and agonistic behavior in terms of the causes, expressions and functions of the behaviors (Hutchinson & Renfrew 1966; Scott 1974). There is no reason to expect that because an animal kills members of another animal species for food, it will necessarily exhibit high rates of "intent to cause physical injury" toward members of its own species. Thus, to state that humans kill or otherwise injure other humans because of a predatory or carnivorous nature is nonsense. The stepping-stones to this erroneous conclusion are a misplaced focus upon the nebulous "physical injury" aspect of conflict situations and the mistaken assumption that all physical injury is the result of similar causes or similar motivation.

Scott's term, agonistic behavior, would still combine intragroup and intergroup conflicts (the first two types listed). It is sometimes argued that in primates the same signals are used in conflicts with members of the group as in conflicts between groups; and that the motivations of the disputants, a desire to maintain a space around oneself or a desire for control over certain resources, are the same in both types of competitive situations. I am not entirely convinced. As Alison Jolly has so evocatively written:

> But one can hardly conclude that the leaping, great-calling gibbon, displaying in a tree top at his male neighbor is aggressive in anything like the same sense as when he glances at his mate and reaches first for a fig — his feelings quite likely differ not only in intensity but also in kind (1972: 179).

Indeed, several primate species have vocalizations and other behaviors which are exhibited only in territorial encounters. Territoriality in primates is not all that common and its presence in mammals generally speaking is rather "spotty" (Hamburg 1973; Scott 1974). The concept was developed from bird studies, and in avian species territoriality can involve either site attachment and/or hostility to rivals wherever they are met. However, in mammals territoriality is more defined as *defense* of a specific *area* (Burt 1943). Most primates (and most animals) habitually maintain a certain amount of space around their bodies which is broached only by relatives, friends or mates; and thus the suggestion is sometimes made that territoriality is just a group extension of personal spacing. But it seems possible that the resemblance is analogous, similar because of function, rather than homologous, similar because of common evolutionary or biological origins. That is, both behavioral systems do result in the maintenance of space, but it does not follow necessarily that both systems have similar causes. Thus Crook has said that in primates: "territoriality is a 'group characteristic' arising out of the cohabitation of individuals living in a given locality," and the term therefore, "cannot refer to the intrinsic aspect of the motivations of an individual, but rather it refers to a relationship between two or more animals with respect to location" (1968: 148). I would suggest that just as it is necessary to distinguish predator-prey relationships from intraspecific conflicts, so it may be important to distinguish between conflicts involving different groups or individuals from different groups in defense of a given location, and conflicts occuring within a social group between individuals. When we place behaviors under a common label we focus on the features they have in common and tend to ignore the features that are different, until we unconsciously may treat the behaviors as identical. This can then lead to the false assumption that two behaviors must be identical *because* they have been placed in the same class. I am not arguing for endless atomization in the study of behavior, but I do think that we need to make our cognitive taxonomy as conscious a process as possible. We often make the most progress in understanding behavior when we recognize that what we assumed to be a monocausal, unitary phenomenon, is in fact a heterogeneous category of behavioral patterns with multiple causes and functions.

CAUSES

Ethologists distinguish between proximal or immediate causation of behavior, that is the internal and/or external stimuli which trigger a behavior, and ultimate causation, which refers to the postulated evolutionary reasons for a behavior. In this section we will be concerned only with immediate causation of aggressive behaviors. We have seen that what is called aggression may be expressed in several different types of situations. It is likewise true that the causes of aggression are multiple and diverse. As we saw in Chapter 3, the idea of aggression being principally the result of "instinct", or a spontaneous and internally-triggered mechanism, has now been greatly modified. Aggression is of course mediated through the central nervous system and augmented or inhibited by physiological factors such as hormones. For example the latter may influence the threshold at which aggressive responses will occur. However, the immediate stimulation or trigger for each aggressive act is usually considered to be an external or environmental factor (see, however, a discussion of apparently "spontaneous" aggression in chimpanzees by DeWaal & Hoekstra 1980). Past experiences, inherited factors, and current physiological state may all help to predispose an individual, or a class of individuals (males for example), or a species to react with aggression to certain kinds of stimuli from the environment. However, as Crook (1968) has noted, there is no evidence for a genetically determined "appetite" for aggressive behavior. This is an important point, because it means that individuals do not just go around with a certain amount of aggressive energy that they have to expend in any way possible. Rather they operate in day to day life with a set of proclivities to react aggressively to certain environmental stimuli, should the stimuli occur. The influence of experiential and inherited factors on the development of this "set of proclivities" or predispositions, or the ontogeny of aggressive patterns, especially as it relates to female and male differences, will be the subject of Chapters 11, 12, and 13.

There are a variety of environmental phenomena which might provoke intraspecific aggression, but exactly what sorts of things are we talking about? It has been said that:

The most general cause of agonistic behavior, related to its defensive function, is to react to any sort of noxious stimulation or emotional disturbance as if a

76

nearby animal has caused it and to attack the animal and drive it away (Scott 1974: 423).

Thus some ethologists refer to simple proximity (closeness) of another animal as a major cause of aggression. However one has to be very careful to avoid circularity in discussing proximity, as Marler (1976) recognizes, because no animals, other than humans, can fight unless they are close together, and we must not confuse this *correlation* of aggression and proximity with *causation*. It is best to think of proximity as provoking aggression only in certain contexts and to certain degrees, when the proximal animal is perceived as a noxious stimulation. All social behavior (mating, mothering, etc.) arises with a degree of proximity, but aggressive behavior usually arises when *certain others* become too proximal, that is, when an animal, in a sense, perceives itself as crowded by these other individuals. Data from a wide range of species show that animals have particular distances which they prefer to maintain between themselves and various others, and that increases in population density and crowding correlate with increases in aggressive levels.

As alluded to above, research also shows that animals respond differently to proximity, depending on who the proximal others are. In primatology the best and most extensive set of experiments concerned with this phenomenon have been carried out by Bernstein (1964, 1969, 1971a, 1974a, 1974b; Bernstein & Mason 1963) through a series of literally hundreds of introductions of "intruders" into established social groups, and the introduction of several strangers to each other for the purpose of group formation. The experiments have been carried out with several different species, but most extensively with rhesus monkeys. Like others who have done research on this topic (e.g. Vessey 1971; Kawai 1960), Bernstein found that several features influence the amount of aggression elicited: the age and sex of the intruder or the strangers; the age and sex composition of the group; the previous social history of the intruder; and individual response patterns, or "personalities." In general, individuals direct the most aggression toward adults of their own sex who are strangers to them; and while male aggression in frequency and intensity exceeds female aggression during the initial phase of the introduction, it gradually declines, while female participation in agonistic episodes increases, until some type of social order is established. Infants and juveniles, as strangers or intruders, are generally accepted into social groups with little aggression.

Why do animals perceive certain other individuals of their own species as noxious or disturbing in specific situations? Why should strangers fight, or why should social groups fight with intruders? The major explanation for agonistic behavior is competition for limited resources, such as food, water and mates, and disputes are often seen to arise over access to these resources, especially in captivity where resources are usually more limited. But this does not explain why strangers trigger aggression more often than familiars. One possible explanation, on the level of ultimate or evolutionary causation, is offered by kin selection theory (see Chapter 18), which argues that individuals help their kin, who are usually familiar to them, because relatives share some genetic material in common, and helping one's kin therefore helps oneself. Strangers are less likely to be one's kin. But in this section we are concerned with immediate causes, so we will leave this argument until later.

On the level of immediate causes of aggression then, it has been suggested that strangers and intruders are perceived as disturbances to normal social life. Bernstein and

Gordon, following their experiments on rhesus monkey introductions, formulated the theory that primate aggression arises primarily through: "efforts to preserve established social patterns and to enforce expected patterns of social behavior, rather than resulting from active competition leading to conflict" (1974: 308). They argue that, contrary to what we would expect from competition theories, experiments in reducing food and space do not always increase aggressive behavior, and at best, do so only temporarily. On the other hand, experiments involving introductions of intruders and strangers always result in the most extreme types and levels of aggression until some form of social order is established. The idea that aggression may be caused by a perceived disturbance to the social order or social code has also been put forward by Hall (1968) and by Scott (1974). The latter argues that social disorganization (for example, as a result of crowding or emigration) both in the wild and in laboratory colonies may be a major cause of violence. This social disruption theory may also be expanded to explain agonistic behavior between familiars. Well-known animals can also violate a social code or behave outside of expected norms. For example, agonistic behavior can arise between familiars if one of them frightens an infant (even if it appears to do so accidentally), or if one of them appears to challenge the status of a high ranking animal, or even if one of them becomes so ill that it behaves abnormally (see, for example, Goodall's 1968 description of chimpanzee polio victims being attacked by familiar chimpanzees).

It does not seem that competition theories and social disturbance theories need be mutually exclusive explanations for what causes aggression in primate societies. Sometimes primates fight when two or more of them want to use the same limited resource at the same time and sometimes they fight after one of them has done something out of the ordinary. And finally they sometimes engage in severe aggression when the immediate cause is not discernable (see Goodall 1977; DeWaal & Hoekstra 1980). I have argued that aggression is really a very heterogeneous category, and precisely because of this I think we need not expect the behaviors so classified to have a unitary cause.

FUNCTIONS

Causes and functions of behavior can be closely interwoven, and theories of causation often lead directly to theories of function. Thus people who propose that aggression is caused by various forms of conflict have sometimes suggested that the function of aggression is *conflict resolution*. In a theory which is not contrary to this, but perhaps more specific, Bernstein proposed that because aggression in primates is caused by perceived violations of the social code and the expression of aggression tends to establish and/or restore social order, aggression therefore functions as an adaptive force in the establishment and maintenance of primate societies. Not everyone agrees with this formulation. Hinde, for example, argued that social groups and social stability occur *in spite* of aggressive behavior, rather than as a beneficial consequence of aggression. He saw aggression as a socially disruptive rather than a socially adaptive force.

Southwick (1972: 20) has suggested the following list of functions for primate aggression:

1. more effective defense against predators;

78

2. better chances of survival in competitive communities;
3. spacing of social groups and regulation of population density through dispersal;
4. home-range extension and forced utilization of new habitats;
5. sexual selection and leadership determination;

Lists such as this one are useful for summarizing succinctly the various suggested functions of aggression. However the drawback to this type of listing is that it leads us to aggregate in our minds the different types of conflict situations: predator-prey relationships (Southwick's #1); intragroup competition (#2); and between-group conflicts (#3). And it tends to mix the question of whether a trait is valuable to, or functional for, an individual, as suggested by #2; or whether it is functional for a social group or species, as suggested by #3. The two questions are not the same and should be carefully separated. Hinde showed this very nicely in a discussion of the value of aggression in human social life. It has often been argued that because aggression arises through natural selection, it must be valuable for a species, even our own. However, natural selection works mainly, if not solely, through individuals and it can select for a set of proclivities to react with aggression to certain environmental stimuli, if individuals that possess these proclivities are more likely to survive and contribute to the future gene pool. But it does not select for aggression as a unitary quality or in a direct sense, and it does not necessarily follow that what benefits the individual also benefits the social group or the species. Indeed, using the restricted or technical definition of aggression (directed towards physical harm), Hinde argues that society's major concern should be to reduce such behaviors.

Does aggression benefit the individual, as Southwick's function #2 would suggest? Surely it depends on our definition of aggression, and the context and degree of aggression under study. Most people would see self-assertive behavior as necessarily self-beneficial and leading to enhanced reproductive success. However excessive amounts of behaviour directed towards causing physical harm or even just self-assertive behavior also can have negative results for the aggressor. Such an individual can frighten away potential mates, neglect the rearing of young, or be injured during repeated violent interaction (see Hinde 1974). In a species where the individual depends on continual social life for survival, as do most of the primate species, skill in developing social alliances may be an equally important behavioral trait. Controlled and carefully directed self-assertive behaviors can be part of the process or skill of developing social alliances (see Chapter 7), but, as we have seen, aggression is not a unitary quality or trait of an individual such that we can say "the more aggressive the better". Indeed, unlimited selection for simple aggressiveness cannot lead elsewhere but to extinction for sexually-reproducing species. This is part of the problem with the oft-repeated argument, stemming from competition theories of causation, that aggression functions to ensure that the "fitter" individuals have priority of access to important resources, such as food and mates. The assumption of this argument is that the most aggressive animals are also the most fit, yet we do not have evidence for this. It is based on an intuitive and subjective supposition that we can recognize the fit because they are strong and the strong because they are aggressive. Deag (1978) has discussed how poorly "fitness" has been conceptualized and measured in primate studies, and how speculating upon the adaptive significance of behavior has become little more than a game. Fitness is a topic that will be discussed further in Chapters 7 and 17, but I would like to point out here that it often appears in the literature as poorly defined and as equally vague as the term aggression.

In summary we can say that aggression has been understood by some researchers as an adaptive mechanism which enables individuals to improve their reproductive success and by others as a mechanism which enables primates to resolve conflicts, to establish and maintain social order, to disperse themselves over resource areas, to defend themselves against predators, or to survive in competitive communities. Still other ethologists, such as Hinde, tend to emphasize the dysfunctional aspects of aggression from the point of view of the other members in a social group, even though these same aggressive behaviors are beneficial to the individual performing them.

AMOUNTS AND COMPARISONS: JUST HOW AGGRESSIVE ARE PRIMATES?

According to Southwick (1972), until twenty-five years ago monkeys and apes were considered to be extremely vicious and dangerous animals. Although primates were generally observed in zoos then, where limited space, boredom, and unnatural group composition led to high levels of physical injury, this was believed to be representative of all primates, wild-living or captive. The classic and most oft-quoted case was an introduction of a few females to many hamadryas baboon males (all strangers to one another) at the London Zoo in 1932. The resultant fighting was severe, with males competing for females to the extent that they killed each other and the females, sometimes continuing to drag female corpses about in pieces. Many extrapolations about competition in primate society were drawn from this incident and nothing was then known about the behavior of hamadryas baboons in the wild to counter the generalizations. Today we know that hamadryas baboons live in multi-female, uni-male societies, with very nearly the opposite sex ratio to that which was set up in the London Zoo; and that hamadryas social groups are established and maintained through elaborate mechanisms involving the development of bonds while females are still young, and through inhibitions between familiar adult males against stealing mates (see Chapter 16). The London Zoo introduction is a very good example of Scott's principle of social disorganization as a cause of violence. Unfortunately it is not the only example of scientists mistaking stress-related violence for normal behavioral processes in animals (Scott 1974).

Primate field studies are so new, really only beginning in any abundance after World War II, that until recently there was little contrary information to the assumption that primates behave in the wild just the way they do in captivity. Until well into this century what little information there was about wild primates still came from European adventurers' stories of encounters; travelers' tales that now appear to have been greatly exaggerated. For what imaginative stories many of them were! Gorillas were believed to scoop up hapless victims into the trees, squeeze the life out of them and drop them back onto the trail in front of their terrified companions; males of several species were believed to be so relentlessly virile that human females in their vicinity must fear abduction and a "fate worse than death"; and all primate social groups were believed to be ruled by despotic overlords, who were, of course, males. Clearly, the traits which different peoples and cultures see in animals, and/or project onto animals, especially those animals most closely related to us, the monkeys and apes, can tell us a lot about the peoples and cultures themselves.

Indeed, along with European travelers' tales, another common source of information about wild primates has been the knowledge of the local people who live in the same area as the animals. One of the problems for every field primatologist is the assessment of stories concerning primate behavior told by these people who have lived alongside the animals far longer than the primatologist. Some of these individuals are obviously knowledgeable about animals, and some of the information available in the primate literature today owes a not-always acknowledged debt to the indigenous people who have acted as guides, advisors and even sources of data. On the other hand, stories about animals make up significant aspects of the mythology of most cultures, and such stories are passed on from individual to individual, taking on cultural meaning and elaboration. Such story-tellers have no reason or directive to remain "factual" in the sense that primatologists would want them to, in order to incorporate such tales into scientific reports. A distinctive line between "fact and fiction" is not easy to draw here, and one cannot base one's decision simply on the tale's plausibility or lack of it. For example, chimpanzees have been reported by the people of tropical Africa to steal human babies and eat them, and Goodall (1968b, 1977) thinks that at least two of these reports have been adequately verified by medical and court examinations. However, baboons have been reported to engage in battles and tactical warfare by some of the peoples of Africa, but primatologists have never observed this nor any indication that it might occur even as a rare event. (Crook & Aldrich-Blake, 1968). On the island of St. Kitts we were told that vervet monkeys post "sentinels" when raiding crops, that they crack whelks (shellfish) with a hammerstone and that they bury their dead. The first description is somewhat confirmed by Poirier's (1972b) observation of vervet crop raiding on St. Kitts, and given that African vervets often manipulate objects and that other nonhuman primates are known to crack open hard fruit with hammerstones, the second is not inconceivable. But what is a primatologist to make of the third? In order to catch spider monkeys in Guatemala, some local Indians repeat a maxim, handed down from ancestral Mayans, that one need only to build a large fire because monkeys like moths are fatally attracted to the flames, and will put down their infants and approach the fire without caution. Even though these latter descriptions may not strike us as particularly credible, one is intrigued as to their possible origins, or the behaviors in the monkeys that might have given rise to the tales.

Everywhere that primatologists work in the field they encounter local stories about their subject animals and must evaluate the tales' acceptibility as "scientific" information. Some descriptions are patently mythical or at least mythologized, but scientists cannot be so arrogant as to dismiss all local knowledge out of hand, nor can they claim that their own observations and descriptions of primates will be totally free of cultural biases and cultural myths.

When reports began to come in from the first extensive scientific field studies (e.g. Jay on langurs, 1965; Schaller on the mountain gorilla, 1965; Goodall on chimpanzees, 1965) they strongly emphasized the peaceful nature of primate societies, in contradistinction to earlier views. Also, major reviews and studies of aggression in primates, such as those described earlier by Bernstein and Southwick, tended to emphasize that aggression is controlled and inhibited even in the most stressful conflict situations. More recently, I would suggest, the pendulum is beginning to swing back the other way, with some theorists emphasizing that primates in the wild are far more aggressive than we formerly thought (see for example, Goodall 1977; Hamburg & McCown 1979; Marler 1977).

Why do these swings in scientific opinion about the prevalence of aggression occur? One of the problems, as we have seen, lies with the concept of aggression itself. It is such a heterogeneous category that one primatologist could legitimately study a monkey's behavior and conclude it to be aggressive while another could study the same animal and conclude that it is not. When we are dealing with a topic this poorly defined and yet, because of our intense concern about aggression in human societies, so socially and politically sensitive, scientific objectivity can be difficult at best, and general conclusions tend to be so vague as to be useless. "Primates are more aggressive than we thought they were" is actually a meaningless statement. Which primates? What does aggressive mean here? What is more aggressive? How aggressive did we think they were?

The recent emphasis on the prevalence of aggression in primates may be related to the increasing popularity of new theories of animal behavior which stress the overriding importance of self-maximization and continual competition rather than cooperation in social animals. Certain types of agonistic encounters between members of socially living animals observed by primatologists over the past twenty years had been largely interpreted as being part of a ritualized and controlled method of resource allocation, benefitting the entire group and the species. These same agonistic encounters now are often seen as wholehearted attempts by the participants to assert themselves and to harm others in the group as a necessary part of an individualistic and competitive "game" of life.

Finally, a major stimulus for the currently popular discussions of violence in primate societies has been the data concerning intraspecific killing in chimpanzees and Hanuman langurs, information which has only become widely available through primatological studies in the past decade and only widely discussed in the past five years. The same species which were described in the 1960's as extremely peaceful and tolerant in social life are now reported to regularly kill members of their own and other groups. Because these reports on chimpanzees and langurs have been contradictory, controversial, and very important to our general assessment of aggressiveness as a part of "primate nature," they should be considered here. Both species are further described in later chapters.

Killing in Chimpanzees

No other primate field study has received so much publicity or provided so many exciting and unexpected discoveries as the research on chimpanzees conducted since 1963 at the Gombe National Park in Tanzania (see also Chapter 14). Although these apes were originally described as living quite amicably with one another and being frugivorous in diet (Goodall 1965), they have since then been found to prey on other primate species and other mammals quite regularly for meat (Teleki 1973a,b). More recently they have been seen to kill and eat chimpanzee infants, as well as to gang-attack and beat to death adult chimpanzees. The killings fall into two categories: attacks on individuals of neighboring communities by adult males, and attacks on small infants of the same community by one female and her daughter.

In 1963, Goodall began to provision chimpanzees with bananas at a feeding site in order to better observe them. Although there were definite benefits in terms of information gained from the provisioning, it is now recognized that extremely high levels of aggression between chimpanzees, and the killing and eating of baboons who were competing for bananas, plus an influx of chimpanzees who normally ranged far away, all were less desirable results of the artificial feeding (Wrangham 1974; Goodall et al 1979;

Pusey 1979). Teleki et al (1976) have also shown that peaks in mortality and slumps in natality and emigration corresponded to the timing of artificial provisioning during the 1963-1973 decade. In the past ten years the program of banana provisioning has been altered and diminished and the number of baboons killed by chimpanzees went from nine in 1968 to only one from 1969-1972 and none since then, although other mammalian prey such as bushbucks and red colobus monkeys are still caught and eaten (Wrangham 1974; Busse 1978).

In 1971 the study community of chimpanzees at Gombe appeared to begin the process of fissioning, as a small group of males and females moved to the south of the larger community. It is also possible that these individuals might have been not leaving their community, but rather returning to a range they previously occupied before being attracted to the banana provisioning grounds. Until 1968 chimpanzees were almost only watched at the feeding grounds, but as researchers began to regularly and increasingly follow them on their trips into the bush, it was found, beginning in 1973, that groups of males regularly made "patrols," characterized by tense and silent searching for chimpanzees of neighboring communities. If large groups of neighbors were encountered, noisy and vigorous mutual displays by males followed. If lone males or lone older females were encountered they were attacked and beaten by all the males in the patrol and death or disappearance often followed. Ten severe attacks on neighboring mothers or old females have been observed and five males and one female of the small "break-away" community were beaten to death, effectively exterminating the smaller group.

The second spate of killings began when one female and her daughter in the Gombe community began to kill and eat the infants of other females in the community with whom they regularly associate. Thus far they have definitely killed and consumed three infants, possibly six, and have been seen to try for three more, unsuccessfully. The infant mortality rate at Gombe in a twelve year period has gone from 33.3% in the first two-year block, to 0% for six years, to 16.7% and 83.3% respectively in the last two, two-year blocks (Goodall 1977). Adult males have been seen to partially eat chimpanzee infants killed during violent attacks on "stranger" or neighbor females, and to behave bizarrely with the dead or nearly-dead bodies, alternately beating, grooming, eating, playing with, or even cradling and carrying the fatally injured infants. Goodall thinks these infants are injured in attacks which are largely directed at their mothers. Adult males have not been seen to so attack familiar females of the home community, and in fact females are believed to seek the support of these males in defending themselves against the one adult female and her daughter who have sought to kill and eat their infants. Goodall suggests this particular female has a "glazed, mad look," that her daughter and other offspring have been very badly mothered, and have learned their grisly behavior from their mother (1979).

That chimpanzees should be sometimes extremely violent and sometimes kill each other is probably no more surprising than that they should make tools, recognize themselves in mirrors, suckle their young for five years, die of apparent grief, and learn American Sign Language readily in captivity, all of which they do. In spite of some claims to the contrary (e.g. Popp & DeVore 1979; Hrdy 1977), intraspecific killing in primates still is rarely (or only sporadically) reported from field studies and cannot be considered the "norm," but then this is also true of many of the other characteristic behaviors and capacities of the chimpanzees. Every year of research reveals to us that chimpanzees are "higher" on our very anthropomorphic scale of behavioral capacities, or as some have said, more like ourselves, than we thought. And, as Hamburg & McCown (1979) note, having expended much effort ridding ourselves of the nineteenth century notions of the totally vicious ape, it is difficult to accept or admit that chimpanzees and gorillas are not as completely pacific as the field studies of the early 1960's had indicated. However, having accepted the proposition that chimpanzees, like humans, have the capacity for intense agonism as well as intense affiliation, what still remains to be adequately explained is the extraordinarily high level of aggression, especially the killing of other chimpanzees, which is now witnessed at Gombe, and why so much of it is only recently witnessed. As Lancaster (1978) has noted, the Gombe community has a higher per capita "homicide" rate than New York City. Individual chimpanzees, who had been seen to associate peacefully for a decade, are now seen to kill each other at a remarkable rate. Similar individuals and events sometimes elicit extreme violence and other times a peaceful response. If there is a key or common underlying trigger to these aggressive episodes, researchers have not yet detected it.

As this information on killings at Gombe was disseminated to the public, newspaper and magazine headlines across North America blazed with phrases like: "warfare and cannibalism found in chimpanzees." In addition to the caution expressed in Chapter 2 that one should be critical in acceptance of such emotion-laden terms when applied to animals, it is also important to note, as Reynolds (1975) has done, that although chimpanzees have also been studied elsewhere, everything that chimpanzees do at Gombe "tends to be seized on rather uncritically by those eager to prove something (what?) about early man" (1975: 125). Indeed, although many researchers at Gombe have been careful in their extrapolations to all chimpanzees and all primates, others — laymen and scientists alike — have treated the Gombe data as a definitive vision into the crystal ball of our own ancient and irrevocable nature, rather than as the study of a limited population experiencing certain unique historical events, which it is.

Chimpanzees have also been the subjects of long-term studies in the Budongo Forest of Uganda (Sugiyama 1968, 1969, 1972, 1973), and in the Mahali Mountains south of Gombe National Park (Nishida 1979a, b). At the latter site they also were provisioned, and males also have been observed to be antagonistic to males of neighboring communities. "Patrols" and mutual displays, but no killings or beatings of other adults have been reported from the Mahali Mountains study area. One infant of a female from a different community was killed and eaten by males, and another infant is suspected of meeting the same fate (Nishida 1979b). In the Budongo Forest, chimpanzees have not been provisioned, and reports still stress the low levels of daily agonism or dominance encounters within and between communities, and the peaceful nature of chimpanzee life. However, a group of males in the Budongo Forest were once found carrying and eating a newborn infant chimpanzee. It is true that neither of these two sites has been studied for quite as long or quite as intensively as the Gombe site, and perhaps these two research projects will require more time in order to answer adequately our questions about the generalizability of Gombe data. Nishida (1979a) defends the provisioning technique of study, which he has used in the Mahali Mountains, as only changing the *rates* of behavior, such as aggression, and merely bringing out the species traits which potentially exist. Others have suggested that provisioning may alter the *nature* of chimpanzee social bonds (e.g. Pusey 1979). In fact, it is not difficult to envision the development of a

84

new and learned cultural tradition, such as regular killing of conspecifics, in a species as behaviorally flexible and quick-to-learn as the chimpanzee. Is the data from Gombe typical or a-typical of chimpanzees? Why were the killings not observed earlier? Have the accuracy and completeness of observations improved, as some have suggested, or have the inter-community interactions changed? Are the chimpanzees being crowded into the park by human disturbances of the land surrounding Gombe, as suggested by Goodall et al (1979)? When a chimpanzee female produces only one infant every five or six years and only about five offspring in her lifetime, can the Gombe community continue to exist given the current infant mortality rate? These questions all await further study.

Killing in Hanuman Langurs

As with chimpanzees, the field reports on Hanuman langurs (*Presbytis entellus*) first stressed the extreme peacefulness of the animals and their society, and it was only later that studies began to report high levels of male-male fighting and the killing of infants. However in this case there seem to be well-established differences between the behavior of the same species of langur in different research areas. At three sites in India — Jodhpur, Abu and Dharwar — Hanuman langurs live in uni-male, multi-female groups which experience periodic attacks and invasions by all-male groups. During these male invasions, and even after a new male has established himself as the one male of the breeding group (a process which is called a male "take-over"), there is much fighting between males. There is also some fighting and some mating between males and females, and many of the unweaned infants of the attacked group disappear. Of some forty infants whose disappearance has been recorded during such disturbances, three have been seen to be killed at Jodhpur, and one seriously wounded, by "usurper" males (Mohnot 1971, Makwana 1979); one was seen to be wounded during a male-attack on a mother-infant pair at Dharwar and later died (Sugiyama 1965); and several were seen to be attacked at Abu with one being seriously wounded (Hrdy 1974, 1977a, b). Local inhabitants in the Abu area also reported to Hrdy that males were killing infants. It is inferred by the researchers at these three sites that all of the forty infants disappeared and died after being attacked by invader males. One assumes that the infant mortality rate is much greater during these periods of male invasions, however I have not seen comparative data concerning the rate of infant death or disappearance outside of the male-invasion context. While there is no doubt that these infants are disappearing in the context of male aggression, there is considerable disagreement over whether their deaths are the results of goal-directed attacks on the part of invading males, or are the results of infant-vulnerability during periods of social upheaval. Are infants caught up in the general violence and "accidently" wounded or killed, or are the invading males "trying" to kill infants?

At other research sites such as Orcha and Kaukori in India (Jay 1965; Curtin and Dolhinow 1978), Junbesi and Melemchi in Nepal (Boggess 1979; Curtin 1977), Hanuman langurs have been found to live in multi-male, multi-female groups and no "male invasions" or infant killings are reported. At the Junbesi site, males were reported to be very competitive during mating seasons and to leave and enter groups in the context of male-male skirmishes. However neither extreme violence and total defeat, permanent exclusion of a male from a group, nor attacks on mother-infant pairs have been reported from these latter sites.

Two major explanations or "theories" have been offered for langur infant killing. One is

that the phenomenon is a pathological result of extreme crowding and habitat disturbance at the three sites in question — Jodhpur, Dharwar and Abu (Curtin 1977; Boggess
1979; Curtin and Dolhinow 1978). The other is that infant killing is a male reproductive
strategy, whereby individuals who have short tenure as breeding males, kill the infants of
rivals, thereby hastening their mothers into estrus. The infanticidal males then mate with
the females and produce their own young (Hrdy 1974, 1977a, 1977b, 1979). The major
proponent of the latter explanation, Sarah Blaffer Hrdy, agrees that there are higher
langur population densities at the sites of infant killings, but argues that the resulting male behavior is not pathological, but an adaptive strategy for successful reproduction
in this high density context. There are numerous ramifications of these arguments, some
of which will be discussed in Chapter 18, which concerns sociobiological theories.

For the moment I simply would like to consider these data in light of the question:
how prevalent and how important is extreme violence in primate societies? Hrdy (1979)
argues that the langur data on infanticide has stimulated interest in, and concern for,
information on killing in free-ranging primates, and as a result, more and more reports of
such killings are being made. Indeed more and more examples of infants disappearing in
the context of male aggression are being published (e.g., Rudran 1973; Wolf & Fleagle
1977; Oates 1978; Struhsaker 1977; Fossey 1979; Galat-Luong & Galat 1979; Marsh
1979). There seems to be an emerging pattern of infanticide in polygynous societies when
new males invade and attack unfamiliar mother-infant pairs. However, it is also true that
most of these are isolated instances and/ or inferred cases of killing. Such inferences
should always be approached with the greatest possible caution. The practice, somewhat
popular now, of reanalysing old data and attributing past infant disappearances to male
killing in the absence of any evidence, is somewhat less than rigorous scientific practice.
For infants, as well as adults, do die and disappear for any number of reasons, including
disease, predation and accidents. Indeed, Sugiyama and Parthasarathy (1977) reported that
the Dharwar population of langurs in 1976 had decreased to 54.5% of its 1961 size,
and they attributed this significant decline to human disturbance. Conspecific killing *is*
difficult to observe in the wild, and if we are to understand its pattern of occurrence, then
actual observations need to be carefully sifted from assumptions and inferences. Also, the
relative importance of infant killings would need to be considered in the context of all
demographic data for the species (especially emigration, mortality and reproductive rates),
before the value of a suggested "infanticidal adaptive strategy" could be determined. As
just three examples, if males join and leave groups too rapidly to ensure the successful
development of their own infants, or if females only produce one infant every two years,
whatever the circumstances, or if adult males in the same area are closely related, then
killing of infants would be unlikely to be advantageous to the males. It would seem that
the life-history approach to primate behavior advocated by Altmann and Altmann (1979),
and the first attempt at a predictive model which takes into account male tenure length
and female reproductive patterns (Hausfater 1977; Chapman & Hausfater 1979), represent such ways of understanding possible patterns of infant-killing in the whole context
of the animals' life-way (see also Ripley 1980).

Perhaps in few other areas of animal research as the study of aggression have the
reported field data been so colored by the prevailing cultural attitudes of the period and
by the advocacy positions of researchers. There can be no simple, unitary answer to the
question of "just how aggressive are primates?" Killing of conspecifics would seem to be

in the "primate repertoire" or at least a potential behavior in certain contexts, but to paraphrase Hrdy (1979: 20): 'in no species is killing a common event.' The vast majority of agonistic interactions in primates consist of threats, displays, retreats, submissive gestures and counter-threats. Obviously the killing of conspecifics in the context of overt fighting is an important topic for further study. However this relatively "uncommon" phenomenon, or still rarely reported event in free-ranging primates, should not preoccupy us so that we focus upon it to the exclusion of other more subtle, but equally important and more frequent patterns of agonism. The killing of "helpless" infants by male canine puncture wounds may catch our horrified attention, but the suppression of ovulation and reproduction in subordinate females by dominant females (see Chapter 16) is no less significant in its biological impact. And being harried slowly, through the use of subtle but continued threats, into a state of depression, or exile from the group, or even death, is no less an extreme form of violence and pain than the more headline-grabbing "gang attacks" of chimpanzee males. That most primate agonism consists of threats, rather than physical confronations, is no reason to underestimate their significance.

"Aggression" not only refers to a loose constellation of self-assertive behaviors with different motivations, patterns and functions in any given individual, but its manifestation is highly variable from species to species. Attempts at comparative reviews of aggression in different primate species, such as those by A. Jolly (1972) and Southwick (1972) conclude that aggressive acts as they occur in different individuals, species, or situations, are really not comparable in any quantitative sense. Macaques, for example, are often said to be more aggressive than Cercopithecus monkeys. However in stressful captive conditions macaques threaten each other far more often, while Cercopithecus monkeys wound each other more often (Rowell 1971; Kaplan 1977). We cannot conclude from this that one species is more aggressive than the other, only that they exhibit different agonistic patterns and react differently to the stress of captivity.

SEX DIFFERENCES IN AGGRESSION

Almost all reviews of sex differences in aggression begin with the statement that males are more aggressive than females (e.g. Moyer 1974; Gray 1971; Maccoby & Jacklin 1974; Mitchell 1979; and Martin & Voorhies 1975). Almost all reviews then proceed to demonstrate that male aggressive patterns are *different* from female aggressive pattern — different in motivation, neural and hormonal bases, pattern of expression, and context. The male and female patterns of aggression have been characterized in various ways, but one frequent theme is to link physical contact aggression more to males, and agonism involving threat and counter-threat ("squabbles") more to females. Many psychologists have argued that in humans the underlying motivation of aggression — "to hurt others" — is the same for males and females, but is channeled into different forms for boys (overt physical expression) than girls (disguised, verbal or repressed form). Bardwick (1971) has argued that physical aggression is more characteristic of boys, while psychological aggression (e.g. verbal manipulation, withdrawal, gossip) is more characteristic of girls, and numerous papers in the psychological literature suggest that socializing agents (parents, teachers, etc.) stimulate and reinforce different aggressive patterns for boys and girls.

Maccoby and Jacklin (1974) contended that socialization hypotheses are really not adequate to explain the nearly universal, or at least highly prevalent, pattern of primate males at a very early age exhibiting more aggressive responses in most categories of aggression. They argued that both sexes are not equal in their initial tendencies to exhibit aggression. Males are believed to be predisposed to learning and using aggressive patterns due to the effects of prenatal androgens, which also predispose them to greater expression of gross motor energy (see Chapter 11).

Maccoby and Jacklin also argued that primate males aggress primarily against each other and less often against females, and indeed discussions of male-male aggression usually form considerable portions of review papers on aggression. However the same may be true of females, that is, they may aggress primarily against each other and less often against males. For example, several researchers have demonstrated that macaque females are more highly aggressive towards each other than towards males in many circumstances (Kawai 1960; Vessey 1971; Erwin et al 1975; Hendricks et al 1975) and may exhibit different levels of aggression depending on the presence or absence of males (B. Anderson et al 1977). The near-impossibility of accounting for all the contextual factors which may affect levels of aggression, as well as the heterogeneous nature of the concept "aggression", may help to explain contradictory findings for sex differences in even one species. For example, rhesus macaque females are said to exhibit more threats in some studies (Mason et al 1960; Chance 1956; Sackett 1974) and rhesus males in others (Altmann 1968; Drickamer 1974). Bernstein (1972) found that pigtail macaque females spend more time in agonistic interactions than do males, but then, as he noted, pigtail females spend more time in all social interactions than do males. There are prominent exceptions to what most reviewers would call the general rule of greater male aggressiveness in primates. Firstly there are certain contexts or environmental stimulus situations, such as attacks on their infants or introduction of stranger females, in which females exhibit more frequent or more severe aggressive responses than males. Secondly, there exist species in all the major taxonomic categories of primates for which females are said to be usually "more aggressive" than males — e.g. Indri and ring-tailed lemurs for prosimians (Jolly 1966; Pollock 1979) callitrichids for New World monkeys; talapoins and patas for Old World monkeys; and hylobatids for apes. Burton (1977), in a review of gender identity in nonhuman primates, noted that females in several species may be the source of most aggressive *and* most socially adhesive behaviors. There are also exceptions to the rule that males direct most of their aggressive behaviors towards other males, such as spider monkeys, in which most agonism occurs between the sexes.

A more productive approach to sex differences in aggression than attempting to compare some overall "amount", is to describe how males and females differ or exhibit commonalities in their response to particular situations. As just one example, in the rhesus macaque "introduction-of-strangers" experiments by Bernstein, described earlier, it was found that both males and females respond agonistically to strangers. However, males participate most in the initial severe fighting and they fight most with other males, but as their agonistic rates decline, female agonistic rates increase. Female rhesus macaques direct their attacks more toward stranger females and while they do not have the large slashing canines of the males, they are capable of smaller pressure wounds with a deadly cumulative effect. Most importantly, Bernstein found that females do not cease to attack a helpless, defeated animal as the males do. The most severe male fights often occur when a female enlists one or more males against another. It seems likely then that at least for rhesus monkeys in this context — and, one might argue that for many species in many contexts — the "set of proclivities" toward aggression and the exact environmental triggers can differ between sexes, and the pattern in which aggression is expressed can be different.

Returning to the question posed at the beginning of Chapter 3 — do males dominate females because males are instinctively *more aggressive* than females — I think that it can be seen to be essentially unanswerable because of the way it is posed. "Aggression" is such a heterogeneous category of behaviors and interactions that "amount of aggression" is not a very useful concept, and relative amounts of aggression are not the most relevant comparisons to make between males and females in order to understand the nature of the phenomena. If we wish to achieve some understanding of the nature of sex differences in aggressive patterns in the primate world, it is surely more to the point to direct our research toward determining when, how and why male and female primates respond agonistically and differently, or similarly, to given environmental circumstances. If we ask — do males dominate females because males exhibit *different* aggressive patterns than females — the question is becoming more answerable, although the concept of "dominance" still needs to be clarified. In the following chapter we shall turn to a consideration of the concept of dominance.

Chapter *SEVEN*
Dominance and Alliance

Vervet

Cercopithecus aethiops

The adaptable and semi-terrestrial vervet ranges widely in Africa. The postures and threat-faces of the female (right front) and subadult male, in their coalition against the defensive adult male (left) are typical of this species.

Almost all people may imagine the troop of monkeys as a society ruled by force. They will think it right to imagine so, when they see monkeys gathered at the center of the feeding ground. How odious is that big male monkey sitting impudently at the center of the feeding ground! He is driving away other male monkeys from females in order to keep them to himself. Moreover, he tries to get every orange, never to share with others. He seems to be the leader but he is no better than a tyrant. Everyone would get such an impression. Now we must discuss the role of female monkeys in the troop . . . Although the society of monkeys may appear to be under the dictatorship of a mighty male monkey, females actually have great influence in the society.

Imanishi 1960: 40,41

We saw in the last chapter that when individuals of a group come into conflict, the interactions involved are referred to as agonism. The study of dominance shifts our focus to the *outcome* of such conflicts, for the nature of the resolution as well as the conflict itself, is thought to tell us something about the relationships of the group members. If one individual clearly physically defeats another, or by some criterion appears to "win" a dispute, the first individual is said to be dominant *in that situation*. If the same individual consistently wins in this type of dispute, it can be said to be dominant over the other *in that type of situation*. Further, if it always wins disputes of any kind, it could be said to be, simply, "dominant." The latter, a generalized attribute, often attributed to physical strength, is the way many of us tend to think of dominance. Yet it is the rarest type of dominance in the animal world. The ability to win a conflict is usually specific to certain kinds of situations and to certain "others" and not generalizable to an overall power over others of the social group. The total despot or tyrannical dictator is less characteristic of animal than of human societies, and even in the latter despots seldom if ever rule by virtue of physical stength alone.

More often than not we learn something about the dominance relationships between members of a primate social group because they *do not* enter into overt disputes when we expect they should. If an important commodity, such as food, water, space, or sexual partners is limited, it is expected that individual group members will compete for these. If, instead of fighting over the limited resource, one individual takes and uses it first, while while the others wait and may or may not have a turn at all, we infer that the "taker" does so because it is dominant over the others in this case and could win the fight over the resource, although it is not necessary to prove so in each instance. There are, of course, always a number of other reasons why one individual should have the first turn at using a resource (for example, the taker is simply closest and therefore in a sense "posesses" it through proximity, see Kummer 1973). This is one of the serious confounding variables with which dominance studies must always contend.

It should be obvious that one aspect of the relationships between members of a social group concerns the ability to exert control, influence, or power over others in given situations. Further, as scientists, we may wish to study and measure how conflicts or potential conflicts of individual interests reveal one aspect of social power, and we may call this the ability to "dominate" others. But how does power and influence over others arise? Is it almost solely due to superior physical strength and physical coercion as is often stated or implied? How is social power expressed? Can we recognize it only in disputes? How does it function? And finally, just how important is the asymmetry or imbalance of power in particular interactions to the overall relationship between those individuals? How important is the social structure of dominance hierarchies to social life in general? These are some of the questions around which current scientific debate over the dominance concept revolves. This chapter will examine these questions as well as the definitions and measurements of dominance. I will argue that we often over-simplify the phenomena categorized together as dominance, as well as overestimating the importance of physical coercion in day-to-day primate life. The ability to exert control and influence over others in given situations is a prominent social reality and probably a sophisticated social skill. An additional focus on *alliances* based on kinship, friendship, consortship and roles, and on social power revealed in phenomena such as leadership, attention-structure, social facilitation and inhibition, may help us to better understand the dynamics of primate social interaction. Also it may help us to place competition and cooperation among social primates in proper perspective as intertwined rather than opposing forces, and female as well as male primates in their proper perspective as playing major roles in primate "politics" through their participation in alliance systems.

DEFINITIONS

What does it mean then to say that one individual dominates another? The dictionary defines it as: "rule or control over others by superior power or influence," and that sort of general, pervasive power over others is what most people think of as dominance. Thus, to dominate others is widely considered to be the ability to rule, control, or influence others. In contrast, ethologists who study dominance in animals have used several definitions, all of which are operational and so more limited in scope than the popular understanding of the term. That is, scientific definitions refer to dominance as special forms of power which are established through conflict and self-assertion or potential for "aggression." Likewise dominance is *measured* through studying the outcome of conflicts and/or the patterns of conflict avoidance. Hinde (1978) and Wade (1970) have also distinguished between two conceptual levels in the scientific study of dominance — the asymmetrical relationship between two individuals (dyadic level) being the first, and the social structure of dominance in an *entire* group (hierarchy at the social level) being the second.

Many definitions and many studies of dominance refer to Schjelderup-Ebbe's now classic studies of peck order in domestic chickens (1922), which showed that strange hens placed together in an enclosure fight by pecking each other until a straight-line order or linear hierarchy is established, such that chickens can be clearly ranked according to their ability to peck others — the alpha chicken can peck all others, more than it is pecked itself, the beta chicken can peck all but the alpha, etc. Schjelderup-Ebbe, who is in a sense the "father" of dominance studies, generalized his findings to a theory of despotism as the fundamental structure of the universe. Whether or not one now accepts his generalizations, a peck order definition of dominance implies that animals organize themselves according to the ability to physically defeat, or intimidate others in conflicts.

Later ethologists were more interested in focusing upon the functional benefits of being dominant, and they came to define and measure dominance as expressed through "priority of access to incentives," which means having precedence over others in using desired and needed resources such as food, water, space, sexual partners and other social companions. If desired objects are limited and not everyone can use them at the same time, those who have first access to them are said to be dominant. The "power" over others inherent in this definition of dominance then, is the ability to "go first," which may or may not be the same as the ability to physically defeat others, depending on how this ability to go first is established. Syme (1974) has referred to the "priority of access" definition of dominance as describing "competitive rank orders," whereas the "peck order" definition describes "aggressive rank orders." Some researchers have argued that competitive rank orders *reduce* agonistic interactions and, in a sense, allocate resources. However, others have rejected this as a group selectionist argument (see Chapter 18) or countered that all rank orders are only evident because of *high* rates of agonistic interactions in the first place (Rowell 1974; Gartlan 1968).

93

The words "dominance," "rank," "status" and "hierarchy" are often used interchangeably, but they do have different meanings. Dominance as we have seen, can take on different operational definitions, that is, it is usually defined by scientists according to how it is measured, but dominance basically refers to some form of power over others established through intimidation. Rank means a relative position in a row, line or series. Individuals can be ranked according to myriad criteria, such as how large they are, or how many points they score on intelligence tests. "Dominance rank" refers to the relative amount of power over others in conflicts and conflict avoidance which an animal can exhibit. Status is often a more general term meaning a state, condition or position, which is not always a relative position, in a ranked order. Thus an individual monkey's status could be "healthy," "old," or "pregnant," none of which are direct statements about the individual's dominance rank or power over others. Although many ethologists use "status" synonymously with, or in place of dominance rank, I will avoid the term in this chapter to help simplify an already complex topic. A hierarchy is a group of things which are arranged in order of rank or grade according to some value system or set of criteria; and a linear dominance hierarchy is a straight-line rank ordering of animals drawn up by the researcher according to what are considered to be the animals' relative abilities to intimidate each other and thus win conflicts or use resources first.

Two very important concepts which are occasionally mentioned, but until recently seldom adequately explored in connection with dominance studies, are "alliance" and "coalition." Although cultural anthropologists often use the former term to refer specifically to unions through marriage, alliances are also defined as semi-permanent or stable associations between individuals for a common purpose or purposes. Coalitions are temporary alliances for some specific purpose. As I suggest later in this chapter, alliances and coalitions may be a key to understanding how an individual primate comes to exhibit power over others in its social group.

MEASUREMENTS

Measures of dominance have been even more variable than definitions of dominance, and since many scientific definitions are operational, knowing something about the various ways in which it has been measured is a considerable help in understanding how the concept is used.

Peck Order
Most simply, one may observe a social group and record all the winners and losers of physical fights. If animal A defeats all others in fights, it is said to be first ranking, most dominant, or the "alpha" individual. This is of course a measurement based on the peck order definition of dominance and is dependent on social interactions in which fights occur and an observer can clearly decide who has won and who has lost. This is not always possible with primates, and the researcher may have to decide upon some fairly arbitrary criterion of winning or losing.

Direction of Agonistic Signals
Many species of primates seldom engage in actual physical fights. Rather, they may

94

exhibit behavioral signals which threaten physical contact and/or they may use signals to indicate they will submit should there be physical contact, so that actual fighting is forestalled. This is probably one of the reasons many studies have shown that physical prowess is not the fundamental determinant of dominance rank in primates, because so often relative physical strength does not come to the test. Since a great deal more of primate agonistic behavior consists of threats and submission than of physical confrontations, primatologists often have turned to the direction of agonistic signals as a measure of dominance. As a result, animals may be ranked according to how many and to whom else in their social group they direct threats, more than they receive threats from. One factor which must be kept in mind with this type of ranking is that individual primates not uncommonly threaten others in their group, who then respond by ignoring them or threatening them in return. It may be difficult to evaluate relative dominance rank in such an interaction. Also, it has been found that frequency of threatening signals, or "amount of aggressiveness," if one wishes to call it that, is not a very good indicator of dominance rank (e.g. Eaton & Resko 1974; Hinde 1978; Bernstein 1970; Bramblett 1976; Rayor & Chizar 1978). Individuals judged by other criteria to be "alpha" (the most dominant) animals of their groups, are sometimes found to exhibit a very low level of threats, and vice versa. We must not confuse the search for an asymmetrical pattern of threat and submission with the overall *amount* of threatening.

Then too, some primatologists have begun to suggest that the *response* to a threat tells us more about a dominance relationship, and that only submissive signals are clear indicators of subordination, rather than just "being threatened." So animals may be ranked according to who they direct submissive signals toward, and the whole group may be ordered into a "subordination" or "subordinacy" hierarchy (Deag 1977). This has been suggested by Rowell (1966, 1974) to be a more useful concept than a dominance hierarchy. Others (e.g. Deag 1977; Wade 1979) disagree.

Direction of Approach — Retreat Interactions

There are some species of primates and some situations in all species in which observers not only seldom see fighting, but where even conflicts in which threats and submissive gestures are exchanged are rare. In such species and situations individuals may avoid certain impending interactions with others by leaving the scene or simply turning away. These are known as approach-retreat or approach-withdrawal interactions.

Observers have in some cases used approach-retreat interactions as a measure of dominance, on the assumption that animals may be ranked according to who will move away or who will avoid interactions. The most common example is when one group member approaches another who is grooming, eating, mating or simply sitting, and the approached individual hurries away before any overt exchange of signals occurs. This form of approach-retreat is variously termed "supplantation" or "avoidance" depending on the decision of the researcher as to whether the surrender of space or goods was forced or voluntary. This particular measure of dominance may also be known as priority of access to space (see below).

In many primates, animals so seldom engage in physical agonism or even threat of physical agonism, that approach-retreat interactions are the only sufficiently frequent clues to the dominance relationship between members of a social group available to the researcher. At least one study has suggested that approach-retreat is a reasonably good measure of dominance rank (Rowell 1966), but most primatologists who have used this measure, including myself, are aware of the rather large assumption about individual motivation that it requires. If an individual avoids an impending interaction can we always assume the avoidance is motivated by fear or subordinate rank, or recognition of the approacher's higher dominance rank? Might it not sometimes be motivated by a simple desire to avoid social interactions at that time? We do know that we can not always assume the approaching individual intends to intimidate, because what an approaching animal does after arriving equally often indicates its intention to simply join the grooming group or to enjoy some other form of social companionship. In general, we are probably fairly safe in assuming that dominant individuals *ignore* or threaten away approaches and invitations to social interactions when they do not wish to be so engaged, while subordinate individuals *retreat* from such approaches, and so we can use this as a measure of dominance rank. But we should be aware of the assumption involved, and the very fact that we are in a sense, *forced* to use such a measure, shows just how subtle dominance relationships in primates can be.

Priority of Access to Incentives

The reader can see that in adjusting our concept of dominance to what we can actually measure in normally occuring social behavior, we have moved quite some distance from a picture of individuals pecking each other or physically defeating each other in fights. Yet, there are species in which even approach-retreat interactions do not occur often enough to be used as a measure of dominance, and some researchers have concluded that there is no dominance hierarchy at all in these species. However, other researchers have not been satisfied with such a conclusion and have developed another way of addressing the problem, which is to artificially create competitive situations such that animals are compelled to compete. This has been a more popular method of studying dominance than simply waiting for conflict situations to arise on their own.

1. Food Tests

The most common method of artificially creating competition is to reduce food or water to just one source, and then study the order in which individuals of a hungry or thirsty group eat or drink. This method is used quite often in the laboratory (for example, one water spigot is offered to a group of thirsty animals). Japanese primatologists have also used a version of this in their studies of dominance in free-ranging Japanese macaques, in the form of the "peanut test" or the "caramel test." One piece of food (a peanut, a candy or a raisin, etc.) is dropped between two individuals, and the one who picks it up and eats it is labeled dominant. These tests are almost always conducted on animals two at a time, because dominance has traditionally been studied in terms of "dyadic" interactions. Even linear hierarchies, although Hinde (1978) and Wade (1977) distinguish them from dyadic dominance, are actually ranked lists built up from comparing pairs of individuals. The reasons for, and drawbacks to, the dyadic approach will be discussed in a section below.

It does seem important at this point to comment on the difficulties of artificial food tests of dominance. Most primatologists who conduct food tests are aware of these problems, but continue to believe the method can be used despite them. When the experimenter drops one piece of food between two primates, a whole array of factors may determine which of the two test animals will pick it up. Many of these factors are believed to obscure the true or "basic" dominance which is being tested, and thus researchers often try to control for these "confounding variables." Dominance rank may vary with social context such that, for example, a monkey will usually avoid taking food from the infant of a dominant female if the mother is present. If however the mother is not present, the test monkey may sometimes treat the infant as though it had its mother's rank, while at other times it may attempt to "dominate" the presently isolated infant. Thus the dyadic rankings recorded between infants and other monkeys will be variable and make little sense without a knowledge of the social context. Even when two monkeys are taken from their group and placed alone, a third monkey's influence may be invisible but significant. Most researchers have avoided the problem of social history by giving food tests to pairs of monkeys who are strangers to each other. In this case, it has been found that the two monkeys should be of equal size, sex, health and age, in order to control for these effects; they must be introduced to a strange cage at the same time so neither is more familiar with the surroundings; they must be equally hungry; they must like the food item equally well (candy, peanut, raisin); and the food must be dropped exactly between them so we do not mistake proximity for priority of access.

To my knowledge all of these conditions have never been met in a food test study of dominance. In addition, given that in normal social life primates never meet in isolated pairs, strangers to each other, in unfamiliar areas, arriving in equivalent physical conditions at the same time to battle over one piece of food exactly between them, one may legitimately ask why researchers would want to create and measure such an interaction? People who have performed these tests generally think that repeated trials reveal either an underlying unitary attribute of the individual primate (dominance), or the underlying foundation of social relations (power conflicts), without the "noise" of confounding factors. In fact, it has been found that food tests do not generalize to other conflict situations in any consistent way. Not only does the test-situation only exist in the artificial laboratory-test setting, it has been found that priority to food does not necessarily correlate with priority to other incentives (see below), and that dominance determined through dyadic tests does not generalize to dominance relationships for the same individuals within the group as a whole (e.g., Hansen & Mason 1962; Rowell 1974; Maslow 1936; Hinde 1978).

In this sense then, much as I.Q. is often defined as "that which I.Q. Tests measure," food test dominance could be defined as that which food tests measure. It is a self-limited finding of unknown value outside of the context in which it was tested. The reason is that dominance relationships, when they exist, do not exist in a social vacuum; rather, they are interrelated with many of the very factors which experiments have attempted to eliminate as "noise" or confounding variables. Rather than peeling away the layers of the behavioral onion to arrive at the core of an underlying "real" dominance rank or dominance relationship, it can be argued that the experimenter has in fact *created* the dominance relationship seen in dyadic food tests (Rowell 1974; Gartlan 1964).

In their critiques of dominance as it has been conceptualized and studied in primates, Rowell and Gartlan also brought together evidence to indicate that the expression of dominance, and the existence of dominance hierarchies, are much more prevalent in captivity than in the wild. They suggested that dominance hierarchies are a pathological result of competition and stress and, rather than controlling agonism, dominance hierarchies are *produced* by abnormally high rates of agonistic interactions. Others, such as Wade (1976) and Deag (1977), countered this sort of argument is "throwing out the baby with the bathwater." They agreed that there have been problems with, and oversimplifications in, the manner in which dominance has been studied, but argued that dominance-related behaviors do occur and are important in many primates, captive and feral.

So, we are left with this important problem concerning food tests of dominance: if researchers find that for certain species they have to artificially create situations of competition and stress in order to observe and demonstrate dominance, does this mean that dominance relationships are (1) real underlying structures even when not ordinarily apparent? (2) one of the potential responses of primate groups to extreme conditions, and therefore of interest in understanding social behavior (3) an artificial phenomena of our own making, of little important in understanding the social behavior of these animals?

I have tried to show that the answers to these questions are as contradictory as are the questions themselves and, in the face of conflicting data, depend in good part on the conceptualization of dominance favored by the researcher.

2.   Sexual Partners

As well as access to preferred locations and to food and water, individuals may compete for access to sexual partners. The theory of sexual selection suggests that females are generally the limiting resource in animal societies and that males will compete for access to fertile females with which to mate. A model has been developed in primate studies, which predicts that males have access to estrous females in direct relation to male ranks in a dominance hierarchy; that is, the alpha or top-ranking male has first access to all estrous females, the beta or second-ranking male has second access, and so on. Many descriptions of the model further suggest that the highest ranking males carry the "best genes," and will be the most reproductively successful. Thus these males will perpetuate the "best genes" and the genetically based aspects of high dominance rank.

This "priority of access to estrous females" model will be considered at some length in Chapter 17, but it should be mentioned here that there are weaknesses in its every premise. Males do compete for access to estrous females, but females also compete for access to preferred male partners. In fact, in groups of geladas, vervets, Sykes' and Japanese macaques I have observed, threats and physical confrontations between females involving who would sit closest to a male or who would solicit sexual interaction first, were not an uncommon sight. Qualities successful in male-male competition, resulting in potential "prior access" to females, may not be those qualities chosen by females. There is no necessary correlation between qualities leading to successful access and qualities leading to successful choice. Although the highest-ranking male may be able to drive away competing males, it does not automatically follow that he can form a successful consort with every female, or that he will have the greatest reproductive succes. It should go without saying that estrous female primates are not equivalent to passive resources, such as water and peanuts. Finally, as I will argue shortly, the more dominant males at any given time period in the history of a social group are not necessarily the possessors of "better genes" than the more subordinate males. Although priority of access to sexual partners, is usually listed as one of the incentives or resources involved in dominance, it has not in fact been used as a way of determining dominance. Studies thus far have looked for correlations between dominance as determined by some other means (e.g., direction of agonistic signals, food tests, and male reproductive success). It is this thorny problem of correlations between different expressions and measures of dominance to which I shall turn next.

CORRELATIONS BETWEEN MEASUREMENTS

When one primatologist is learning from another primatologist about a group of primates, the question that often comes up first, or at least early on in the acquaintance process, is, what is the dominance hierarchy? This is almost a way of asking, "What is the social organization?"; although more and more primatologists are becoming aware that the best use of the dominance concept is to restrict it to its proper domain of specified forms of power, control or influence, in specified social interactions. If a researcher answers, "A is dominant over B and B over C," we have seen from the discussion of different measures used that the researcher may mean A can physically defeat B in a fight, *or* A threatens B more often than B threatens A, *or* B directs submissive gestures to A, *or* A takes the raisin in food tests, *or* A has first access to space or sexual partners. In addition it might mean that A is always the mounter in non-sexual mounts, and is groomed more often, *or* A leads B out to forage, *or* B pays more attention to A.

If the researcher could say that all of these aspects of the relationship between A and B were true, we would have an enormously useful concept in dominance because it would be the most valuable and predictive model for understanding the interactions between A and B. Unfortunately, dominance relationships in animals never have this absolute, unitary quality. A number of studies have now been conducted which examine the question of correlations between different measures of dominance in the same group of primates (Rowell 1966; McKenna 1978; Bernstein 1970; Drew 1973; Eisenberg & Kuehn 1966; Gartlan 1968; Saayman 1971). Although one study in particular (Richards 1974; and also see Loy 1975) reported high correlations between some measures, most studies found consistently low correlations between dominance rankings based on different measures.

Hinde (1974) said that the dominance concept would be useful only if it had the consistent or correlational quality of an "intervening variable"; that is, if the different ways in which dominance can be measured and assessed agreed with each other, and if the asymmetrical relationship of dominance between two individuals could be manifested in a variety of ways. An intervening variable is a characteristic such as "age and sex" or a postulated characteristic such as "hunger" or "aggressiveness" which has explanatory power because it can successfully predict a variety of behaviors and interactions. Dominance rank might be such an intervening variable which could predict how an individual or a dyad would behave in a variety of situations. It is probably fair to say that the consensus of opinion today is that dominance rank does not have this unitary quality or universal predictive value and explanatory power in primate species. Measures of dominance as closely related as water tests and food tests may not correlate. An individual may go first to the water spigot, but not to the peanut (Syme 1974). However, Hinde has recently altered his opinion (1978). He now suggests that dominance is useful both for the description of asymmetrical interactions in which one individual bosses the other, and for the description of a particular type of social structure in which individuals are rank ordered into a hierarchy according to some criterion of ability to boss others related to aggression. These two concepts of dominance, he thinks, are not necessarily related.

I would suggest that the frequent failure to find correlations between different measures of dominance indicates that it is incorrect to think of dominance rank as a unitary characteristic possessed by an individual, and incorrect to think of dominance relations in dyads, and dominance hierarchies in groups, as a single phenomenon generalizable and applicable to all social interactions. This does not mean that dominance is not a useful concept in understanding primate interactions. The ability to win in conflicts and to intimidate others is an observable and important social skill: it is, however, situation- and context-dependent (see Bernstein & Gordon 1980; Bernstein et al. 1979). Individuals win in certain conflicts and go first in certain competitive situations, but not in others. Even where correlations do exist, they are still not high enough to suggest a unitary phenomenon being measured by several independent techniques (Syme 1974).

DOMINANCE, MOUNTING, AND GROOMING

Grooming

One example of just how all-pervasive the effect of dominance is considered to be, is the common assumption that dominant individuals receive more grooming and more attention and more "presentations" (an interaction in which the perianal or anogenital region of one individual is placed close to the face of another), while subordinate individuals receive more mounts in non-sexual contexts. Both McKenna (1978) and Bernstein (1970) pointed out that grooming is in part a social strategy which may be employed in a variety of ways (to approach the mother of a new infant, to reduce the struggles of a youngster being weaned, to court a potential sexual partner), and has no necessary relationship to dominance rank (however, see Seyfarth 1977). We do not even have evidence that to *be* groomed is somehow more desirable than to groom (Altmann and Walters 1978).

Similarly, presenting the anogenital region and mounting the presenter are known to occur in several contexts: in greeting, in sexual interaction, during agonistic encounters or tense situations, during grooming, and in approaches to infants. It has been assumed widely that presenting and being mounted are submissive or appeasement gestures, ritualized from female copulatory behavior, while mounting is ritualized from male copulatory behavior and represents a dominant or even an aggressive gesture: "The more dominant individuals play the masculine role, while the subordinate monkeys usually enact the part which corresponds to the female role" (Carpenter 1964: 329). Thus, there is a persistent idea in the primate literature that non-sexual presenting and mounting, especially between males, may be labelled dominance behavior, and that dominant animals mount subordinates while subordinates present and are mounted (for example, see Wickler 1967; Altmann 1961; Chance 1963; Tokuda 1961). On the other hand, several studies have reported no relationship between rank and the direction or frequency of male-male mounting (Bernstein 1970; Jay 1965; Kaufmann 1965; Simonds 1965; Rowell 1972b).

Hinde and Stevenson-Hinde (1976) suggested that presentations might be gestures of "social approval," and Rowell similarly referred to presentations as a gesture of "politeness" used by all ages and both sexes in a number of primate species. Yet a third suggestion was made by Andrew (1964), who noted that presenting might have evolved from anogenital scent-marking rather than copulation, as presentations often bring some of the scent-glands into close proximity with another individual's face. (Simonds 1965; Rowell 1972b)

Hanby (1974) published a very detailed analysis of male-male mounting and presentations in Japanese monkeys, and a review of past studies, including many of the references listed above. She found little support, not only for the motivational basis of the "mounting equals dominance" theory, but also for the assertion that dominant animals do indeed always mount less dominant ones. Her study and review led her to point out that even when mounting occurs in an agonistic or tense context, the mounting animal might just as easily be motivated by fear, or need for reassurance, or the tendency to revert to infantile clinging, rather than by feeling "aggressive."

She also found that presenting in an agonistic context is not a straightforward indicator of subordinate rank in a dominance interaction, as had been assumed. Her data do not support the widespread idea that presenting is a device used by subordinates to deflect attack from high-ranking monkeys (see exposition of the theory in Wickler 1967), since monkeys of all ranks present. Thus, it also seems that presenting, even in tense situations, is not necessarily motivated by either submissive feelings or a desire to appease. If motivation is our interest, it is more in line with the data thus far available to suggest that presenting is motivated by a desire for affiliation. The other contexts for presenting are all affiliative (grooming, copulation, infant interactions, greeting), and it seems reasonable that cohesive gestures would appear also in times of social tension. There is no *a priori* reason why attempts at affiliation in times of stress should be interpreted as signs of submission or inferiority. Researchers have tended to assume intuitively that high-ranking animals have no need to affiliate; I would suggest that skills in enlisting support may be important components of high dominance rank. And, to paraphrase a statement by Hinde (1979), individuals may address gestures of social approval and social affiliation to each other for reasons quite unrelated to the dominance hierarchy: the most popular individuals and the most dominant may not be the same.

MISCONCEPTIONS OF DOMINANCE

Having described definitions and measures of dominance and their correlations or lack thereof, I would now like to discuss what I consider to be rather widespread misconceptions about how dominance is determined, and its general importance and applicability to social interaction. These are first listed, then explained. Studies of primates generally show that dominance is not:

1. determined primarily by the physical attributes of size and strength;
2. determined primarily by amount of "aggressiveness";
3. a permanent trait of an individual over time;
4. a trait "possessed" by an individual apart from a social context;
5. a genetically inherited trait;
6. more characteristic of males than females;
7. the basis of primate social organization;
8. equivalent to all forms of social control, influence or power.

1. The primate literature is replete with examples of individuals who are either old, ill, toothless, physically weak, or even female, exercising various important forms of dominance over other members of their group (e.g. Itani 1961; Southwick & Siddiqi 1967; Alexander and Hughes 1971; Hall & DeVore 1965; Bernstein 1969, 1978; Bolwig 1959; Kawamura 1965; Chalmers and Rowell 1972; Neville 1968). Macaques and baboons are the two types of primates in which researchers most often see overt conflicts. Probably one of the most striking interactions for a new researcher to observe is a larger, apparently better-equipped physical specimen showing submissive behavior to a smaller, or older, or physically inferior individual. Obviously it helps an individual in a general sense to be physically well and strong, but the power of social learning and social tradition is so pervasive in primates that it often overrides the importance of physical factors in the determination of dominance rank.

2. Aggressiveness is usually measured by frequency of threatening behaviors or frequency of fights. However, it has been found repeatedly that individuals exercising the most control over others in the group in terms of their ability to *win* fights or of their priority to incentives, are not necessarily the ones exhibiting the most frequent aggression in daily life. A low-ranking or middle-ranking animal, as measured by the direction of agonistic signals or priority of access to some resource, may spend its entire day engaging in conflicts or threatening the few individuals below it in the hierarchy, while a high-ranking individual may engage in conflicts only rarely. Again, we would not say that aggressiveness in the sense of self-assertion is irrelevant to dominance rank, partly because an animal usually would have difficulty becoming or remaining high-ranking if it avoided all conflicts or was timid. The mistake, or oversimplification, occurs in assuming that the amount of aggressiveness is the major cause of the dominance rank in the first place. Bernstein (1976), Riss and Goodall (1977), Rowell (1974) and Mason (1978) have suggested that higher ranking animals exhibit more skills in the astute manipulation of aggressive patterns — they exhibit self-assertion in contexts most appropriate and not in others. So, if anything, the direction of causality may be the reverse (rank leads to characteristic levels of aggression), although this too may be an oversimplification. We may not be able to generalize about causation or even correlation between the features of dominance rank and aggressiveness.

3. Neither is dominance rank a permanent trait, although changes are usually slow, sometimes requiring years to occur, and thus years of observation to document. Although both individuals and hierarchies tend to be conservative in terms of dominance, the longitudinal studies that have been conducted — Goodall's on the Gombe chimpanzees, Sade's and others' on the Cayo Santiago rhesus macaques, Japanese studies of macaques, the Altmann's team on Kenyan baboons (Ris & Goodall 1977; Sade 1972; Koyama 1971; Hausfater 1975) — all have discussed how dominance rankings of individuals in study groups change over the years. A number of studies show that age may correlate with rank, and maturity with high rank, more than any other factor (Hanby et al. 1971; Rowell 1974; Drickamer 1975; Hausfater 1975; Stephenson 1975). For males, length of time spent in the group, or "tenure," may correlate significantly with rank (Drickamer and Vessey 1973; Norikoshi & Koyama 1975). Finally, for rhesus and Japanese macaques, we know that entire matrilineages may rise or fall in rank (Koyama 1971; Sade 1967; Gouzoules 1980a).

4. It helps in understanding these longitudinal data to recognize that dominance is not an attribute that is possessed by an individual apart from a social context (e.g. Bernstein & Gordon 1980). As we saw with "aggression," an individual does not go about its life possessing a certain amount of dominance like a certain amount of body hair. The various forms of dominance which we recognize are actually inferences about aspects of relationships *between* individuals and so do not properly reside *within* individuals. As social contexts change over time, or simply space, so dominance relationships may change. The "alpha" male in a macaque group may leave and join another group in which he is the lowest-ranking male, at least initially.

5. It follows from the above that dominance cannot be a directly genetically inherited trait because it is, in fact, either a relationship between two individuals or a network of social relationships, *not* an individual trait. Nonetheless, there are surely behavioral and physical factors which contribute to an individual's dominance rank in certain times and places and which in turn have been influenced by the individual's genotype. Thus far we know so little about dominance and so little about the genetics of behavior, that it is difficult to hypothesize how these genetic influences might operate, other than the very general observation that a well-functioning, healthy body and a reasonably self-assertive temperament probably result in an active, confident animal. One does not have to participate in a false nature-nurture dichotomy, as Wade (1976) accuses the dominance critics Rowell and Gartlan of doing, in order to suggest that genetic influences on dominance ranks are probably very general and indirect.

6. Arising in tandem with the notion that male primates are more aggressive than female primates (see Chapter 6) is the idea that males have more obvious and stable hierarchies. Some early field studies reported stable hierarchies in the males but not the females, and although many studies have since reported otherwise, this remains a pervasive assumption in the literature. In particular, females have been believed to temporarily elevate their rank during estrous periods by depending on a high ranking male. In some groups this may be true, although I rather suspect that sometimes heightened irritability in estrous females is mistaken for rank changes (see Rowell 1972). In fact, long-term studies of primates have shown that because male primates are more prone to mobility between groups, and because dominance is dependent upon social context, males are more likely to have unstable rank relationships. Neither female nor male hierarchies are as permanent as we once thought, but in many species female primates are as likely as males to form stable dominance hierarchies. We might speculate that the emphasis in the literature on male dominance results from the observation that in some species males are more apt to engage in severe fights while females engage in other, less easily observed expressions of conflict (see Chapter 6). The unsubstantiated idea that dominance and dominance hierarchies are more important and more stable in males than females appears to stem from an overfocus on males and overemphasis on physical strength, as opposed to other means of winning conflicts. It may also stem from the skewed picture of primate social life we tend to receive from short-term studies. As Sade (1972) has said:

> Short-term studies have emphasized that the dominance relations among males are obvious and stable, while those of females are not obvious or are unstable. These assumptions are part of a larger set of assumptions about

macaques in which investigators have by and large emphasized the presence of adult males, especially the dominant male, as occupying the central position or performing the most important roles in the organization of the monkey society. These conclusions are in fact reversed by consideration of longitudinal data (1972: 388).

7. Some early treatises on primates suggested that dominance was so pervasive and important to primates that it actually determined all aspects or the whole organization of primate social life (e.g. Chance 1963; Zuckerman 1932). It is highly doubtful that any primatologist, including the original authors of this suggestion, would defend this notion today, yet the idea is so entrenched it continues to enjoy wide currency, especially in popular writings. And ethologists themselves continue to disagree over just how important are the various forms of dominance to social life in general. How important it is seen to be depends in part on the species studied, and, in part, on the researchers' own concept of dominance. Nevertheless, we can say that dominance rank in the oversimplified or even misleading sense of physical prowess, is not a very adequate predictor of individual primate behavior, or a good predictor of the various power relationships in a social group, or a good predictor of how individuals will relate to each other as relatives, friends and mates (e.g. Harding 1980).

8. This leads to the final point that while dominance in the sense of winning conflicts through individual effort is important in primate social life, it does not offer an explanation for all social behavior, interactions and relationships. Nor does it represent the *only* form of social control, influence and power found in animal societies. Social power is defined as the ability to influence other individuals to do or conform to one's will (Beattie 1964; Loy 1975). Although we readily recognize that humans may make others do or conform to their will through means other than individual patterns of intimidation, we have been slow to recognize other possible sources of social power in animals. Yet individual primates do rally others to their side in on-going conflicts, either as a result of stable alliances based on phenomena such as kinship, or as a result of their social skills in forming coalitions. Further, some forms of social power may not even relate to conflict as does dominance. Examples are leadership in group movement, attention-structure, and patterns of social facilitation and inhibition.

ALLIANCE AND COALITION

In 1958 Kawai distinguished between the concepts of independent (basic) and dependent rank in Japanese monkeys . Basic rank, he defined as a more or less permanent property of the individual based on such features as strength, weight and early psychological development of the personality. Dependent rank, he defined as situational or context-related, varying with the proximity and motivation of potential allies in a conflict. Since I think that dominance is always context-related, and not a unitary or permanent trait of an individual, I have never been comfortable with the concept of basic rank. Not only does individual dominance rank change over the years in primate groups (e.g. Leonard 1979; Chikazawa et al. 1979), but experimental evidence has shown that it is possible to contrive experiences of victory and defeat in laboratory groups such that individuals exhibit sharp changes in confidence (Kawai's "personality" trait). However, the concept of

104

dependent rank was an important milestone, being an early recognition of the significance of alliances and coalitions in determining the outcome of conflicts. Alliances tend
to be based on long-standing social networks such as kinship and friendship, while coalitions may be based on more immediate considerations, such as temporarily-perceived
common interest. "Persuading" others that they have a common interest with oneself in
an on-going conflict, that is, rousing others to one's support, is an important manipulative
skill found in primates.

Primatologists have recognized for some time that alliances or shorter term coalitions may strongly influence dominance (Maslow 1936; Carpenter 1950; Hall and DeVore
1965; Altmann 1962; Bramblett 1970) and yet, until recently, good studies focusing on
the dynamics of alliance have been rare. We have preferred to think of dominance rank
as an individual quality or achievement, of interactions as always dyadic, and of hierarchies as linear. None of these are necessarily true, but they are methodologically much
easier to research. Thus we have found ourselves in the odd position of viewing alliances
and coalitions as "tangles" in the simple line of dominance — when in fact dominance is
by and large the *product* of alliance in normal primate social life. The "tangle" actually
occurs when we insist on continuing to apply simplified formulations, such as "linear
hierarchies" to complex support systems.

Recently several good studies of the nature of alliances and coalitions (e.g. Cheney
1978; Kaplan 1977, 1978; Watanabe 1979; Mason 1978; Anderson & Mason 1978; de
Waal et al. 1976; Walters 1980; Leonard 1979) have been published, and they demonstrate that there are individual species and age and sex differences in the patterns of giving
aid and support during conflicts. Kinship is one of the primary sources of alliance networks (see Chapter 9), although alliances with non-kin are also a common phenomenon.
Watanabe (1979) suggested that Japanese macaque males ally most frequently with
recipients of aggression, and attack the aggressors (similar to description of "control"
role, see Chapter 8). However Kaplan (1977, 1978) suggested that rhesus males ally in
large part with adult females, regardless of the female's role as victim or attacker in fights,
or her relatedness to them. It is his suggestion that female rhesus give aid in order to assist
their kin and to reinforce bonds with peers, while males give aid to adult females as part
of their acceptance and integration into the group, leading to ultimate reproductive
success. Cheney (1977) found in common baboons that immature individuals attempt to
form alliances with members of other and higher ranking matrilineages, while adult
females form alliances with related individuals. On the other hand, Walters (1980) also
studying common baboons, found that alliances generally were formed to reinforce a
dominance hierarchy based on birth rank, and kinship was not an important predictor of
which individuals intervened in conflicts. It is clear that the formation and maintenance
of alliances is very important to assessments of dominance, and to a larger concept of
social power. However it is also clear that alliances will prove to be multi-faceted and
multi-functional social phenomena which must be carefully studied in the context of
different social systems.

LEADERSHIP

Leadership is not a well-defined or well-researched topic in primatology. As part of
the assumption that dominance was a unitary phenomenon applicable to all aspects of

social life, the most dominant individuals of the group have sometimes been referred to as "leaders." Individuals who will defend their group against outside attack, who will interfere with and in some sense suppress, fighting in others, and who lead group movement have all been called leaders. The problem is that being dominant, defending the group, interfering in fights, and leading group movement do not always occur in the same individuals. They are all however important forms of social influence and power.

Most primatologists speak of leadership only in the very restricted sense of leading group movement. Yet it seems that even this restricted interpretation may be difficult to apply, as the individual at the head of a progression (first) is not necessarily the individual determining the route chosen. Kummer (1968) distinguished between the "initiator" and the "decider" of group movement (the "I.D. system") in hamadryas baboons and this concept has been applied since to a number of primate species. The initiator is the first individual to move out and to head the progression. It is at times quite difficult to decide exactly who is determining group movement, and one must often rely on assessments of which individual can bring about the halt or the redirection of group movement through its own actions. In a sense this is the individual to whom the others are "paying attention" and I shall return to this concept shortly.

Determining the patterns of group movement is a very significant form of social influence since individual primates (with the exception of chimpanzees and spider monkeys) do not wander off on their own to find adequate sources of food, water and shelter. However it is not a form of social power based on conflict and intimidation. With the possible exception of hamadryas baboon males who "herd" their females, primate leaders do not compel or force others to follow them. On the contrary the followers in a sense create the leader. Thus there is no reason why the attributes that contribute to high dominance rank in an individual should also be the qualities found in leaders of group movement. Rowell (1969, 1974) has argued that the oldest individuals, those who have been in the group for the longest time and who have the best knowledge of the group's range (usually females), have the most influence on group movement. In Chapters 14, 15 and 16, I shall discuss leadership of group movement and dominance ranking in nine species of primates. For most of these examples the highest ranked individuals, if any, are not the leaders of group movement.

ATTENTION-STRUCTURE

Chance (1967; Chance & Larson 1976) has developed an approach to primate behavior based on the manner in which individuals pay attention to the other members of their group. Chance and Jolly (1970) suggested that dominance might be based on an individual's attention-attracting qualities rather than its physical prowess and ability to win conflicts. This view has been soundly criticized by Hinde (1974) who argued that attention-catching displays in chimpanzees for example may influence dominance (Reynolds and Luscombe 1976) precisely because they are intimidating displays which show off physical prowess.

I do not think that attention-structure is the basis for dominance, and while the concept is exceedingly difficult to research in many situations (how can we objectively decide who is paying attention to whom without extensive experimentation?) I do think that attention-structure might be a clue to forms of social influence and power which are not dependent on conflicts. Primates do not observe or focus upon all group members equally. For example, the attention-catching and social-binding properties of a female with a new infant have been described repeatedly. Individuals determining group movement are in essence those to whom the others are paying the most attention at that time. Non-human primates do not appear to be active teachers, therefore new behaviors are acquired largely through observation. But innovations are learned more quickly when pioneered by some individuals than others. Receiving attention from others is more significant than just "being popular," since it can result in various forms of social influence. So, "attention" does seem to be an observable phenomenon in many instances, but, like other measures of social influence, it does not appear to be a single power vested only in a few constant individuals.

SOCIAL FACILITATION, COERCION AND INHIBITION

Facilitation is a behavioral mechanism whereby a socially-living animal tends to do as its group members do. If others eat food items, it eats likewise, if they sleep, it sleeps, if they groom, it grooms and so forth. Individuals are not only strongly stimulated to conform to the general patterns of group behavior, they may be at times compelled to do so. We recognize this "social coercion" in immature individuals, where we refer to it as the process of socialization, but adult primates are also strongly influenced and constrained in their behavior by the behavior of other group members. The last chapter illustrated that one major trigger of agonistic attack is disruption of the social order by unusual or abnormal behavior. Adult non-human primates, and in a sense, "primate groups," are highly resistant to change. Established patterns, for example, of foraging or interacting, may become "traditional" and conservative. This may be another explanation of why dominance hierarchies tend toward stability, if not permanence. An individual macaque male attempting to enter a new group and rise in rank, for example, does so against the resistance of females in particular who react negatively to newcomers (Packer & Pusey 1979), and also of males who endeavor to maintain their present ranks. Kummer (1971a) has postulated that social inhibition is also a behavioral mechanism which reduces an individual's tendency to perform certain activities (courting a sexual partner, leading the group), if another individual is already doing so. This concept is discussed further in Chapter 8.

DOMINANCE AND SOCIAL POWER

Dominance appears to be a significant if complex aspect of social power. Although it has been variably defined and measured, the concept essentially refers to power established in conflicts through agonism or the potential for agonism. It may be conceptualized at the level of dyads or hierarchies. However, the winners of different types of conflicts are not necessarily the same individuals, and alliances and coalitions have a strong, even overriding, impact on conflict outcome. Further, dominance is not necessarily an ubiquitous aspect of primate social life; its significance varies with the form of interaction, with the species studied, and even with the perceptions of the observer.

Most importantly, dominance as studied by primatologists is not the same as total influence, power, or control over others. Thus to return for the last time to the question which sparked this and two other chapters: "do males dominate females" . . , it should be apparent now why the question as such cannot be answered. It could mean: do males drink first at a water spigot when thirsty? do they defeat females in fights? do they get the peanut? and so forth. The only honest answer to these questions is: some males, in some cases, in some species — but not in others. The following examples illustrate my point. Fully adult male chimpanzees are considered to rank over all adult females in terms of aggressive *or* competitive orders. In terms of first access to food in free-ranging groups, adult female ring-tailed lemurs and Indri rank over the adult males. In peanut tests, the first-ranking male in a Japanese macaque group usually ranks over all the adult females; however the second-ranking male usually ranks *under* the top-ranking female. In terms of direction of agonistic signals, adult talapoin females rank over adult talapoin males. In siamangs the adult female and male do not engage in any interactions which can be assessed in terms of dominance.

But this array of answers might be wasted breath and ink, because I suspect that what most people mean by this question is: do males have pervasive social power over females in nonhuman primate societies? Or, is male behavior "unlimited" (Deag 1977) by the females? The obvious short answer is no, it is not, but I hope that by now neither the question nor the answer will be considered appropriate.

What I have attempted to convey in this chapter is not only that dominance is multi-faceted, but that it is only one form of social power. Nonetheless, from the very inception of the modern study of social behavior in animals, a major focus has been the investigation of overt, conflict situations, or readily visible contests for limited prizes, on the assumption that it is these patently combative encounters which will reveal life's true winners and losers, as well as the power relationships which some have believed to be the real basis of all social organization. If, as often happened, animals would not enter into these contests two by two in the normal course of day-to-day life, scientists created situations in which space, social companions, and resources were artificially limited, and animals were forced into dyadic competitions. Unfortunately out of this was born what is officially known as the study of dominance, but what is loosely considered as the study of social power. "Unfortunately," because in our haste to document one aspect of social life, that is, the manner in which individuals in a society exert influence over each other, we may have stripped the process of its regular vestments, and may fail even to recognize the Joseph's coat of social power as it occurs in the normal, on-going social life of animals.

Chapter EIGHT

Roles

Cotton-top tamarins
Saguinus oedipus
Small, monogamous monkeys from the New World, tamarins commonly bear twins which are then often carried by the male (right). They show little or no dimorphism.

Ecologically, the position of the dominant animal is of little significance unless it is associated with leading or protective functions. Consequently, thinking in terms of dominance is currently being replaced by thinking in terms of roles.

Kummer 1971a: 61

We used to speak of linear dominance hierarchies — now it is a concept more like role-playing, because different measures of 'dominance' rarely rank the same individual in quite the same way.

A. Jolly 1972: 2

Statements such as those quoted at the beginning of this chapter occurred frequently in the primate literature about ten years ago. More recently the term and the concept of "role" have been criticized (or found unhelpful) by several writers, and role studies have become fewer. In this chapter I trace the history of roles as the concept has been thought of and applied in primate studies. In so doing I hope to show that some criticisms of primate role studies are justified, whilst others appear to stem from misapprehensions about what roles are. The term itself has brought some confusion, for like "aggression" and "dominance," it was borrowed from the sphere of human behavior and developed as an analytic tool by ethologists in their attempts to understand animal social life.

When we observe the members of a social group interacting, and even more importantly, when we compare social interactions across groups of the same species, it quickly becomes apparent that some behaviors are common to nearly everyone in all groups and can perhaps be called "species-specific behaviors." Examples are characteristic facial expressions or vocalizations which may be exhibited by anyone in the group. On the other hand, some behavior patterns are unique to only one individual in the group and may be labeled "idiosyncratic behaviors." For example, in a group of Japanese monkeys I studied, one monkey alone underlined her threats with a demonstrative bite of her own wrist; another, suffering perhaps from visual impairment, often carefully removed apparently-invisible specks from the air in front of her; and a third would vigorously rub her ventrally-clinging infant up and down on her stomach like laundry on a washboard. Each of these behaviors was unique to one individual in the group. Lastly, certain behavioral patterns are exhibited largely by particular individuals of the group who share one or more attributes, such as having an infant, or being of a certain age, rank and sex. In spider monkeys it is mainly mothers who use their bodies to "bridge" the gap between branches of adjacent trees so that older infants and juveniles may cross, and mainly adult males who engage in vigorous vocal and branch-shaking territorial displays (Baxter 1979). It is these sorts of behaviors which occur regularly in given situations and are attributable to subgroups or classes of individuals within the group, which have sometimes been called roles.

Scientists have recognized from almost the first studies of social behavior in primates (Carpenter 1934, 1942a), that some behavior patterns occur regularly but not universally in primate groups, and that different subgroups within a society exhibit characteristic behaviors or sets of behaviors. The primate literature is liberally sprinkled with references to mothering behavior, leader male behavior, peripheral male behavior and so forth. The basis for this recognition and categorization has been the *observed regularities* in behavior, in certain contexts, such that there is normative as well as idiosyncratic differentiation of behaviors within a group. Beginning in the mid-sixties a number of primatologists referred to these normative patterns as "roles." Several, but not all, of these primatologists directly contrasted the concept of role to the concept of dominance, and thus role analysis for animals was conceived, born and developed partly in the context of dominance criticism. Although, as I argued in Chapter 7, criticism of the traditional dominance paradigm was warranted in part, this may not have been the best developmental environment for the role concept, if only because some of the criticism now leveled at role studies may be part of a "backlash" on the part of dominance advocates. This is unfortunate because rather than setting up the two concepts in opposition, dominance and roles might well be viewed as interrelated social structures, as I will suggest in this chapter. However, it must be admitted that some of the criticism of role studies in animals results not from a false dominance-roles dichotomy, but from the multiple and confusing usages of the concept "roles" by different primatologists.

WHAT IS A ROLE?

There are many definitions of the term "role," differing from discipline to discipline and sometimes from researcher to researcher, but the term is a *methaphor*, derived from the theatre. We borrowed the word so long ago and we use it so often in everyday conversation, that we find it difficult to remember that "role" is a figure of speech, designed to communicate a likeness between some types of social behavior and the part an actor plays in a theatrical production. A role (French: rôle; Latin: rotula) originally meant the name for a piece of parchment on which was written the actor's part of the script for a play. Later it came to mean the part itself in the play, and still later it was developed as a metaphor for any behavior or conduct which "adheres to certain parts (or positions) rather than to the players" (Sarbin & Allen 1968: 489).

Therefore, a role is not just behavior, or behavior seen in one category of individuals; it is a repeated, regular, and predictable pattern of behavior, observed in particular situations, in certain subgroups of a society which occupy given positions in relation to other subgroups. It is a generalized or abstracted normative pattern of social behavior, which we may call a social part or position, independent of the individuals who play the part or fill the role. As Nadel (1957) described it, role operates in that strategic area where individual behavior becomes social conduct. Or as Talcott Parsons defined it: "It should be made quite clear that statuses and roles . . . are not, in general, attributes of the actor, but are units of the social system" (1951: 25).

Finally, and most directly relevant to primate studies, is Reynold's operational definition of monkey roles: "Statistically probable behaviors in a given interactional and ecologicial setting" (1970: 450). Thus the role concept can be used as an analytic tool whereby normative patterns are derived from observed behavior. This tool aids us in the prediction of social behavior, interactions and relationships.

Social scientists have developed several models of human societies as networks of interacting roles (e.g. Banton 1965; Parsons 1951), some of which are extremely complicated models, and only portions of which are applicable to other than human animals. It has been suggested that the concept of roles is too founded upon human traits, particularly human cognition about roles, to be very useful in studies of animals. However, most of the terms used to describe animal behavior are borrowed from the realm of human behavior, and many carry with them connotations inappropriate to animals (e.g., play is "nonfunctional," or dominance is "odious" — see Chapter 7). However all borrowed terms are attempts to better explain the unfamiliar through likening it to the familiar, and all classifications inherently carry implications. Sarbin and Allen (1968) suggested that most scientific controversies are actually arguments as to which set of metaphors more nearly represents "truth." Indeed, a disagreement over metaphors might be one explanation for the suggestions quoted at the beginning of the chapter, that thinking in terms of dominance be replaced by thinking in terms of roles. More recently the role metaphor has seemed to some to connote group selection, and has thus fallen into disfavor.

111

WHAT IS ROLE ANALYSIS?

Role analysis is one way of approaching the study of social structures and normative patterns in social behavior. As such it is an abstraction or model of social interaction and is most unlikely to correspond to an individual animal's motivation for behaving in a certain manner. Hinde (1974) believed this to be one of the shortcomings of role studies. For animals, role analysis does not explain on an immediate level or in proximate terms, what directly causes or triggers an individual's behavior. It does attempt to identify some regular, repeated elements of behavior, and may attempt to explain how these elements function in the individual's life, or in the ongoing social life of the group. Role analysis may also be used towards an adaptive or ultimate level of explanation of animal behavior, as Gartlan (1968) has attempted to do.

Social scientists have made much of human "role expectations," "role demands," "role conflict," and "rights and duties." Their studies have focused upon the *conscious* enactment and enforcement of normative behavior, and emphasized cognition and verbalization about social norms in human society. These scientists largely have studied people's ideas about what individuals in certain roles *ought* to do, and human beings very clearly do operate in social life according to their ideas about roles. Obviously monkeys cannot give us their ideas on appropriate "leadership behavior," for example, as humans can, and what is more, we do not have substantial data to suggest that nonhuman primates have such opinions or that their behavior is directed by conscious awareness of role behavior. Actually, humans are not always consciously aware of role-playing either, because it is a term that we as observers simply ascribe to the behavior of those who are interacting — playing the fool, the villain or the hero for example. Nonetheless, humans are often aware that their behavior does and ought to conform to certain social norms (e.g., being a "good father") so that awareness of the role is one cause of their behavior. We need to make it very clear that animal roles are our labels for certain regularities in observed behavior, not conscious ideas within the mind of the animal.

If animals are not conscious of performing a role, how do they come to do so? Probably through a complex of biological, psychological and social influences. Of course we cannot totally demarcate certain factors as biological only, others as psychological, and yet others as purely social, but the attempt to delineate different sources of influence on regular patterns of behavior may help us to understand how they arise. The primary biological attributes influencing normative differentiation of behavior within a social group are age and sex differences. That is to say, adult males of a given species usually show certain regular patterns of behavior which distinguish them, for example, from the juvenile males or the adult females of their group. Some species of course exhibit more strongly differentiated age and sex patterns than do others, and I will return to a discussion of these differences at the end of the chapter, and in Chapters 14, 15, and 16.

In addition, psychological factors may be very influential for role-playing, especially the many attributes often lumped under the vague term "temperament," attributes such as confidence, assertiveness, placidity, nervousness, etc. Thus a confident, assertive adult female may behave differently, and play a rather different role than a fearful, nervous adult female or a placid, slow-to-react adult female. General perceptual-motor skills are known to be very important in human role-playing, and it will be argued later in this chapter that these skills may be influential in animal role-behavior also.

112

Individuals do move through different roles in their lifetimes as they mature, and as the social dynamics of their groups change. Crook (1974) considered this to be the most significant aspect of animal roles. We can say that any individual could *potentially* fill all the roles found in their social group except those directly related to biological, corporeal structures, such as "lactating mother" or "inseminating male." Yet specific individuals only fill some of the total constellation of roles in their society and this is probably due to social factors, our third and final influence. These perhaps may best be explained through reference to learning, especially through social facilitation and social inhibition.

Social facilitation demonstrates that members of a society have a strong effect on each other's behavior, such that one individual may do as others do, and activities are thus coordinated. Social learning also involves observation and imitation. Imanishi (1965) suggested a long-lasting effect of this type for which he borrowed the term "identification" and by which he meant that young monkeys essentially do as they see certain adults do. Thus they learn and internalize a role model from observing the adults.

This idea has never been well-tested and most ethologists would probably consider it too mentalistic and too anthropomorphic. That role learning is important to primates can be seen from the findings of deprivation studies which indicate that inadequate mothering and inappropriate sexual behavior are found in isolate-reared monkeys (see Chapter 13). Any kind of role learning surely involves the interaction of biological (especially maturational) factors, and social (role models, facilitation) factors. While we have said that role theory does *not* assume animals are consciously aware of roles, and while role "expectation" would not be an appropriate causal explanation for an animal's behavior, nevertheless nonhuman primates are responsive to regularities in behavior, and might well learn roles as predictable patterns of interactions. They may reinforce, positively or negatively, behavior which concurs with or deviates from a normative pattern.

It has been shown that as well as facilitating behavior, individuals may inhibit one another from performing certain activities if some member(s) of the group is already doing so. For example, a male hamadryas baboon usually will not attempt to form a bond with a female who is already bonded with a familiar male (Kummer 1973; Bachmann & Kummer 1980); an adult female gorilla may not lead a group out to forage as long as a silverback male is performing that behavior (Fossey, cited in Hinde 1974); and an individual may not act as a defender of its group if someone else is already doing so (Bernstein 1964b, 1966a). As Kummer stated: "social order provides a complete scale of coordinating mechanisms, ranging from strong encouragement to violent discouragement of 'doing likewise', depending, it seems, on the number of animals that a task requires" (1971a: 58). A role always implies relationships between members of a group (Parsons 1951; Sarbin & Allen 1968). That is, leaders imply followers, mothers imply offspring etc., and thus the role an individual enacts is always dependent on the possibilities offered by the social group or by the social circumstances. Social relationships, the context of the social system, as well as normative behavior patterns, are the three fundamental elements of role theory in the social sciences and these can all be applied to animal role studies.

In summary, I have suggested that an animal performs a role in its social group, not because it is conscious of doing so, but because its behavior is influenced by factors such as age, sex, perceptual abilities, temperament, social models, and social facilitation or inhibition. This may explain how behavior is circumscribed, but may not explain completely why certain behavior patterns occur with such regularity within groups over time, across groups of the same species, and even across different species of the order primates.

Consider Bernstein's experiments on the "control" animal role (1964b; 1966a). He compared a number of groups of different primate species in similar captive settings and found that each group contained one animal who controlled intragroup disturbance and defended the group against external challenges. Interestingly, "control" behaviors occurred even in the apparent absence of a dominance hierarchy. Most importantly, removal of this individual resulted in another group member immediately exhibiting the "control" behaviors never before observed in this individual. Return of the original control animal resulted in again inhibiting the control behaviors of the substitute.

For reasons not entirely clear, no other role in primate societies has emerged so definitively. The point of the example here is to emphasize that there may be parts or positions or roles in primate societies which must and will be performed whatever the age, sex or temperament of the individual who finally fills them. Males may normally act as watchful defenders in macaque groups, but if there are no males in the group, females will perform this role (Kawamura 1965; Koyama 1970; Neville 1968). Adults may normally respond to external challenges, but in a captive group formed only of juveniles, a two-year old will defend its group (personal observation). Adult females in certain species may normally be the primary caregivers for infants, but on the death of the mother, a male or a young female without an infant of her own may adopt and raise the infant.

This strongly suggests that at least some roles are essential to primate social life (Bernstein 1966b). Ethologists are often concerned with the adaptive aspects of behavior; their major interest may be in how roles *function* to promote the fitness of the individual, and perhaps also of the group. According to this view then, primates perform roles because these roles are an essential part of the primate adaptive complex.

Although a number of primatologists have used this functional approach to role analysis, either explicitly or implicitly (Gartlan 1968; Rowell 1974b; Eisenberg et al 1972; Lancaster 1976; Bramblett 1973), and although this argument would explain the striking regularity with which behavior patterns reoccur in a group over time and across groups, still there is a certain "just so-ness" about the reasoning, and a suggestion of group selection about the adaptive significance or functions of some role behaviors in animals, that has disturbed a number of commentators on animal role studies (e.g. Wilson 1975; Hinde 1974). A control role or mother role may appear clearly functional for individuals and for the group, but a juvenile male role may not (Bernstein & Sharpe 1966; Fedigan 1976). The fact that some primatologists have focused upon normative aspects of roles and others upon functional aspects of roles (neither of which *necessarily* implies group selection) has created some confusion and perhaps has been partially the cause of some of the criticism of role studies in animals. However, the differing approaches, which we might describe as structural-functional differences, have also characterized human role studies. Critics have argued that clusters of behaviors simply associated with age/sex classes do not need the label of "role," and indeed they might not (although anthropologists refer to age/sex class behavior in humans as roles). But primate roles are usually defined by more than just age and sex (e.g. rank, kinship, spatial status), and I do feel that the concept of roles fulfills some of Hinde's criteria for being a useful intervening (predictor) variable (see p. 99). But if simply removing the term would alleviate the confusion and objections, then it might be well to do so. In any event, although structural role studies are now infrequent, the analytical procedures they use are becoming more common. The search for clusters of behaviors statistically associated with particular subgroups or classes of individuals continues, and can be found in a variety of studies, including recent research on kin effects (see Chapter 9).

TAXONOMY OF ROLES

Not surprisingly, the concept of roles has multiple interpretations and usages in the primate literature as well as the social sciences literature. Southwick and Siddiqi (1967) and Gartlan (1968) referred to particular behaviors, such as a territorial display or protection of the group from external threat, as a role, such that each presumed function was called one role. Others referred to the behavioral profiles or a constellation of behavior patterns exhibited by a class of individuals, such as central adult females or peripheral males, as a role (Lancaster 1976; Crook 1971; Rowell 1974b; Fedigan 1972a, 1976). Still others referred to the relative contribution of each individual to a social interaction, as a role; for example Hinde and Atkinson (1970) in a paper titled: "Assessing the roles of social partners in maintaining mutual proximity." None of these different usages were necessarily correct or incorrect. All have appeared in the social science literature, but leave primate role studies open to Wilson's (1975) criticism of lacking firm operational definition.

It may be helpful to consider briefly two schemes of role taxonomy or role classification which have been proposed to deal with the fact that normative social behavior can be analysed on several levels. Linton (1945) proposed that social roles be placed on a continuum from ascribed to achieved roles. Ascribed roles are based largely on biological attributes such as age, sex, and kinship, and would correspond well to previously mentioned monkey roles. One of the characteristics of ascribed roles is that individuals are considered to be performing the role at all times (Sarbin & Allen 1968) and thus the measurement of activity profiles is a common approach to the study of these roles. Achieved roles are determined by distinctive qualities and skills found only in certain individuals, and these roles are more specific in their behavioral scope, so that psychological and social factors are more important than biological factors, and individuals may move in and out of the position of an achieved role. This would correspond to such nonhuman primate roles as control animal (Bernstein 1966) and consort male (Reynolds 1970). An individual may perform several of these roles successively or simultaneously and the role refers to only certain of the individual's interactions with others.

Another taxonomy of roles often used by social scientists is that proposed by Parsons (1951) which distinguishes between what he has called "single-stranded" and "multiplex" roles. Single-stranded roles, as explained by Benedict (1969: 209-210), are characterized by: (1) functional specificity, — they pertain to a single type of transaction; (2) affective neutrality, — they do not involve a strong affective or emotional commitment on the part of actors; (3) a short time span, — they do not last long, though they may be repeated; and (4) a tendency to be stereotyped or impersonal.

Multiplex role relationships on the other hand tend to be: (1) functionally diffuse, — the actors perform many different functions for each other, e.g., family relationships; (2) affectively charged, — actors feeling strongly about each other; (3) continuous over a relatively long time, — actors interacting over and over again; and (4) nonstereotyped or highly personal — it matters very much who the actors are. Most of the characteristics involved in the concept of multiplex roles, like those of ascribed roles, would correspond reasonably

115

well to the approach used by primatologists who study holistic patterns of subgroup behavior ("mothers," "peripheral males," etc.); while the concept of a single-strand role would be more in accord with Southwick and Siddiqi's role of "protection of the group from external threat" for example.

Thus we see that roles can be analyzed at many levels on a gradient from multiplex to single-stranded relationships or from ascribed to achieved roles. Many of these levels have been used by primatologists in their discussions of primate roles, proceeding from the more general to the more specific, and they can be listed as follows: (1) the separation of roles between adults and young or between females and males (for example, Lancaster 1975, 1976); (2) the separation of roles between age/sex/status classes such as adult peripheral female, or adult dominant male (for example, Bernstein and Sharpe 1966; Rowell 1974b; Fedigan 1971a, 1976); (3) the separation of roles in each single, apparently-functional behavior (for example, Gartlan 1968; Southwick and Siddiqi 1967); (4) the contribution (or part played) by each individual in a particular form of interaction, such as consort female (Reynolds 1970).

DOMINANCE AND ROLES

Gartlan (1968) suggested that a dominance rank is not a unitary character or property of an individual, but rather a role that an individual plays, and this was reiterated by A. Jolly (1972) in the passage quoted at the beginning of the chapter. This is a plausible way to approach the controversial topic of dominance for we can see that dominance rank would fit the description of an achieved role very well. It is also possible to think of the relationship between dominance and roles at a more general level and to suggest that rank may be only *one* manifestation of a multiplex role.

> The style adopted by an individual in encounters with each other member of the group defines the animal's social position. Within this perspective, the relative dominance of an individual is seen as an aspect of the individual's social position and not a cause of it. Social position in primate groups may perhaps be best described in terms of roles. (Crook 1971: 150-251).

That is, differential ability to exert control over others in conflict situations may be one aspect of a larger role. Also, speaking in functional terms, we could say that in some cases an individual's prerogatives or priority of access to different resources corresponds to a whole constellation of behavior patterns, some of which have adaptive significance. For example, mothers are often found to have priority of access as far as their own infants are concerned, and this helps to ensure the survival of their offspring.

In a similar vein, Rowell (1974b) compared adult male roles in six species of African monkeys and concluded that defense of the group was the common role feature in these species. In addition she suggested that when adult males are larger and equipped with defense response systems, (e.g. enlarged canines) they may also be more intimidating in their behavior to young and to adult females, especially under stressful environmental conditions. Rowell postulated that in a sense females "tolerate" aggressive behavior from males in direct proportion to the amount of actual protection they receive from those males. Thus what has been seen as the dominance of these males over females can also be seen in terms of the general or multiplex role relationships between females and males in these species. In this sense then, dominance and roles can be viewed as interrelated rather than dichotomous structuring mechanisms in social relationships.

SOCIAL SKILLS AND ROLES

I argued in Chapter 7 that in the socially-living primates, social skills are an important key not only to dominance and differential ability to exert control over others, but also to survival and reproductive success. One way of approaching social skills might be in terms of differential abilities to enact a role. Role analysis necessarily focuses on the existence of normative patterns in animal social behavior, yet individual differences and flexibility in behavior are essential to the primate adaptive pattern. There are many different ways to perform a role and still conform to the general configuration of that role. For example, in macaques some control animals run to mediate conflicts at the first scream, others tend to turn their backs on minor squabbles and only interfere in longer-lasting, more violent fights; some mothers retrieve infants who make the slightest cry, others ignore shuddering, shrieking offspring for long periods of time.

Primatologists may occasionally apply value judgments to role skills, for example when reference is made to "good" or "bad" monkey mothers; however they are also making the inference that there are forms of social and natural selection which may act upon differential role skills and therefore make the ultimate value judgment. As an example, take the case of the "consort male role." Stephenson (1973) has shown that in Japanese monkeys there is great variability in the courtship behavior of males, differing from troop to troop and from male to male, although a general normative pattern of male courtship can be delineated. Social perception and motor skills on the part of the males are probably important to the individual rendition of the elaborate weaving, bobbing, whirling, posing display of the courting male. And if we adhere to the tenets of sexual selection, female choice of male partners, and thus male reproductive success, is at least partially based on their display skills. In addition, because they are "series mounters," Japanese monkey copulation is a synchronized process (see Chapter 10) in which the female and male must stay together and cooperate for an extended period of time, so the consort male role also involves skills in maintaining the proximity and cooperation of the female.

We could go through this example again from the point of view of the consort female role, because females also show differential ability to court, attract and maintain proximity with a male. Or we could look at a different role, such as the "alpha" animal in some form of dominance hierarchy, and suggest that alliance-forming skills, skills at perceiving and predicting social situation, and skills in behaving like a top-ranking animal may all be important to the role.

FEMALES AND ROLES

There is no need to say that evolution has brought about major psychological differences between males and females to explain role differences. The only difference that we need to establish is that females lactate and males do not — all the rest will follow through quite simple and obvious processes of social dynamics reinforced by socialization. (Lancaster 1973: 99)

It was noted earlier in this chapter that any individual primate can potentially fill all the roles in its society except perhaps a very few roles specifically dependent on a biological structure, such as "lactating mother" or "inseminating male." In fact there are many recorded cases of individuals *attempting* to fill even these roles: females will mount other females or males and go through the motions of the "inseminating male" role; young females and males will sometimes allow infants to suckle although they are not producing milk. There are even a few documented cases of non-pregnant, non-lactating females suddenly producing milk after adopting an orphaned infant (Hansen 1966; Harlow et al. 1963). Nonetheless, there are tendencies in each primate society for an individual of a given age, sex, temperament, rank, etc. to exhibit certain normative behaviors and not others.

Detailed descriptions of female and male roles in different primate societies will be the subjects of Chapters 14, 15 and 16, but two commonly-held stereotypes, or perhaps I should even say "fallacies" about primate female roles warrant discussion here. (1) Females are mothers and do little else in society. (2) Females *must* be nurturant mothers and care-givers because of fixed, biologically-based traits.

Since the second can be derived from the first, let me counter it first. Numerous deprivation studies have establshed that in primates the role of mother must be learned in a social context (see Chapter 13 for detailed discussion), and that the biological phenomena of being pregnant, giving birth, and then lactating, are not in a way sufficient in and of themselves to ensure mothering behavor. It will be remembered from the discussion of biological potential versus biological determinism in Chapter 3, that to suggest a behavioral complex or tendency is potentially present in females, and that it is adaptive for these behaviors to occur, is not at all the same as arguing that this behavioral complex must inevitably occur, whatever the context. Lancaster, in the quotation given above, makes the point that a social "division of tasks" such that females are the primary care-givers for infants, can be most parsimoniously explained as the result of social process ("social dynamics reinforced by socialization") based on the simple biological fact that females can lactate and thereby feed young infants, and males cannot. We do not need to postulate selection for elaborate biological mechanisms and psychological differences between the sexes in order to explain why different roles exist.

Even the fact that it is only female primates who are equipped to feed young infants does not always lead to mothers playing the primary care-giving role — in the monogamous species of primates the adult male plays a very large part in raising infants. In some New World species the mother, after giving birth, only takes the infants for brief periods to nurse them, while the male and older offspring do all else concerned with the rearing of the young. This is particularly interesting because, as Burton (1977) has noted, the females in these monogamous species go through the same biological processes of gestation, parturition, and lactation and they have similar "female" hormones (estrogens, progesterone) as do females in other primate species. Yet they do not perform all the same care-giving functions as do mothers in other species. In other words, the significant biological events of being female have stayed fundamentally the same, but the social structures and processes have changed, and the roles have altered with them.

Returning now to the first cliché about primate female roles, that is females function as mothers and little else in society, it should first be noted that with the exception of nursing infants, each individual monkey, ape or prosimian is an independent foraging unit. Every individual must gather food for itself. Our own cultural tradition of males providing economic support for dependent females seems to be such a deeply-held, pervasive model of social life that we tend wrongly to project it not only onto all humans (see Chapter 19 on the evolution of early human social life), but onto other animals as well. But there is no division of tasks in nonhuman primate society which involves the exchange of infant care for economic support. Rather, females raise infants in addition to gathering enough food for themselves.

There has been some discussion in primatology of a division of tasks such that females are infant-raisers and males are group protectors. Males may not gather food for others in the group, but can sustain the group (for an example of this argument, see Crook 1972). This may be true for a few of the ground-dwelling species of Old World monkeys, where the males are substantially larger than the females and are said to perform a specialized defense role. However, ideas about predator defense are largely theoretical as little actual predation on these primates has ever been observed. In a situation where I was able to observe a group of ground-dwelling macaques repeatedly defend themselves against predators, and on one occasion, chase a bobcat running off with an infant monkey in its mouth (Gouzoules, Fedigan and Fedigan 1975), animals of many ages and both sexes participated effectively in the defensive action. This species is one in which adult males only are traditionally thought to perform a specialized defense role.

The common baboons are probably the most widely-publicized example of large males protecting their groups and yet Rowell reports:

> Baboon males are often described as defending their troop, but this I never saw and find difficult to imagine, since Ishasha baboons always reacted to any potential danger by flight . . . the whole troop flees from any major threat, the males with their longer legs at the front, with the females carrying the heaviest infants coming last (1972: 44).

Not many of the primate species are territorial in the strict definition of the term (see Chapter 6), but for those that are it is often reported that females participate actively in territorial defense (Lancaster 1973; Mason 1968b; Charles-Dominique 1974; Ripley 1967; Epple 1970a; Box 1975a; Kleiman 1977). Primatologists have concentrated mainly on the large, sexually-dimorphic, terrestrial species which are easier to observe. However if we wish to make generalizations about nonhuman primate sex roles, we must also consider the many arboreal and/or monomorphic primate species in which sex role differentiation may be less strictly demarcated than the much studied baboons, macaques, and chimpanzees. For example, studies of arboreal Cercopithecus monkeys (Struhsaker 1969), spider monkeys (Baxter 1979) and squirrel monkeys (Candland et al. 1973) have shown that females in these species respond vigorously to potential sources of danger.

Thus as well as raising offspring and feeding themselves, female primates may defend themselves against predators and will help to maintain a territory if that is characteristic of the species. Most important of all, longitudinal studies have shown that it is females who generally remain in their home groups and home areas for life, while males of many species show a tendency to wander. This means, as both Rowell (1972) and Lancaster (1975, 1976) have noted, it is the adult females, especially older ones, who have the best knowledge of the resources and terrain within the group home range. Rowell (1972) has said that even in the common baboon, with its postulated "army model" of group movement with the males out front and leading (DeVore 1964), it is generally the older females and not the males who determine the direction of group movement. Altmann (1979) recently reported nine years of data concerning the group movement patterns of common baboons, and was unable to find any clear pattern to the order in which individuals led or moved within traveling groups ("progressions"). He concluded that the placement of individuals within progressions is either random, or the ordering relationships, if they do exist, are very subtle.

Longitudinal studies have shown us that females are often the social foci of the group, and that female-offspring units form the "core" of many primate societies (e.g. Sade 1972; Eisenberg and Kuehn 1966; Lancaster 1975, 1976). The importance of female kinship will be described in Chapter 9. Finally, as studies of Japanese monkeys and rhesus macaques (Kawamura 1965; Koyama 1970; Neville 1968), and my own study of roles have demonstrated, some females do exhibit patterns of behavior which can be called a female leadership role, and which complements a male leadership role. This female leadership role probably occurs widely, at least in Old World monkeys, and is further described in Chapter 14.

In summary then, roles vary from species to species, and sometimes from group to group, depending on the social organization of the group. While female primates commonly take a major share in caring for infants, this too is dependent on social learning, and there are examples in the primate world of males participating extensively in the raising of infants. The raising of offspring is of course a very important task. Primate young are very dependent for long periods of time, and the survival of offspring is considered *the* major criterion of success in evolutionary theory. In addition to performing what is the most important task for any animal, female primates also feed themselves and they may protect themselves, use their knowledge of their home range, control their own intragroup conflicts, and through their affinitive ties with friends, relatives and mates, establish and maintain bonds which are the cement of primate social life.

Chapter NINE

Kinship: The Ties that Bind

Geladas

Theropithecus gelada

These are very dimorphic, terrestrial monkeys from Ethiopia. The male (left), with the large "cape" shows the characteristic defensive "lip-roll," and the female screams defensively.

KINSHIP — THE TIES THAT BIND

Reports from many of the early, short-term studies of primates stated that mother-infant ties were broken after about one year when the infant was weaned and became independent of the mother. In contradistinction, one of the most significant findings of longitudinal research is an appreciation of the enduring importance of kinship to many aspects of social life. Short term studies, which can reveal a "cross-section" of primate societies have taught us many things; however the significance of long-term studies to our understanding of primate dynamics cannot be overstressed. Primates take a long time to grow up, and most social relationships seen in adults are the result of years of interactions. It is only by collecting data on particular individuals and groups over many years that certain questions, I would dare to say the more interesting and important questions, concerning social and population dynamics, can be answered.

In almost all of the longitudinal studies of intact groups in the field and in the laboratory to date, infants have been found to maintain close ties with their mothers at least until adolescence, and female offspring often do so until the death of the mother. Further, an infant develops special bonds with the group members who are likely to be close to its mother — mother's siblings and mother's mother for example — and these bonds may be maintained even after the death of the mother. For several species, it has been demonstrated that differential behavior is shown between individuals which we, as ethologists, recognize as different kinds of biological relatives; and which the animals may distinguish on the basis of different degrees of familiarity and types of affiliative interactions, or by a particular kind of relationship articulated through the mother. Borrowing a term from social anthropology, this has been called a "kinship system," because it involves differential social patterns based on distinguishing categories of recognized individuals. As both Reynolds (1968) and Fox (1975) have noted, if animals can be shown to direct differential behavior toward categories of individuals known to be differentially related, then the operation of a kinship system is indicated, even if the categories of relatives are not linguistically labelled. It has been pointed out to me that many American anthropologists tend to view kinship as a nomenclatural, semantic system, while many British anthropologists tend to see the semantic system as a by-product of differential behavior in terms of experience and sentiment. It is only from the latter perspective that one might speak of "kinship" in nonhuman primates (B. Blount, personal communication).

However, the borrowing of the term and concept does not of itself imply necessary homologues between human and nonhuman kinship systems. In an area that is extremely contentious, one conclusion agreed upon by social anthropologists is that human kinship systems vary enormously, and while some type of biological grid is always used, social kinship systems do not have to articulate closely with a biological system based on classification according to percentage of genes shared. For example, in some cultures individuals may call several females by the same term they would use for their mother; or they may not distinguish between what we call siblings and cousins; or they may distinguish between two types of cousin whom we would lump into the single category of "first cousin." Kinship systems are taxonomical, or classification, schemes which structure one's interactions with the other members of a society, and which, according to the central anthropological tenet of "cultural relativity," are all equally valid ways of ordering and perceiving social life. Because almost all primatologists come from cultures which share and employ a similar kinship system and a similar understanding of biology, we have tended to project essentially two systems onto our study animals — a biological classification, and our own cultural kinship system. Thus we discuss their interactions either in terms of "aunt," "cousins," and "siblings," or in terms of biological articulation based on birth, mother's mother, mother's brother, etc. This use is justified as an analytic tool to help us organize and attempt to make sense of the data on social bonds, unless we assume that our own cultural system is in fact universal and therefore natural and correct, which would be another good example of the aha! reaction.

In a later chapter, theories of the evolutionary basis of kinship will be presented. Here I would like to discuss the more proximate social dynamics of kinship, or the manner in which a knowledge of kinship helps us to understand the phenomena of social structure, dominance rank, group fissions, and cultural behavior in nonhuman primates. Because longitudinal studies are difficult to conduct and yet are so necessary if we are to determine how primates are related, thus far we only have good documentation of the importance of kinship from the long-term research of teams studying Japanese macaques in Japan, rhesus macaques on Cayo Santiago, and chimpanzees at the Gombe Stream Reserve in Africa. These will be largely the species and the studies to which this chapter refers. Evidence that kinship is important to the social life of common baboons, pigtail macaques, vervets, geladas, spider monkeys, and howlers, is also beginning to accumulate (Nash 1976, 1978a; Rosenblum 1971a; Massey 1977; Lancaster 1975; Bramblett 1978; Ehardt-Seward & Bramblett 1980; Dunbar & Dunbar 1975; Rondinelli & Klein 1976; Eisenberg 1976; Neville 1972). However, one caveat to this may be the suggestion by Altmann and Altmann (1979) that the pervasive influence of kinship on all aspects of social behavior is only possible in rapidly expanding populations (high birth rate, low mortality rate), where individuals are more likely to grow up surrounded by their relatives. Also, the bonnet macaque is one species for which it has been suggested that kinship does not play a large part in structuring social life, largely because mothers and offspring do not form very exclusive bonds (Rosenblum 1971a; Lancaster 1976; Clark 1978; Caine & Mitchell 1979).

This brings me to the point that what has been referred to as a kinship system in these nonhuman primate species is based on the phenomenon of enduring mother-offspring bonds. Indeed S. Gouzoules (in prep., a) found that *only* mother-offspring interaction patterns could be statistically distinguished from those of other kin and non-kin dyads. With the exception of the gelada, the species listed above are generally found in multi-male, multi-female groups where paternity is unknown to us, and as far as we can tell, unknown to the animals in these societies. Females of these species form very exclusive, dyadic bonds with their offspring, while males tend to react to young of the group as a class, or occasionally to form special bonds with the infants of their sisters, mothers or friends (see Chapter 12 for details). In Japanese and rhesus macaques, females stay with their natal troop throughout their lives, while adult males tend to roam. This, plus the fact that females remain closely bonded to their mothers even after they have their own offspring, results in a related group of adult females and young of both sexes who are linked together through descent from a common female ancestor. Such a descent group in nonhuman primates has been variously and rather loosely known by terms, again borrowed from the social sciences, such as uterine group, consanguineal group, maternal genealogical group and matrilineage. All of these terms were chosen to highlight the fact that individuals of such a group are linked through descent from a common female ancestor, called a "primogenetrix" by Japanese primatologists, and a "connector female" by Chepko-Sade and Sade (1979).

In the case of chimpanzees and geladas, although the data are still incomplete, there is some evidence that females may leave their group of origin during adolescence and enter or form another group containing male mates. They may still maintain ties with their mothers and matrilineal kin, either by returning to the natal group at times (chimpanzees), or by occupying a home range contiguous to the range of the mother's group (geladas). It has also been suggested that gelada females may leave their natal group with sisters, or as small groups of mothers and daughters, and form a new neighboring group with a new adult male (Dunbar & Dunbar 1975). (However, it should be noted that Mori 1979a,b argued from his eight month study that female geladas remain permanently in their natal groups, although group fissions and new male take-overs do occur.) Such patterns are certainly different from the macaque system described above (and chimpanzees and geladas have not yet been studied past two generations to determine what happens then), but in all longitudinal studies where relations are documented thus far, the mother is the hub of the system. Because of this, and also because all other kin relations may be seen as ramifications of the original mother-infant bond (Lancaster 1975), primatologists have also referred to kinship groups as matrifocal units (Lancaster 1975), or matrifocal families (Reynolds 1968).

Since, even in longitudinal studies, researchers "cut into" an on-going social group, all of the adult females of a particular society may be related at some level unknown to us (for example, elderly females of the group may be sisters) and thus assigning individuals of the group as relatives versus non-relatives may be somewhat misleading. However, given the nature of the data to be presented in this chapter, a distinction between at least two levels of kin relations may prove useful: the "matrifocal unit," defined here as a mother and her immediate or first generation of progeny; and the "matrilineage," defined here as descent group of several generations traced back to the first female ancestor recognized as such by researchers. An example of part of the genealogical records for the Arashiyama West group of Japanese macaques, arranged according to matrilineage, is presented in Figure 3. Two matrilineages, named the "Matsu" and the "Pelka" lineages are shown, and the names are taken from the names of the first female ancestors known to the Japanese primatologists. Many matrifocal units are represented; one example is Wania and her first generation of offspring; Wa65, Wa66, Wa67, Wa70 and Wa72.

DIFFERENTIAL BEHAVIOR AND SOCIAL DYNAMICS

The basic element of what has been called primate kinship is discriminatory behavior toward certain individuals, or groups of individuals, which we recognize to be different classes of biological relatives through female lines. What exactly are we referring to as discriminatory behavior? First of all, numerous studies have found that many positive or affiliative behaviors seem to be directed preferentially toward other members of the matrifocal unit, and only somewhat less often toward other members of the matrilineage. For example, grooming, "aiding" or supporting in agonistic encounters, sitting in bodily contact (e.g. "huddling"), and alloparenting (care of other's offspring), have all been

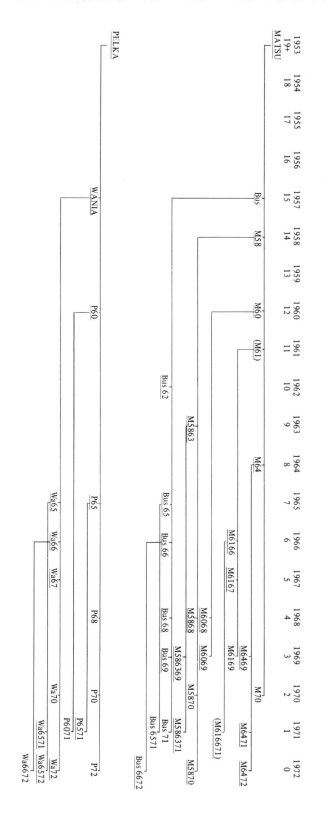

Figure 3. Two Matrilineages of the Arashiyama West Group of
Japanese Macaques, 1953-1972*

KEY:

NAMES UNDERLINED = FEMALES
(NAMES) IN PARENTHESES = DEAD

x⌐
 └y   = x GAVE BIRTH TO y

*Format courtesy of Koyama (1967). Data courtesy of Koyama (1967, 1970) and Mano (pers. comm.).
NAMING SYSTEM:
Originally all troop members were given individual names. Beginning 1958 offspring were given name of mother plus year of birth. For example, MATSU58 was born to MATSU in 1958; MATSU5863 was born

shown to be directed more often toward matrilineal kin than to members of other matri-lineages (Kaplan 1977, 1978; Loy & Loy 1974; Goodall 1968b, 1971b; Nash 1978; Missakian 1972, 1973; Massey 1977; Koyama 1967, 1970; Yamada 1963; Sade 1965, 1972; Dunbar 1980; Miller et al 1973; but see also S. Gouzoules, in prep., a). Individuals from the same matrilineage are also more likely to rest, sleep, travel and forage together (S. Gouzoules, in prep., b). Thus, in terms of their distribution within the larger troop, relatives are more likely to be found in close proximity and to be interacting frequently. On the other hand, some behaviors tend to be directed proportionately more often toward "non-relatives," or between matrilineages: these are agonistic, play, courtship and mating behaviors.

Kinship is such an important key to many interactions in Japanese and rhesus macaques that a veritable fog seems to lift from the observer's eyes as a chaotic and random cuffing, screaming and feinting match between an assortment of animals crystallizes into an encounter between two matrilineages in which various relatives are enlisting and giving support to each side. A chart of the clustered distribution of these monkeys during a grooming period, a resting period, or during the night's sleep, takes on meaning when the matrilineages are colored in, like countries on a map.

It is easy to imagine how a young rhesus or Japanese macaque infant, born into this kin-structured world, learns that certain individuals are always likely to be close at hand and willing to support the infant in, or rescue it from, its first endeavors away from the mother's body. Older siblings may still be so close to the mother that they huddle up against her ventrum behind the nursing infant, and thus the infant sleeps in what must be a warm and secure "family pocket." Mother's sisters and mother's mother are also often nearby and they may provide care as allomothers (see Chapter 12), or should they have infants of their own, such "cousins," "aunts" and "uncles" may be the first peers the infant encounters.

Female kinship ties are so important to social life in these macaques and such a sizeable proportion of the interactions between matrilineages tend to be agonistic, that I have occasionally found myself wondering how and why different matrilineages live together at all. Indeed, groups do often split along consanguineal lines, and some cases of one matrilineage alone establishing its own new group are known. However there are several factors which could contribute to group integration. To begin with, it may well be that all of the genealogical descent units in a troop are themselves related through some more distant primogenetrix, and they will certainly unite in opposition to a neighboring troop. Many affiliative interactions, such as grooming, do occur between individuals from different matrilineages, though not as frequently as within these kinship groups, and thus individuals from different genealogical subunits should not be viewed as constantly interacting agonistically. In fact, because matrilineal kin spend much more time together, they have higher absolute frequencies of interactions of all types with each other, including agonistic ones. Play and other forms of peer interactions, which seem important to social learning in Japanese and rhesus macaques, tend to occur in large peer groups made up of young from a cross-section of consanguineal groups. Juvenile and pubertal males, in particular, may spend as much time in friendly interactions with peers as with relatives. Older males tend to be mobile between troops, and adults are often found living in a troop with few or no matrilineal relatives. Adult males usually form alliances and "friend-ships" with certain females or groups of females, which cross-cut genealogical lines (e.g. Grewal 1980). They also court and mate with females who are more likely to be non-relatives, and during the mating season the females spend more time away from their matrilineal kin and close to their consorts.

Thus, although matrilineal kinship is a strong structuring principle in the social life of these species, there are also factors other than genealogical relatedness which integrate different matrilineages within the troop. The larger group could be characterized as consisting of relatives, mates and friends.

KINSHIP AND DOMINANCE

Rather early in the history of primate social behavior studies, Japanese researchers made the remarkable discovery of "rules of dominance," based on kinship in Japanese macaques. Largely using food tests and recording the direction of agonistic signals between monkeys on the provisioning grounds, it was found that offspring rank according to the rank of their mothers, and that, at least until adolescence, siblings rank in reverse order of their ages, and that mothers usually rank over their offspring. Further, whole matrilineages may be ranked against one another in large groups such as those found in provisioned Japanese and rhesus macaques (Kawamura 1958; Kawai 1965; Koyama 1967). There are of course exceptions to all of these rules, but they do seem to hold true in general for this species; and at least some of them may be generalizable to rhesus macaques (Loy & Loy 1974; Missakian 1972), pigtail macaques (Massey 1977), and to a lesser degree, vervets (Ehardt-Seward and Bramblett 1980).

If kinship is the key to many aspects of dominance in these species, the behavioral phenomena variously known as "alliance," "agonistic aid," "support," and "fight interference" constitute the mechanism by which it operates. Studies such as those by Massey (1977) and Kaplan (1977, 1978) have begun to demonstrate that monkeys, females especially, come to the aid of their matrilineal relatives during agonistic encounters. Mothers, in particular, respond so readily to threats to their offspring that other group members treat the offspring from infancy as if they held the mother's rank. Bramblett (1976) refers to this as falling in the mother's dominance shadow. Marsden (1968) has suggested that it is meaningful to think of a mother and her immediate offspring as constituting a social unit which shows aggression towards, or receives aggression from, other matrifocal units in maintaining a dominance hierarchy. This is not to say that monkeys do not occasionally take advantage of a high ranking mother's absence to tweak a young offspring found alone, for primates are ever opportunists, but in general the mere possibility of the mother's aid maintains the rank of her young at her own level.

Further, in Japanese and rhesus macaques, mothers tend to support their younger over their older offspring when conflicts between them arise, and although some have felt this to be an inadequate explanation for why siblings, especially females, rank in reverse order of age, I believe it conforms with the other known results of support systems. It is an expression of how important the original configuration of mother-offspring bonds was, that elderly sisters whose mother is long dead, are usually found still to rank in reverse order of their ages. Brothers tend to conform less to this rule, and especially at puberty, when ties with the mother loosen for most males, older brothers tend to rise in rank over younger ones. Adult male rank in general is probably less affected by kinship, especially as males usually leave their group-of-origin and live with non-relatives. Although there are cases of sons of high-ranking females growing up to be high-ranking males in their home troop and even in non-natal groups, as was originally hypothesized by Japanese primatologists, there are also cases of sons of low-ranking females becoming alpha or first ranking males in their troops of adoption. Age, and length of time, or "tenure" in the group seem to be the primary factors determining adult male rank in a number of primate species (Norikoshi & Koyama 1975; Lancaster 1978; Rowell 1972; Packer 1979a; Kummer 1968). When males change groups they often begin with very low rank in their new groups, whatever their rank in the natal troop.

Japanese primatologists have also found that whole matrilineages may be ranked as higher or lower in dominance than others, although such rankings usually do change slowly over time. However there may be conspicuous exceptions to unit ranking, as certain individuals can be scored, using the technique of recording the direction of agonistic signals, as much lower or higher ranking than the other members of the matrilineage. In the history of the Arashiyama West troop, one female from a low-ranking lineage consolidated an alliance with the unrelated alpha male through dint of her own efforts, and managed to rise to the top of the female hierarchy. Later, some, but not all, of her matrilineal kin used her support, actual and potential, to effect similar rises in rank, while the entire lineage of the former alpha female dropped in rank (H. Gouzoules 1980a). Four years later when the alpha male walked away and left the troop permanently, the alpha female and her matriline were able to maintain their rank in the absence of any alliances with top ranking males and in the face of challenges from other females.

Earlier in the history of the Arashiyama troop in Japan, another matrilineage had maintained very high rank while the eldest son of the primogenetrix was a leader in the troop, but after a troop fission which separated the male from his matrilineage (see Koyama 1970), and later, after the death of the primogenetrix, this consanguineal group declined in rank. It occurs to me that while many primatologists have conceived of the "politics" of monkey social life and leadership in terms of "male despots," others might argue that longitudinal studies are now indicating matrilineal "dynasties" would be a more apt, if equally anthropomorphic, conceptualization. In fact, studies such as those of the Arashiyama troop have shown that long-term female kinship relations articulate with shorter-term male dominance relations to produce a far more dynamic and sophisticated social system than we had ever imagined (see Chapter 14 for more on this topic).

MALE EMIGRATION AND TROOP FISSION

Many early primate field studies suggested that social groups were "closed," as members did not appear to transfer from group to group. The more data we accumulate, the more it becomes evident that male primates of many species move out of their groups of origin and may even switch groups several times in their lifetimes (see review of male transfer data in Demarest 1977; Lancaster 1978a). Chimpanzees, gorillas and red colobus form prominent exceptions. In these species more females than males leave their group of origin and move to other groups (Nishida & Kawanaka 1972; Harcourt et al. 1976; Harcourt 1979; Marsh 1979). No one really knows what causes males in a number of species to "peripheralize" from their natal group and move to another, or in some cases to "solitarize" or literally live alone for a period of time. However we can suggest beneficial or functional results of this phenomenon, such as outbreeding, and the spread of cultural traits (see below).

Many researchers have stated or implied that young males are pushed out of home groups through aggression directed towards them but such expulsion has only been documented for some monogamous species, in which case it is directed at female and male offspring alike. Some explanations have focused on puberty as a time when internal changes bring about a propensity in males to wander (Nishida 1966), and reports do indicate that many males change groups around this time in their lives. However in macaque society, adult males are also prone to suddenly leave social groups, even when they have struggled to join them only a few years before (Sugiyama 1976). Other researchers have suggested that low-ranking males leave to seek better mating opportunities (e.g. Drickamer & Vessey 1973; Packer 1979a), and mating may indeed have something to do with group changes at least in macaques, since "strange" males often appear during breeding seasons. However low-ranking males do not always have less opportunity to mate in their groups and, in any event, they would likely begin with very low rank in a new group. Further, studies indicate they do not leave more often than high ranking males and indeed, alpha males are known to leave their groups.

Both Sugiyama (1976) and Itoigawa (1974) have concluded that male Japanese macaques are simply more mobile than previous studies had indicated, and that these monkeys may wander for any number of reasons, such as following the example of a playmate or elder brother, because of a weak bond with the mother and matrilineal kin, or maturational changes, or simply to seek a rich food distribution outside the troop's range. For whatever reasons, most males in the species documented thus far do not spend their adult lives with their matrilineages.

As Lancaster (1976) noted, females may leave their groups also, but with the exception of chimpanzees, they do not do so alone as do many males, rather their matrilineal kin leave with them, provoking what is called a group "fission". The process of group division and/or the natural formation of new social groups is a phenomenon which primatologists believe offers a significant insight into the ontogeny and the dynamics of social groups, and is one which, owing to the "cross-sectional" nature of most studies of free-ranging groups, rarely is fully documented. In all of the cases where kinship has been adequately known at the time of fission, it has been shown, for rhesus macaques, common baboons and Japanese monkeys, that groups fission along female genealogical lines (Missakian 1973; Chepko-Sade 1974; Chepko-Sade & Sade 1979; Chepko-Sade & Olivier 1979; Nash 1976; Furuya 1969; Koyama 1970). The actual descriptions of fissions are quite involved, as group divisions sometimes take long periods to be completed. The arguments concerning causes of fission are very complicated and usually involve changes in rank and alliance patterns between males and between females and males. However, I would like to reiterate the point made so nicely by Koyama, that when group fission does occur, it is not a disorderly or chaotic process (a "falling apart" of society), it is rather the departure of one or more matrilineal groups and their allied males for a new home range.

KINSHIP AND MATING PATTERNS

One of the undoubted *results* of the frequent transfer of primate males from the natal group to another group is "outbreeding," or the avoidance of mating with close relatives. Whether or not "inbreeding avoidance" is one of the *causes* of male mobility is another question entirely, and a controversial one. Firstly, evolutionary causes, which refer to the adaptiveness of a pattern, need to be separated from proximate causes, which refer to immediate stimuli such as internal processes (motivation, etc.) and social pressures. This means it is one thing to argue that natural selection has promoted a tendency for males to leave their groups of origin because outbreeding results in greater reproductive success, and another to argue that males leave their home groups because they are psychologically and/or socially inhibited from copulating with their kin and must seek mates elsewhere. Both mechanisms, natural selection for outbreeding, and "inbreeding avoidance," may be operating, but the existence of one does not prove the other.

Sade (1968), Tokuda (1961-62) and Immanishi (1961) first commented on the lack of mother-son matings in rhesus and Japanese macaques. This is especially striking in species where mothers and sons may be very close in other respects (recall that father-daughter relations are unknown). Sade hypothesized that males who remain in their natal group are inhibited from mating with their mothers by the reverberance of the role of infant in their adult relations with their mothers, and by the superior dominance of the mothers. His is therefore a theory of psychological and social inhibitions. As Lancaster (1978) and Fox (1975) noted, Sade's suggestion was picked up quickly and widely in anthropological literature, caused considerable controversy, and "achieved almost the status of a law in some circles" (Fox 1975: 22).

The study of the "incest taboo" in humans has a long and controversial history, and for many people is intimately linked to their concept of "human nature." Some form of proscription against mating with close relatives is found in all human cultures, and although the particular kin included in the taboo vary from culture to culture, mother-son avoidance is the most common. Theories concerning the incest taboo in humans are many and varied, and range from biologically-based theories concerning inherent human *preferences* or desires for mating with relatives (Freud 1935) to inherent *aversions* to doing so (Westermarck 1922). In contrast are sociocultural theories that incest avoidance rules are uniquely human social inventions conceived and established in deliberate contradistinction to the prevailing promiscuity of animals (Levi-Strauss 1968, 1970; White 1959; also see discussion in V. Reynolds 1976).

Unfortunately, it has seemed of paramount importance to some scientists to prove that the human species is exempt from all biological influences and is in every way unique, while to others it has seemed essential to prove that all human and animal behavior can be reduced to the same biological mechanisms. This has resulted in long and unproductive arguments between those who set up rubicons for human uniqueness and those who seek only to cross them. The studies of "language" or communication skills in nonhuman primates, as well as "proto-cultural" or learned and shared group behavior, and now kinship and mating patterns, all have been mobilized in such battles, whose emotional style often does little to further our understanding either of human or nonhuman social behavior patterns.

Several attempts have been made to develop an ethological and/or biosocial theory of incest avoidance (Bischof 1975; Demarest 1977; Parker 1976; Kortmulder 1968; Murray 1979). As I noted earlier, the existence of a psychological incest avoidance mechanism, especially for mother-son matings in non-human primates, cannot be proven using only the data of male emigration or group transfers, unless it can be documented that an inhibition against mating or an unwillingness to mate with matrilineal kin is the proximate cause of male emigration. The only evidence for this point is circumstantial; the finding that males often leave one group and attempt to join another at the beginning of mating season in rhesus and Japanese macaques, and that adolescent female chimpanzees often change groups when they are in estrus. However, a psychologically and socially based avoidance of mating with kin, or a preference for non-kin, could be examined in cases where individual males do remain with their natal groups when sexually mature. Is there evidence that such males avoid mating with matrilineal kin?

One such study by Missakian (1973) found that some rhesus macaque males around the time of puberty do mate with their mothers and sisters (14.8% of sexual behavior observed involved genealogical relatives and 5.4% involved mothers and sons). In a few cases, the mother was even the preferred sex partner, although such incestuous matings only occurred in the first active mating season for the male, and other elements of normal consortship were absent. Seven out of Missakian's ten cases of mother-son mating involved mothers who were still dominant over their sons. On the other hand, for Japanese macaques, Imanishi (1961) and Tokuda (1961-2) reported no cases of mother-son matings, and Enomoto (1974, 1978) found that males who remain with their natal groups do not mate with genealogical kin. He attributed this pattern to female "psychological avoidance" or an unwillingness to mate with kin.

For the Arashiyama West troop of Japanese macaques, my colleague, Harold Gouzoules, and I have been collecting mating and consorting data for five years and out of a total of more than 1000 matings, we have never observed a mother-son copulation, only one brother-sister mating, and one grandmother-grandson mating. The Arashiyama West troop is presently isolated from other groups and almost all males remain in the natal troop so that the opportunities for mother-son and within-matriline matings are constantly increasing. In spite of this, cases of within family mating continue to be rare to non-existent. Some of these sons rank over their mothers, while others do not. Data analysis is still in progress, but calculations from one of the years, presented in Baxter and Fedigan (1979), shows that avoidance of mating with relatives extends to the entire matriline, although to a lesser degree than to the matrifocal family.

For chimpanzees, Goodall (1968, 1971) found that mothers and mature sons in the same community do not copulate, and brother-sister copulation is rare and discouraged by the female. Therefore the data we do have thus far on avoidance when close relatives are actively mating and soliciting partners in the same group, are limited to a few groups and are contradictory. As more information becomes available from longitudinal studies on kinship and its effects, we hope to learn more about how kinship relates to mating patterns. It may well prove, as Lancaster has suggested, that: "The frequent observation that some males prefer or avoid mating with specific subgroups in baboon and macaque societies may indeed reflect preferences for mating between individuals from unrelated genealogies" (1978: 72).

Should we find that nonhuman primates generally avoid mating with close kin, this would certainly not reduce the elaborate human cultural beliefs and institutions concerning incest to a mere biological instinct. It has been noted a number of times that similarities in social patterns do not prove "biological determinations," and that social learning and biological propensities are not easily distinguished, nor meaningfully dichotomized. As more complex and elaborate social and cognitive processes are recognized in nonhuman primates (indeed, in all mammals), it is disheartening and unnecessary for so many people to take this to mean they must reduce their evaluation of human abilities, rather than to increase their appreciation of primate capacities. If it is found that humans should have shaped and constructed their social worlds through a symbolic elaboration and transformation of primate patterns, this need not be surprising or degrading.

PROTOCULTURE

A great deal of the behavior of primates can be called cultural in the sense that it is transmitted by learning from generation to generation. This is true not only of social behavior, but of behavior toward the environment, as simple a thing as the home range of a group (Jolly 1972: 350).

The concept of culture obviously loses a great deal when accommodated to the dimensions of biology. What we can gain from the operation is the wider context of evolution from which culture emerged as one possible way of life . . . (Kummer 1971a: 13).

Yet another controversial topic has been the study in Japanese monkeys of the invention and transmission through learning of new behaviors, whereby traditions are established which have been referred to as "culture" (see extensive review in Itani & Nishimura 1973). The latter began when, in the process of provisioning monkeys with foods in order to accustom them to human presence, it was discovered that only some monkeys, young ones in particular, would accept and try new foods quickly; others would watch and perhaps try the foods afterward (Itani 1965). The most famous case, a story which has now been told many times, involved the feeding of sweet potatoes and later wheat, to the Koshima troop (Kawamura 1959; Kawai 1965). The food was spread on the sand of the ocean shore and one young female innovated the technique of washing the potatoes, first in a freshwater brook, and later in the sea. Others watched her and many came to use this technique of food preparation, even seeming to develop a preference for the taste of salt from sea water on the food. Two years later the same female dropped handfuls of sand and wheat into the water and picked off the sand-free wheat which floated to the top; again others learned from her invention. After a decade of observation, it was found that the majority of the troop performed the potato-washing and wheat "placer-mining," and that infants learned the behaviors readily from their mothers as part of the group feeding pattern. Individuals of one particular matrilineage however, failed almost entirely to learn these new behaviors.

From this and other studies on different troops of Japanese Macaques, primatologists documented several patterns of innovation and transmission of newly-acquired behaviors. The young of the group are more likely to invent new methods of food preparation than the adults, and their peers and matrilineal relatives (even their mothers) are most likely to learn from them. Adult males, who often have no relatives in the group, are considered the most "conservative" or least likely to take up a new behavior spreading through the group. Yamada (1957) reported on a different sort of experiment in the propagation of new traits. A high-ranking central male (peanut tests) left the Minoo-B troop and acquired the habit of wheat eating while away. After his return wheat was offered to the troop for the first time, and the eating of wheat was observed to spread from that male to adult females and then to their offspring within a matter of hours, rather than the weeks or years recorded in other studies. Thus, learning seems to spread more quickly from high-ranking individuals to others, from older to younger monkeys, and again, along kinship lines.

When Kawamura reported his findings on the origin and spread of group-specific traditions, there arose what Itani & Nishimura nicely referred to as "heated discussion and even derision over whether or not monkeys have culture" (1973: 31). Japanese primatologists then attempted to avoid stepping on the ubiquitous sore toe of human uniqueness, by referring to what they studied as "protoculture," "preculture" or "subculture." These studies have also been criticised as simply showing the results of artificial provisioning, although provisioning was simply the mechanism which allowed researchers to document the invention and transmission of new cultural behaviors. (The effects of provisioning are yet another subject of controversy.) There is ample documentation of variability from group to group in other behaviors, such as courtship, prevalence of bisexual and homosexual behavior, communication patterns, paternal care, swimming, etc., and these studies will be referred to in upcoming chapters. The important point for us here is the significance of matrilineal relations for the transmission of newly-acquired and group-specific behaviors.

COMPARATIVE PRIMATE KINSHIP SYSTEMS

Fox (1975) argued that the uniqueness of human kinship systems lies not in the discrimination and categorization of different classes of relatives, but in the particular combination of "alliance" and "descent" which lies at the base of all human kinship systems. He defines alliance as the allocation of mates, or patterns concerning who mates with and/ or marries whom, while descent refers to membership in a group because of common ancestry. Fox suggested that some nonhuman primate societies such as the multi-male, multi-female species whose societies are structured by matrilineages exhibit descent. Other primates, such as the polygynous-living geladas and hamadryas baboons, do not have enduring descent groups, but rather have alliances or permanent assignment of mates. Only human kinship systems have both. Fox's theory is a useful stimulus and basis for generating discussion and working toward a cross-species understanding of kinship systems, and it is the only paper that I know of which attempts a comparative perspective on the structure and social process of kinship systems. In addition to any reservations cultural anthropoligists might have about this comparison, I would like to offer several comments and criticisms from the primatologist's point of view.

Firstly, the androcentrism of this paper is remarkable, although I recognize that several of the examples only follow in the tradition of male bias in the analysis of human kinship systems and in the study of primate societies. The primacy of mother-offspring bonds in the nonhuman kinship systems thus far reported is missed entirely, and different primate social groups are defined and distinguished according to the number of adult males present in the group. Further, in multi-male, multi-female societies, females are said to "belong" to all the males (why not say the males "belong" to the females, or why use this concept of ownership at all?) and it is asserted emphatically that dominant males have first choice of estrous females (the " 'Big Male' chestnut" again, see discussion in Chapters 7 and 17).

Most surprisingly, polygynous societies (Fox uses the term, "harem") are said to be avowedly patriarchal; social groups existing only as long as the male exists, and after the death of the male, females are said to be "reallocated"! Polygynous societies will be the subject of Chapter 15, but it should be pointed out here that many studies show the females in such systems to be closely bonded to one another, to form coalitions against the one adult male of the group, and new groups to result from females choosing to leave and attach themselves to a new male (e.g. Dunbar & Dunbar 1975; Hall 1967; Poirier 1969a; Crook 1977).

Further, as I mentioned at the beginning of this chapter, the Dunbars' study found that matrilineal *descent* was a probable structuring principle in the polygynous societies of geladas, yet much of Fox's paper rests on the argument that geladas and other polygynous species exhibit alliance but not *descent*. Fox does recognize that the importance of matrilineages was missed in earlier studies of macaques because longitudinal data were lacking, and I would suggest that the same may prove to be true of many polygynous species. At least we should reserve judgment concerning the presence or absence of descent systems until we have the necessary genealogical information.

A female descent system as well as quite regular patterns concerning who can or does mate with whom (alliance) may indeed exist in geladas. Furthermore, studies of mating partner choice in multi-female, multi-male societies are in their infancy, and it may well prove that there are regularities, if not rules or "determination" in the mating patterns of these supposedly promiscuous species. While I would not argue that this approximates in any way the complexity of human exogamy systems, which have elaborate rules for the exchange of marriage partners, I cannot agree with Fox's devaluation of the evidence for mother-son mating avoidance and regular patterns of outbreeding in Japanese and rhesus macaques. Fox himself points out that some data on Japanese monkeys shows them to be little more than a "naming system" away from "exogamy plus matrilocal residence," as two groups may be found regularly to "exchange" males.

The comparison of human and nonhuman kinship systems is controversial for several reasons, some of which have been previously mentioned in this chapter. Not least are historical controversies in the discipline of anthropology concerning the manner in which humans became socially and behaviorally distinct from other primates during the course of human evolution. In the nineteenth century early anthropologists or "social philosophers" often suggested that all human societies had, or would, evolve through several universal stages of culture. One prominent school of such thought proposed an initial stage of promiscuity, followed by a matriarchal phase, followed by male rebellion and the establishment of patriarchy (the final stage), while another proposed a fourth and final stage of the monogamous or nuclear family.

In the early twentieth century as anthropology became empirically oriented, a major effort was made to disprove these earlier theories of universal stages in the evolution of human social life, and the greatest effort of all was devoted to the discrediting of the idea of a universal matriarchal stage. Although some feminists do return to such a theory apparently in an attempt to buttress arguments concerning the possibilities of female social power, this promiscuity-matriarchy-patriarchy scheme of social evolution has been considered, by other female anthropologists, as an attempt to show patriarchy as the "logical culmination of civilization" (Webster 1975; Martin & Voorhies 1975).

Most of the theories and arguments of the nineteenth century social philosophers are considered specious, or at best non-empirical, and are often labelled "arm-chair theories," by contemporary cultural anthropologists. However, as no truly matriarchal societies have ever been documented (Gough 1979), a patriarchal system seems to many to be a better choice for the prototypical pattern of human society. It is therefore not surprising that the primacy of male-ruled systems is an assumption underlying much of the writing of twentieth century anthropologsts. Nor is it surprising, given this history, that a certain heavy silence hangs in the air when the importance of matrilineal descent and matrifocal units in nonhuman primates is described to cultural anthropologists. Not only has their terminology of kinship been borrowed, but the phoenix of a "feminist Golden Age" (Gough 1979) may seem to be threatening to rise from ashes they thought cold. (Of course, as Gough has noted, belief in such a "myth" is really not necessary in order for humans to plan for parity of the sexes in the future.)

Power over others may take several different forms, and the question of "who is dominant in this society, females or males?" is essentially unanswerable in that form (see Chapter 7 on dominance). Since the terminology of human kinship systems is being applied to nonhuman primate social relations, care must be taken with the terms. For example, it should be noted that the term, "matrilineal" which refers to a descent system reckoned through female lines, is decidedly not the same as "matriarchal," a term referring to rule by females, specifically a mother-figure. While it is true that human females are generally found to have more freedom, independence, and prestige in matrilineal societies, and while it is true that female kinship systems may structure female dominance relations in some nonhuman primate species, nonetheless, we need to distinguish between statements concerning female-male power relationships and statements concerning kin relationships.

In nonhuman primate species, such as the Japanese and rhesus macaque and the chimpanzee, where we consider matrilineal kinship ties to be a very important structuring principle; it would be impossible and meaningless to suggest that either females or males "ruled" the societies. In the last chapter of this book, I will describe a theory of early human social life, based on chimpanzee analogies and homologies, which argues for the first human kinship system as matrilineal. Here I should note the caveat that many primate species have yet to be studied; perhaps some may not be structured by kinship at all, and others, such as monogamous societies, may operate according to very different principles.

However, it remains significant that our understanding of the importance of females and the enduring matrifocal unit to many fundamental processes of primate social life, has been the one consistent finding of longitudinal research on primates of different species. It is to be hoped that our growing awareness of the female's role in social processes, especially through kinship bonds, will help bring a better balance to the perception of sex roles, or the parts that females, as well as males, play in the cohesion and continuity of primate societies.

Chapter TEN

Sociosexual Behavior

Mandrill

Mandrillus sphinx

Shy and forest dwelling, these largely terrestrial monkeys of Central Africa are difficult to study. The female (right) presents her perineal region, with its swollen estrus skin, to the much larger male as a sexual initiation signal.

We are accustomed to thinking of sexual attractivity primarily in terms of the female as the "attractor" and the male as the "attractee," whereas in fact the phenomenon is mutual and reciprocal . . .

Beach 1976b: 472

When I first studied a population of geladas living on a large zoo island, sexual behavior was an obvious if brief affair. A female and a male from one of the two heterosexual groups living on the island would engage in a short but vigorous copulation, accompanied by noisy grunting vocalizations. Often this would occur up on one of the rocky promontories of the island and everyone nearby, animals and people alike, would be able to take note. Visiting parents were frequently seen to suddenly gather up their fascinated youngsters and hurry them on to the next exhibit.

Later, when I began to observe vervet monkeys, one of my primary goals was to witness mating so I could obtain information on conception, gestation and birth. Months passed, and the closest behavior I ever saw to mating were low-keyed invitations from the one fully-adult male of the group, who would occasionally approach females and touch them gently but lingeringly on the hips. Usually the females responded by turning quickly and threatening the male away. Slowly it became obvious that the abdomens of all the adult females were expanding, and despite my many hours of observation, I still had not seen the behaviors which gave rise to the now-apparent pregnancies. When I finally did see mating in vervets, the interaction was over so quickly that I understood why it had been so difficult to observe. After one of the male's "polite enquiries" the female would stand holding her body stiffly, or after a female invitation through a sudden and brief presentation of her posterior, the male would mount from behind, thrust several times and ejaculate. Then they would separate quickly. After some copulations the female would turn and slap or scream briefly at the male. A moment later there would be no behavioral cues that copulation had taken place. Not until my research with these vervets was complete did I learn that they are called the "sexless monkey" in parts of Africa.

Neither of these two patterns of sexual behavior prepared me for the mating of Japanese monkeys, which occurs only during the breeding season of the year, and sometimes appears to be the predominant activity of that time of year. The two early clues to the on-coming mating season in these macaques are conspicuous reddening of facial and perineal skin, and a conspicuous rise in the number and the intensity of agonistic interactions, particularly those initiated by adult males. The most serious wounding of both females and younger males occurs at this time of the year, and both sexes could be said to undergo distinct "personality" changes. Some (but not all) of the normally quiet, sociable males become excitable and aggressive, as do certain of the females in estrus. Young females, in estrus for the first time, acquire an odd, fixed, staring facial expression and unusual vocalization patterns. They often move away from other group members and sit alone in the brush screeching loudly, or they may appear to run quickly through their repertoire of Japanese macaque vocalizations, from weaning sounds to alarm calls. Although this may be a form of sexual advertisement, these females often run in apparent terror when approached by a potential mating partner. At other times they actively approach males and other females of the group for mating, often backing right into them. Potential mating partners usually appear nervous around these young, excitable females, and their mating efforts are not very successful. Older adult females generally appear to have calmer, though active, estrus periods, but they too may occasionally go into a frenzy of mating activities and behave in ways not otherwise seen, but which can be characterized as hyperactive. Young males may court older females and if the object of their courtship does not respond, may stage tantrums highly reminiscent of infant weaning tantrums. Adult males may launch lightning-speed attacks on females and run off before the startled female has time to catch her breath, or they may attract her attention with long-range tree shaking displays, or close range weaving-and-strutting displays, occasionally involving branch-dragging. As the breeding season progresses and aggression levels drop, more and more individuals become involved in mating, even prepubertal monkeys, and normal social and foraging patterns appear to be disrupted. It is certain that a researcher would not fail to notice mating in a Japanese monkey group.

Indeed, one of the most striking aspects of sexual behavior in the primate order is its extraordinary variability. The range is from species like vervets, in which the act of copulation is rapid and "unemotional" (Lancaster 1978a) or unaccompanied by elaborate preparatory and subsequent behaviors, to species like Japanese macaques, in which mating involves extensive courtship, special sexual bonding, and even temporary shifts in the whole tenor of social life. It also varies from species that will mate throughout the year, such as chimpanzees, to those like rhesus, that will only do so during sharply demarcated breeding seasons. This variability naturally renders generalizations about primate sexual behavior (and chapters such as this), difficult, and, in addition, may help to explain some long-standing controversies in primatology concerning the importance and significance of sexual behavior to primate social life. Earlier studies of nonhuman primates in captivity tended to emphasize the "continuous" sexual nature of all primates as the prime factor in the cohesion of social groups (Zuckerman 1932), while the first field studies in the 1950's and 1960's documented the presence of a distinct mating season in many primate species. Laboratory studies at about the same time were able to demonstrate monthly peaks in female receptivity corresponding to the concept of "estrus" or "heat" in nonprimate mammals. It was then pointed out by many researchers that sexual behavior is periodic rather than continuous in nonhuman species, and even rare in some cases. For example, A. Jolly (1972) has noted that female primates are likely to be mating for no more than twenty weeks in a twenty year lifespan. Yet one of the more striking characteristics of primates is the stable group of adult females and adult males, a group which persists throughout the year and over the years. So, in primatology, the role of sexual behavior and sexual bonding in social life was de-emphasized for a period, and it was even suggested that sexual behavior might be a disruptive rather than a cohesive factor in social life (Washburn & DeVore 1961). More recently, research is again emphasizing the relative emancipation of all primate females from limited periods of hormonally determined estrus, and therefore once again stressing the far-reaching impact on social life of sexual bonds, and the promotion of friendly social relationships through sexual behavior (see Lancaster 1978a; Saayman 1975, discussion of estrus in Chapter 11). It seems, as was noted with the concept of aggression, that the significance of sexual behavior depends to some extent on the perspective of the primatologist, and to a large extent on the particular species studied.

Thus research on sexual behavior in primates has a long, and somewhat controversial history, and has accumulated a body of findings which cannot be adequately reviewed here. Certain aspects of sexual behavior have been, or will be, discussed in Chapters 6, 11 & 17 and reviews can be found in Michael and Zumpe 1971; Lancaster 1978a; and Rowell 1972b. In this chapter I will describe briefly some of the concepts and behaviors fundamental to our discussion of primate sexuality. I will look at the roles females and males play in the sexual behavior of several species, and the nature of sexual bonding, particularly consort bonding in those species.

139

BREEDING SEASONS

Field studies of many primate species have noted the birth of infants at one time of year, indicating a birth season or birth peak, and also indicating a corresponding mating or breeding season. Some observers have also documented the occurrence of copulations during only one period of the year. Although there are certain prominent exceptions to seasonal mating, such as the chimpanzee, and there are still species for which we do not yet have adequate data, the documentation of an annual reproductive cycle in many primates by Lancaster & Lee (1965) and their argumentation that seasonal breeding is a general characteristic of primates, is widely accepted.

Although much research remains to be done, at least two mechanisms are believed to be at work in the seasonality of primate mating: physiological and behavioral responses to environmental changes, such as annual daylength and rainfall patterns; and synchronization of breeding through interactive social stimulation. It is particularly interesting that in seasonal breeders, such as squirrel monkeys, rhesus monkeys and Japanese macaques, both females and males show extensive changes in endocrine conditions from non-breeding to breeding periods (see review in Michael and Zumpe 1971). That females play a key role in the coordination of mating in these species is suggested by studies demonstrating that the presence of sexually-active females brings about changes in the reproductive conditions of males (Rose et al 1972) even in the non-mating season (Vandenbergh & Drickamer 1974), while the reverse — male induced reproductive coordination of females — does not occur (Vandenbergh & Post 1976). That female-female interactive stimulation, perhaps through pheromones, coordinates or synchronizes their ovarian cycles is one tentative, but intriguing, suggestion.

ESTRUS

Much of the contention in the study of primate sexual behavior revolves around the concept of estrus. A history of the use and understanding of this concept in primatology is summarized in Chapter 11; here I will simply present a definition of estrus as the term is used in this book: estrus is female motivation (proceptivity) and/or willingness (receptivity) to mate. It refers to a behavioral phenomenon which does not always occur on a regular cyclic basis, or in correlation with known endocrine states, but always does imply periodic rather than continual willingness to mate. It should also be noted that primatologists vary, from those who consider estrus to be a discrete, measurable behavioral and physical condition of the female corresponding to ovulation, to those who view estrus as a concept of little value in monkeys and apes because of extreme variability in patterns of female sexual behavior and lack of correlation with known endocrine factors.

COPULATION

Beach (1976) has said that the species-specific copulatory pattern for most mammals is relatively stereotyped and unremarkable. Essentially males mount, thrust and ejaculate, while females assume a posture which facilitates intromission of the penis into the vagina, and maintain or readjust this position as the male thrusts. The occurrence of sexual climax in female macaques has been described by several researchers (Zumpe & Michael 1968; Burton 1971; Chevalier Skolnikoff 1974; Michael et al 1974; Goldfoot et al 1980). From these reports orgasm in the female primate is characterized by a clutching reaction and rhythmic vocalizations, muscular body spasms and contractions of the vagina. Indeed, according to Burton's experimental study (1971), the rhesus female exhibits several of the copulatory phases described by Masters and Johnson for the human female.

In most nonhuman primate species the female presents or stands with firmly raised rump and the male then mounts from the rear, often by grasping the back of her legs with his feet, or standing nearly upright behind her, with his hands pressing down against her hips for balance and support while he thrusts. Although it is true that a standard copulatory pattern can be delineated for each nonhuman primate species, nonetheless there exists considerable variability between species, and occasionally between groups and individuals of the same species, in the number of mounts, the number of thrusts and the length of time from first intromission until ejaculation or cessation of copulation. There is also evidence of alternative copulatory positions, such as ventral-ventral positions, mating with the male in a sitting posture, mating while both partners hang from branches, etc. (Hanby et al 1971; Chevalier-Skolnikoff 1974; Goodall 1968b; Baxter 1979; Galdikas 1979).

While the interactions preceding copulatory behavior, and the social and environmental factors related to copulatory patterns, are more multiplex and offer more research possibilities than the consummatory act itself, one very important but little explored variable in copulatory patterns is the type of mounting; that is, either a single mount with thrusting leading to ejaculation, or a series of mounts. In some species of primates the male mounts, thrusts a few times, dismounts and sits behind the female for awhile, and then mounts again. He repeats the pattern several or many times before reaching ejaculation. In series-mounting rodents, it has been demonstrated (Adler 1969; Adler et al 1970) that females exhibit a lower rate of successful pregnancies if experimentally deprived of series mounts while still receiving an ejaculatory mount with viable sperm. It is hypothesized that in these species, series mounting helps to stimulate a neoendocrine response in the female in which progesterone is secreted in amounts sufficient to permit successful implantation of the fertilized egg. A similar mechanism might be operating in series-mounting primate species. Series-mounting also has several implications for sociosexual behaviors, not the least of which is the increased difficulty of successfully completing a copulation to ejaculation, and thus the need for sustained cooperation between a mating pair. Perhaps it is not coincidence that, with few exceptions, the species which copulate through series-mounts also exhibit a sociosexual pattern known as "consorting," and those which single-mount generally do not. This correspondance between series-mounting and consorting has been noted by Michael & Zumpe (1971) and Saayman (1970).

CONSORTS

Carpenter (1934, 1942b) coined the term "consort bond" to describe the sexual relationship in multi-female, multi-male primate species, whereby "a female and a male form an association with a high degree of reciprocally interactive behavior and rapport"

(1942: 196). Since then consort bonds have been described in at least twelve primate species, but few studies have focused directly upon the phenomenon, exceptions being studies by Bernstein (1963), Reynolds (1970), Michael et al (1978) Fedigan & Gouzoules (1978). The references to consorts are plagued by a lack of standardization so that on the one hand, some researchers seem to regard a consort as little more than a copulating pair, and on the other, some seem to view it as little different from any stable heterosexual friendship bond. Consorts will be further described later in this chapter, but it should be noted here that consort relations are a unique form of pair bonding in primates. They involve temporary but very intense and exclusive attachments between two individuals, based on sexual attraction, but implicating and influencing many other social patterns. As such they provide an excellent opportunity for the study of both bond-formation and mate selection in primates.

BISEXUAL BEHAVIOR AND HOMOSEXUAL BEHAVIOR

It has been noted for many species of primates (and indeed many species of mammals) that females may mount and thrust and that males may present their perianal region and stand properly to assist another to mount them (see reviews of these behaviors in Beach 1968 and Michael et al 1974). This performance by females or males, of behavior more often seen in the other sex, has often been called "sex reversal" or "sex inversion," implying that it is an abnormal pattern. However both Beach and Michael, who focused on what they call the "bisexual behavior" of females, have stressed that it is so common for females to mount as well as to be mounted, that it must be considered part of the normal behavioral repertoire. Further, Michael has suggested that female mounting occurs most often when females take the initiative in sexual interactions.

Homosexual behavior has occasionally been reported in the primate literature, but is usually considered to result from stressful and abnormal conditions of captivity, or from misapplication of the term "homosexual" to sex-derived ritualized patterns such as presenting and single mounting between two males in a greeting or an agonistic context. We cannot demarcate in any absolute sense sexual and copulatory behaviors from such sex-ritualized patterns. However homosexual behavior which can be called primarily "sexual" in nature and context, because series mounting and/or directed stimulation of the genitals and/or ejaculation and/or consort bonding occur, has now been repeatedly documented between females and somewhat less often between males in three species of macaques, some groups of which are not in captive conditions. These are the rhesus macaque, the Japanese macaque and the stumptail macaque (Carpenter 1942; Altmann 1962; Akers & Conoway 1978; Erwin & Maple 1976; Hanby et al 1971; Eaton 1978; Fedigan & Gouzoules 1978; Chevalier-Skolnikoff 1974, 1976; Kling & Dunne 1976). In addition, female-female sexual behavior is occasionally reported for other primate species, such as the squirrel monkey (Talmage-Riggs & Anschel 1973), the vervet (Struhsaker 1967a) and the talapoin (Wolfheim & Rowell 1972).

In Japanese macaques homosexual and bisexual behavior occurs in some troops but not others, and may be considered to be another example of precultural behavior (Stephenson 1973; Itani & Nishimura 1973). In our own study (Fedigan & Gouzoules 1978) of female homosexual behavior in the Arashiyama West troop, a group of monkeys which has been studied for twenty-six years and which has always exhibited high levels of such behavior (Koyama, personal communication), we found homosexual behavior, which took place in the context of definite consort bonding, to be part of a larger pattern of female sexual initiative. Females who had not recently given birth exhibited very sexually active patterns of behavior during the mating season — soliciting partners of both sexes, copulating and consorting on many days, mounting both their female and male partners, and engaging in high levels of homosexual activity. No females were exclusively homosexual and those females who formed homosexual as well as heterosexual consort bonds did not exhibit lowered fecundity. Rather, giving birth resulted in a sharp drop in all sexual activities in the following year, including bisexual and homosexual behaviors, in formerly active females. We could find no other variable (age, rank, availability of males, etc.) which correlated so well with the pattern of homosexual and bisexual behavior as recentness of giving birth. Although both female mounting and female-female homosexual relationships in animals *can* occur as a result of stressful captive conditions, we would suggest that all such behavior should not be dismissed as pathological or dysfunctional, a practice which results in "explaining it away" rather than explaining it. While homosexual behavior is not directly functional in the sense of procreation, it may be part of a larger sexual pattern which *is* adaptive in terms of reproduction, or it may have other significance for social living. For example, in our study we found that females who had engaged in homosexual consorts during the mating season were likely to remain affinitively bonded (friends) throughout the year in contrast to male-female consort pairs which were not translated into year-round bonds. Since sexual partners are almost always unrelated, these friendships cross-cut matrilineal lines and are a potential source of alliance and bonding in addition to kin ties. It is hopefully evident from this description that the phenomenon of homosexual behavior may have entirely different causes, patterns of manifestation, and results, in different species and different primate societies. As Beach (1976) noted, people often draw quick conclusions about human homosexuality from the animal data, but surface similitude in behavior does not in itself justify theoretical inference.

FEMALE AND MALE ROLES IN SEXUAL BEHAVIOR

There are several reasons why one receives the impression that primate males are sexually active and females are sexually passive or reluctant; an impression which is not supported by the data, when the time is taken to consider them fully. In laboratory studies of sexual behavior, as Czaja & Bielert (1975) have noted, the emphasis usually has been upon passive female cues, such as vaginal odors, which may stimulate the male, rather than upon the actual initiative and motivation of the female to mate. The very terms which have been used in these studies, "female receptivity" and "female attractiveness," connote the female as receiving rather than engaging in sexual behavior. Beach has put it this way:

A great deal has been written about sexual receptivity of females, but of equal importance is the sexual 'proceptivity' which estrous females exhibit when they take the initiative in approaching, investigating and sexually soliciting the male (1976a: 473).

143

The female's tendency to display appetitive responses finds little opportunity for expression in laboratory experiments which focus exclusively on her receptive behavior, or upon the male's execution of his coital pattern. The resulting concept of essentially passive females receiving sexually aggressive males seriously misrepresents the normal mating sequence and encourages a biased concept of female sexuality (1976b: 115).

Another reason for a general misconception of female and male roles in sexual behavior is that some schools of evolutionary theory, particularly sexual selection and parental investment theories, are often interpreted as predicting that males should be sexually ardent and females reluctant. Theoretically, this is so because males may impregnate any number of females and thus pass on their genes to the next generation, while females have a limited number of opportunities to pass on theirs, restricted to the number of ova they produce and the relatively few offspring they can bear. Thus it is often suggested that males will compete for access to as many females as possible and be continually sexually active, while females will be reluctant and very choosy. As will be further discussed in Chapter 17, male competition has been much researched and discussed, whereas female choice until recently has been largely ignored, especially in the study of primates. While female primates are no doubt selective in their choice of partners, they are also initiators of sexual behavior, and there is ample mention in the literature of males also showing preferences for certain sexual partners and not others (e.g. Seyfarth 1978; Saayman 1970; Hausfater 1975; Enomoto 1978; Herbert 1968; Hall 1967). There may be many ways of assuring one's genetic contribution to the next generation, including strong appetitive behavior on the part of the female which helps to ensure that she will conceive, and careful discrimination and selection of the most appropriate or most "likely-to-succeed" mating partners on the part of the male. In addition, sexual behavior in the nonhuman, as well as the human primate, can serve many functions not directly and obviously related to procreation.

Yet another source of the notion that females tend to be reluctant to engage in sexual behavior is the fact that nonhuman female mammals are not continually motivated to mate. In many animals, female motivation and willingness to mate is correlated with the fertile phase of the ovarian cycle, while in primates the correlation is not clear, and as a result, estrus is a subject about which there is much disagreement. All primate sexual behavior seems to be relatively emancipated from direct hormonal controls, and in human primates most of all. Yet, female hormonal cycles do have influences on nonhuman primate mating patterns, and they do interact with social, psychological, neural, environmental and perhaps other variables, to result in the periodic nature of female sexual motivation. The picture is further clouded by variability between individuals and situations, and by the fact that neither male animals nor all humans can be accurately described as "continually" motivated to mate. However, it is clear that during their periods of estrus, female primates often actively seek coitus with males, and their appetite for sexual interactions may be said to at least match that of the males. For example, Saayman has said of chacma baboons: "The roving, appetitive behavior of inflating [early estrus] females from one male to another was conspicuous" (1970: 86)

Whether sexual behavior is initiated by the male or the female is sometimes difficult to determine and has not often been the focus of study. It also varies from species to species, from group to group, and from individual to individual. However, as Lancaster noted: ". . . fieldwork publications are filled with reports of females soliciting copulations and refusing copulation attempts by keeping their hindquarters lowered" (1978: 68). And there are incidental reports for a good number of primate species of the female partner "often," "usually," or even "solely" initiating sexual behavior. Thus females are said to play an active role in the initiation of sexual behavior in lemurs, howlers, langurs, patas, geladas, common baboons, rhesus macaques, Barbary macaques and gorillas (Petter 1956; Carpenter 1934; Jay 1965; Hall 1968a; Rowell & Hartwell 1978; Dunbar 1978; Hall & DeVore 1965; Saayman 1970; Lindburg 1971, 1975, 1980; Taub 1979; Fischer & Nadler 1977).

How female primates initiate and express their motivation for copulation has also been little researched, but Michael & Zumpe (1970) have done one such study on the rhesus female. They documented several behaviors which are clear expressions of the female's sexual motivation state, and which stimulate coitus with the male. In our studies of Japanese macaques (Fedigan & Gouzoules 1978; and see also Wolfe 1979), we found that females not only make genital presentations to males, they also initiate sexual interactions, and stimulate the continuation of series mounts, through particular vocalizations (estrus calls), facial expressions (an intense stare and lip quiver), body movements (hand slaps, head bobs, reaching back), mounting of the partner, and in some cases where the male fails to respond or to begin copulating, by literally pulling him up onto their backs. It is of note that even the motor act of coitus is usually described in active/passive terms (males mount & thrust, females stand), but the female's behavior of standing stiff-legged and rump-raised while orienting her perianum to the proper angle so that intromission may occur, has been shown by laboratory research to be a learned behavior in primates (see Chapter 13) which is essential to successful copulation, and so is, in fact, more than just passive acquiescence to being mounted.

Although male courtship in a few species involves elements of intimidation toward potential female partners (for example, chimpanzees and Japanese macaques) actual forced copulation or "rape" has only been described for one nonhuman primate species — the orangutan. Other studies sometimes specifically note that "rape" does not occur (e.g. Goodall 1968b). In the case of the orangutan, at least three observers (Galdikas 1978; Rijksen 1978; MacKinnon 1979) have described occasional forced copulations as occurring between adolescent (usually low-ranking or migrant) males and non-estrus adult or adolescent females. Rijksen regards this as a form of dominance interaction. Galdikas, on the other hand, regards the forced copulations as sexual in nature, arguing that young males fall back on this secondary reproductive strategy because females always choose fully mature males as sexual partners. All three researchers report that estrous orangutan females actively initiate sexual interactions with the mature resident males and form consorts with them. These adult, resident males do not perform the behaviors which the observers have called rape. The researchers all seem to agree also that these forced copulations are not in fact a successful reproductive stratgey, since no pregnancies are known to have resulted from them. MacKinnon says:

> In view of the small size of the orangutan penis and the inherent difficulties
> of suspensory arboreal copulation, it seems that only with female cooperation
> can intromission be successfully achieved. Such cooperation is usual only

within the context of an established consortship, in which a female is familiar with and shows little shyness toward her partner. The more aggressive forms of copulation with uncooperative females, which I have referred to as rape would seem to be futile reproductively (1979: 262).

SEXUAL BONDS AND SOCIAL STRUCTURE

Although mate selection, particularly the concept of female choice, will be further examined in Chapter 17, it is pertinent here to discuss some aspects of the "sociosexual matrix," or the social factors which form the structure and context for sexual patterns. Nonhuman primates are often said to be sexually promiscuous, and indeed some social theorists have used rule-regulated versus promiscuous mating patterns as yet another demarcating line for the uniqueness of human societies (see Chapter 9). However, in the primate world, sexual behavior is seldom promiscuous in the sense of involving indiscriminate, totally casual, or random matings. With rare exceptions, mating takes place in the context of social bonding, and there may well be regular patterns of mate selection for every species. The term "promiscuous" has often been applied before we have even tried to determine if there are any regularities or rules for the patterns of mate selection. In addition, "promiscuous" may not be the best term to apply to any animal mating patterns because of its pejorative connotations, from the human sphere, of being "oversexed," indiscriminate, or unable to form lasting bonds.

Some species of primates characteristically live in small monogamous or polygynous groups, and most of the sexual behavior occurs between the small number of adults living in such a group. We know little as yet of how mates are chosen and how bonds develop. For the monogamous species it has often been inferred and occasionally reported (e.g. Chivers 1974 on siamangs) that the young of the group will move out of the parental unit around the time of sexual maturity and occupy a home range contiguous to the parents' area, or perhaps travel through neighboring areas until an unattached individual of the other sex is encountered. But exactly how a new monogamous bond and social unit is developed remains to be documented. For at least one polygynous-living species, the hamadryas baboon, it has been reported (Kummer 1968) that young adult males adopt yearling females and care for them in a parental manner until the females reach reproductive maturity and form sexual bonds with the male who has reared them. For the gelada, it has been hypothesized that younger males follow existing heterosexual units and either take over as the group male when the former male becomes old, or that perhaps the follower manages to entice some of the females, especially the younger ones who are probably offspring of the group, to form a new social unit with him. It is clear that we still have much to learn about the formation of social units which structure the patterns of sexual behavior, in monogamous and polygynous species.

The species which might give rise to the notion of promiscuous sexual behavior in primates are those which live in the larger, multi-female multi-male units. Both females and males in these societies mate with multiple partners; however even in these species many regularities in partner choice may occur and temporary, intense pair bonds may be formed. In the case of the chimpanzee (Goodall 1968b) and the Barbary macaque (Taub 1980a) copulation is often rapid and unaccompanied by elaborate bonding procedures, or partners may be exchanged often. The chimpanzee is sometimes reported to form special consort bonds (Tutin 1975), but at other times "line-ups" of males waiting to mate with a female in rapid succession, or groups of males traveling and mating with one estrous female, are reported. According to Taub, Barbary macaque females form not consorts, but "consociations," or transitory and exclusive sexual relations consecutively with all the reproductively mature males of the group, switching partners several times a day. Although it would certainly not be justified, at this point in our understanding of these species, to suggest that their mating is indiscriminate or without regular patterns of partner choice, it does appear more casual and less related to elaborate pair bonding than sexual behavior in some other primate species.

In contrast, many of the multi-female, multi-male species are reported to mate in the context of consort bonding; for example pigtailed, rhesus and Japanese macaques, chacma, olive and yellow baboons, squirrel monkeys, some species of howler monkeys, and the Hanuman or common langur of Northern India (see review in Michael & Zumpe 1971). Hausfater (1975) has also suggested that the one-male unit of hamadryas baboons may represent an elaboration of the consort relationship. Although there is much variability in manifestation, and consorts may last for only a few hours, or may endure for days or weeks, consort bonding is typified by courtship patterns to initiate the relationship and by affinitive interactions in addition to copulation, such as reciprocal grooming, mutual support in agonistic interactions, and simultaneous traveling, foraging, and resting patterns with close physical proximity. Often the consort couple separate themselves from the larger group, perhaps moving to the periphery, and become occupied and exclusive in their interactions, paying a great deal of attention to each other and resisting attempts by any third party to join or to interact with them.

My colleague, Harold Gouzoules, and I have been studying the process of consort bonding over a period of five years and I will outline here some of our preliminary findings. Although these suggested patterns are by no means indicative of mating systems in all primates, or even all consorting species, they may, by example, illuminate certain aspects of bond-formation and mate selection.

INITIATION OF CONSORT BONDS

In the Japanese macaque it is often very difficult to determine if a female or a male initiates a sexual relationship. As Eaton (1978) has noted, a female under observation may approach a male with a solicitation gesture and we assume this is an initiation, but we cannot know if the male directed a courtship signal at her (or as Eaton says: "sent her roses") over a week ago. That females may also send such fleeting but strong sexual invitations long preceding mating, became clear to me when I was once lucky enough to observe it happen. I was sitting on my heels writing data into a notebook resting on my knees, while closely observing an adolescent peripheral male eating leaves from a shrub. A consort pair moved into our vicinity and the female rapidly approached us, positioned herself directly facing us, and abruptly quivered her lips and gave us a "look" or stare of such intensity that I was temporarily nonplussed, as her eyes were directed just over my shoulder. I turned my head in time to see the young male recoil under the almost physical impact of the signal. Then the consort male stepped into the clearing and looked at us;

and the peripheral male hurriedly resumed eating leaves. I also had been observing this consort couple, consisting of a high-ranking female and male, and had noted that while the male courted and followed the female assiduously, she had neither presented to, nor huddled up against, nor groomed him, and did not seem very responsive to his courtship. After the intense, fleeting invitation by the female to the young male, the consort couple moved off, the female now expressionless, and the peripheral male followed at a discreet distance. He followed them thus for three days during which time the high ranking male did not mount the female. On the fourth day I found the female and the young peripheral male copulating in thick brush, while the former consort male noisily shook branches, pant-grunted in display, and repeatedly climbed different trees to peer around, apparently in search of the female. This incident is also a good example of the difference between "access" to an estrous female and successful consortship and mating (see Chapters 7 & 17).

Female Japanese macaques generally initiate consort bonds by using any or all of the many vocalizations, expressions and gestures described earlier. Particularly common, and apparently effective, is the intense stare with lip quiver I described above, and which is perhaps similar to what Saayman (1970) has called the "eyeface" in chacma baboons. Interestingly, it is only between mating couples and mother-and-offspring that direct eye contact does not function as a threat and a distance-increaser in this species. The Japanese macaque males are most elaborate courters, and Stephenson (1973, 1974) has studied the cultural variability in their courtship patterns. Generally speaking, it appears that much of male courtship is designed to draw attention to the courter. Such "attention-seeking" behaviors often begin at a great distance from the female, with displays such as tree-shaking with vocalization, or "bird-dogging" — a stylized, swaggering, tail-raised approach and jogging retreat, with periodic freezes in which the male lifts a leg and juts out his chin. If the female does not leave, the male may begin close-range courtship gestures, such as placing his hand or foot on the middle of her back and peering around into her face, followed by whirling around and bringing his hindquarters into her view. If the female still remains (and females do often walk out on these performances, or if the male is very high ranking she may crouch when touched and fail to respond), and the female has perhaps returned some initiation gestures, such as eye contact and lip quiver, then the couple may initiate body contact, especially through grooming, which in turn often leads to the beginning of a mount series.

MAINTENANCE AND TERMINATION OF A CONSORT BOND

The physiological aspects of series mounting in primates have not been studied, but it is clear in Japanese macaques that if the female does not respond immediately to the male's hip touch by standing correctly, his impetus, and somtimes his erection, are lost, and the couple sit down and "wait" for the next mount attempt. Furthermore, the male must experience more than one of these successful mounts with intromission and thrusting before ejaculation into the female occurs, and thus two willing and cooperative partners are essential. Occasionally one sees a mating couple in which one or the other partner does not seem to be doing their part toward a successful mount; either the female stands and the male does not mount, or the male hip touches and receives no response.

Such individuals often shudder and vocalize in a manner reminiscent of infant weaning tantrums (perhaps these are "frustration" gestures), and if the lack of coordination continues, the consort breaks up. Usually however, a consort couple seem highly motivated to coordinate their activities, and even if copulation is disrupted by the arrival of a third party, the couple will make every attempt to reunite and resume mating. Even when females consort with young, prepubertal males, who typically exhibit erections, mounting and thrusting, but few intromissions and no ejaculations, the female partners will willingly stand and the male partners will mount for literally hours. In younger partners and female homosexual partners in particular, the process of mutual accommodation to conduct a mounting series is evident, and sometimes unusual or unique signals are seen to develop between such pairs.

After a particular mount series has ended, the couple may move slightly apart or one may begin to groom the other. Also after some copulations the female or the male may leave entirely, bringing the consort to an end, although consorts are typified by more than one mount series. If one partner leaves, the other may follow and attempt to maintain the relationship. In this group it could not be said that either females or males are more prone to terminate a consort bond, as it seems to be individually variable. A high ranking male partner (in terms of direction of agonistic signals) often tries to maintain long consorts with a few females. Interestingly, such a male seldom directs aggression toward a consort female who is trying to leave, but rather he tries to keep up with her, to groom her and consolidate his bond with her, and to keep away other potential male partners by his own proximity to her. Females who initially have been highly motivated to mate with such males, may eventually come to make strong efforts to get away from them. An important comment on the socio-environmental aspects of "estrus" in females is that when followed persistently by a male over time, they may come to be very unresponsive and pale-faced, but if they shake off their consort shadow, they can be found immediately thereafter with bright-red faces, copulating vigorously with some other male. Occasionally the reverse occurs, that is, females persistently follow a male who has consorted with them and then left. Most frequently however the termination of a consort bond occurs when a couple simply ceases to engage in mount series, although the other aspects of consorting, such as grooming and close proximity of partners, may last somewhat longer. Perhaps we could say that most consort partners simply drift apart, and one or both begin the process of courting new partners. Because they begin again with new partners, we may think of them as promiscuous; however, as will be described in Chapter 17, the choice of individuals as consort partners does seem to be governed by several variables such as age, kinship, and past experiences.

MATE SELECTION, AGGRESSION AND SEXUAL SELECTION

Occasionally the initiation of sexual behavior in Japanese macaques is associated with aggression directed by the male toward the female (Tokuda 1961-2; Enomoto 1974; Eaton 1978). The male may chase the female in a very bouncing, stylized manner in which he does not catch her, or he may chase her very determinedly and catch and bite her. The male may also frighten the female by springing on her unawares and bounding away quickly, or he may sneak up to her, bite her and then leave. The female invariably

149

screams and shows every sign of being frightened (defecation, etc.) and either runs away or avoids the male. The Japanese primatologists sometimes say that these attacks have a "positive" effect on the female; however what is odd about the attacks is that they are differentially performed by certain males and not others, that the attacker usually leaves immediately, and that we have not been able to show that such patterns lead to consorts being formed either immediately or later. On the contrary the most "popular" males of a mating season are often elaborate courters, but non-aggressive toward potential partners. Females do not seem to respond sexually to males who frighten them excessively, but rather try to get away from them. This does not mean that some males would not try to use intimidation to initiate consorts, and much as female homosexuality in this group is part of a larger pattern of female sexual initiative, male aggression toward females might be part of a larger pattern of males drawing attention to themselves. This seems particularly likely as aggression seldom occurs between females and males who are actively consorting and mating, but tends to occur between individuals who are not consorting. Even harrassment of mating couples in this species seems a form of male "notice-me" display, directed toward the female of the pair, while the male partner may simply sit still during the harrassing display of a higher ranking male and later rejoin his consort partner.

Another possible explanation may lie with the seasonal breeding of this species, as most male aggressive behavior occurs in the earlier portion of the mating season, shortly after extreme physiological changes in both sexes. While it has been shown that testosterone levels do not correlate with aggressive behavior, when males are compared with one another, sharp and extreme testoserone increases or decreases in one male over time do correlate with behavioral changes (See Chapter 11). It is interesting to note that endocrine changes in females immediately prior to and during early estrus are also associated with increases in aggressive levels (Rowell 1972b). In contrast, for a year-round breeder such as the chacma baboon, it is noted explicitly that agonism, either between males or between females and males, is not part of the process of consort bond formation (Saayman 1970).

Finally, aggressive behavior by males toward potential female partners could be a "side-effect" or misdirection of aggression which is more often directed toward competing males as part of intrasexual selection. For example, in the mating season of the herring gull, the males are initially belligerent to both females and males, and only slowly develop more appropriate responses to females (Tinbergen 1960). While the elaborate courtship of the Japanese macaque male and the preferences and advances shown by females to certain males, conform in many respects to the classic sexual principles of male-male competition and female choice, two additional factors are of note: males also discriminate between and choose female partners (Enomoto, 1978), and they seldom confront another male directly in competition over a female. Sub-adult and adult males actively avoid each other in mating season, and the wounding which does occur at the beginning of the mating season, especially to young pre-pubertal males, seems to result from large mixed-sex, mixed-age agonistic encounters in which aggression is redirected to low-ranking and vulnerable but non-reproducing young individuals of both sexes.

CONCLUSION

The outstanding characteristic of primate sexual behavior is its variability in manifestation. This chapter has attempted to describe the relationship of sex roles, social structure and social bonds to mating behavior. In particular, through an extended example of one group with which I am very familiar, I have tried to show that both female and male primates are active, discriminating sexual participants, and that sexual behavior takes place in the context of a mutual bonding process.

The easily seen external genitalia of the male infant Japanese macaque are the only apparent differences between male and female infants at birth. Photo courtesy of Karen Dickey.

Most infants are born with a distinctive black coat, which they later lose. Variability in coat color can be seen in this infant, whose natal coat is very pale. His mother is 25, the oldest recorded Japanese macaque female to give birth.

Related females who give birth in the same season are often found near one another. Thus, infants of related mothers may be each other's first companions and play partners.

Mothers are generally very protective, and actively restrain overly adventurous infants, holding them firmly by the foot or hand and staring at them emphatically.

An unrestrained infant may approach an adult male, whose reaction is unpredictable; he may retrieve, ignore, avoid, or threaten and punish the infant.

Juveniles, like the two in this photograph, frequently show great interest in infants, especially their younger siblings, and try to entice them away from their mothers.

An infant soon moves from clinging to its mother's ventrum to riding on her back, even when she goes swimming.

Long after it has lost its natal coat and is quite independent, a young monkey may occasionally hitch a ride on its mother.

Observational skills, such as watching what to eat, are very important to young macaques, with the mother being the first important focus or model.

Watching mother choose and eat food items is often followed by "muzzling," presumably to catch olfactory clues about the food item.

Food sharing is seldom seen in Japanese monkeys, but the young often forage close to their mothers, and eat from the same source.

While they maintain close ties with their mothers, juvenile macaques quickly become quite independent and adept foragers.

Peer groups become a second focus of attention and activity; learning from one's peers then becomes important.

A major peer group activity is play; the "play face" is a clear signal of this juvenile's intention to play.

Uncertainty or tension, reflected by this juvenile male's "grimace," is sometimes present in peer interactions.

Threats, involving the raising of the eyebrows ("lid"), and sometimes the dropping of the jaw ("gape"), are another sign of tension or conflict.

Juveniles are often protective of their younger siblings, retrieving and carrying them.

Infants seem to enjoy this attention from alloparents, willingly accepting it even when the juvenile is little bigger than the infant.

Rough and tumble play is often said to be characteristic of juvenile males such as these; however, young females can also be found in such play groups.

Adults of both sexes sometimes play with the young. Juvenile males, like the small one on the right, may be particularly drawn to the company of adult males.

Subadult female play may take the form of "play mothering": these yearlings are well beyond the need for protective restraint.

Juvenile and subadult males may form exclusive groups and begin to move to the periphery of the group. They forage, play, groom, and huddle together.

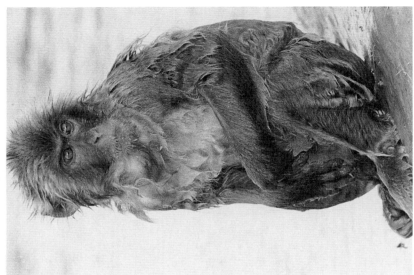

Adult peripheral males immigrating from neighboring groups remain spatially separated while attempting to forge social bonds with the more peripheral females of the troop.

Until they peripheralize or leave their group of origin, young males continue to maintain links with the core of the group through their mothers.

Adult central males often appear vigilant, performing "lookout" duties from good vantage points.

Females, too, can be seen performing vigilant or "lookout" behavior.

At night, or in cold weather, whole family groups (females and males) can be found in tight huddles, which the Japanese primatologists refer to as "dumplings" of monkeys.

During the mating season some adult males develop impressive "capes" and adopt a swaggering courtship walk known as "bird-dogging."

Close bonding of consort couples, who sit, groom, and even forage together, seems to facilitate mating in a series mounting species like the Japanese monkey.

Consort couples are exclusive in their attentions; here the female threatens the photographer, who is a potential disruption.

Here the male shown in the previous photograph is being mounted by his female consort partner; she exhibits the same double-foot clasp mounting style as he did.

The double-foot clasp mount, shown by this male, sometimes has been defined as a uniquely male behavior.

In the winter months, following the autumn mating season, adult males relax and play more, and take a greater interest in the young.

Females maintain their close kin ties, grooming each other and each other's infants.

Wrinkled, white-furred old females remain especially active and influential within their matrilines. These subgroups usually are composed of an elderly female with her daughters, granddaughters, and their immature offspring of both sexes.

Old males, too, can maintain their rank and influence long past their physical prime. This old, arthritic male was still the second ranking male in the Arashiyama West group when he disappeared at an advanced age.

PART III

GROWING UP:
THE ONTOGENY
OF SEX DIFFERENCES

Chapter ELEVEN

Fertility & Virility: Hormones and the Development of Sex Differences in Behavior

Stumptail macaque
Macaca arctoides
These S.E. Asian macaques are moderately dimorphic. The female (left) suckles an infant still with its white natal coat.

The biological factors involved in the sexual differentiation of behavior have been studied largely in terms of endocrine (hormonal) processes and their interrelationship with neural processes, although the workings of the latter system have been more inferred than directly investigated. This chapter describes some of the research concerning the relationship of neuroendocrinology to developing female and male patterns of behavior, especially sexual behavior. Although some research on reproductive physiology and behavior, such as that by Rowell (1963, 1967, 1971), Bramblett et al (1975), and Rose et al (1971, 1972) involves the monitoring and measurement of physiological and behavioral processes in intact monkeys of various species living in social groups, by far the majority of the studies have involved experimental alteration of the physiological and behavioral processes in laboratory rhesus monkeys. As a consequence, this chapter mainly describes experimental research, such as ovariectomies, male castration, and prenatal hormone alteration, on rhesus monkeys maintained in isolation in the laboratory.

The relationship of physiology to behavior is always an extremely challenging and difficult area of research. This is especially true in the case of behavioral and physiological *ontogeny*, where the continual and significant changes in the growing and maturing organism make the determination of causal relationships a complex and uncertain matter. Correlations themselves never prove causation, and this is a warning to be carefully heeded in the study of hormones and behavior, where interpretations of correlational studies often seem to imply uni-directional causation from hormones to behavior. In addition, primate behavior, as compared to the behavior of other mammals, has a far more tenuous and in-direct relationship with hormonal influences, and far more mediation by social learning and cortical control (Beach 1965, 1970), so one can see that this is an area where simple answers can neither be expected, nor accepted.

Endocrinologists working with primates are aware of these difficulties and limitations. However, there is a certain tendency in some of their writings to treat the plasticity of primate behavior and the importance of socio-environmental influences, not as an integral part of the system, but as an impediment to their research, creating difficulties in experimental design — the implication is that once controlled-for or eliminated, the importance of these "difficulties" may be forgotten. Michael, a leading researcher in primate reproductive physiology, has said of rhesus monkeys:

> Of course the complexity of their behavior introduces many variables into any experiment one attempts, but as often happens in science, new methodologies are developed for overcoming the difficulties. To be more specific, when we first began to study the sexual behavior of the female rhesus, it was initially troubling to find that this behavior was not wholly hormone-dependent (1975: 71).

In fact, the dependence of all primate behavior on the social environment has been more than just initially "troubling" to the endocrinologists. The problem these researchers face is that what they consider to be adequate investigation of physiological processes requires a rigorous methodology which necessarily involves extensive manipulation of the animal and control over biological and environmental influences. Yet it is precisely this manipulation and control which may seriously alter the behavior and physiology of the animal under study. For example, we now know that periods of sexual receptivity in the rhesus female and levels of testosterone in the rhesus male may be enormously influenced by environmental events such as social interactions, or by manipulation and handling by experimenters (Rose et al 1972; Rowell 1972b; Saayman 1975). This means that in an effort to increase control over their research variables, experimenters may alter those that they wish to study, and by creating new environmental conditions, may even create new and confounding variables of which they are not aware.

Phoenix, some of whose research will be described below, has expressed this most cogently in a paper on the role of androgens in adult male rhesus sexual behavior:

> Our research animals are kept in what some would describe as unnatural conditions. In many respects laboratory housing resembles the conditions in some of our more enlightened prisons. Therefore, our description of the sexual behavior of the male rhesus may no more resemble his sexual behavior in nature than would a description of an inmate on a week-end pass resemble the sexual behavior of the average human male in nature (1974a: 249-50).

He went on to elucidate the point made in the fourth chapter of this book; that there is no clear-cut distinction between "natural" and "unnatural" conditions, and that rhesus in India sometimes live in considerably stressful conditions. However, his later statement, that environmental and behavioral differences may be irrelevant to the problem he studies is debatable for two reasons: firstly, environmental conditions, and aspects of the organism such as its behavior or its psychological state, are always relevant to questions of endocrine secretion because of the interactive nature of hormone-behavior-environment relationships. Secondly, it is not possible to seal off one's research totally from larger considerations, because the ability to generalize from a specific piece of research is really the whole point of the endeavour, and the temptation to generalize from specific findings is an inclination with which many of us are familiar. We are left with a question analogous to the one we encountered in food tests of dominance: if researchers find that they have to create a totally artificial situation in order to study their problem, what can this work tell us about the behavior and physiology outside of this artifical situation? Or in terms of Phoenix's figure of speech: what can we conclude about the sexual behavior of the average human male from that of a prison inmate on a weekend pass? Actually, given the conditions in the experiments I am about to describe, a more accurate description would be that of a prison inmate in permanent solitary confinement, allowed only brief, periodic visits, still in a cell, with a surgically-treated female prison inmate.

When we wish to demonstrate the complex interaction of hormones, behavior and environment, and even the possible effects of the research environment itself upon primate reproductive physiology, a good cautionary example is the study of "estrus" in primate females,the history of which is outlined briefly here. The earliest studies, based on observations of baboons in zoos (Zuckerman 1932) and rhesus macaques in laboratories (Hartman 1932) stressed continual receptivity in all primate females. In captivity copulation often *does* occur throughout the female menstrual cycle and throughout the year. However it seems likely that the conditions of captivity themselves contribute to this pattern, because, beginning in the 1950's, field studies established definite mating (and non-mating) seasons in many of these same species of Old World primates, including the rhesus monkey (Lancaster and Lee 1965). Spurred on partly by the growing awareness that these species would only mate at certain times, and partly by the direct correlation between "heat" and ovulation in many nonprimate mammals, some primatologists explicitly or implicitly embraced the idea that female nonhuman primates have regular cyclic patterns of receptivity directly associated with the specific stage of ovulation in the monthly cycle. Thus receptive behavior, or estrus, came to be considered perfectly synchronized with, and a mere indicator of, the significant biological event of ovulation. Both the theory of continual receptivity, and that of limited estrus during ovulation, are now known to be at best oversimplifications for nonhuman primate females, whose periods of receptivity are variable from species to species, from female to female, and even from estrus to estrus in one individual over time. It is variable because estrus is a behavioral phenomenon which is integrally related to environmental circumstances as well as to physiological factors (Hanby et al, 1971; Rowell 1972b; Kaufman 1965; Saayman 1975).

This discussion is not intended as a rebuttal of the importance of endocrine factors to behavior, nor a denial of the difficulties with which researchers in this area must contend. It is intended as a caution against hasty extrapolation from very specialized laboratory studies, and against an unquestioning acceptance that endocrine research, any more than deprivation research or socialization studies, offers the definitive explanations of the causes of sexual differentiation of behavior.

OVARIECTOMIES

Interactions between hormones and behavior in primates most often have been studied in terms of those hormones secreted by the gonads — progesterones, estrogens, testosterone. A frequent technique has been to remove the structure (ovaries, or testes, and occasionally, the adrenals) which normally secretes the hormone under study, and then to administer known amounts of the hormone at specified times. Subjects thus treated are paired with a member of the other sex under controlled conditions and their behavior observed. This experimental technique has been used, on females especially, in order to research many aspects of the reproductive behavior and physiology of the monkey such as: rhythmic changes in male-female sexual, affinitive and aggressive interactions, female receptivity, female attractiveness, and the importance of sexual pheromones to heterosexual interactions (see studies such as Michael & Herbert 1963; Michael et al 1968; Michael & Keverne 1968; Michael 1975; Michael & Welegalla 1967, 1968; Michael et al, 1974b; Herbert 1967, 1968, 1970; Herbert & Trimble 1967). To summarize the results of these studies, it can be said that while removal of the ovaries in the rhesus monkey apparently has little effect on female differentiated patterns from birth until puberty (Goy 1968), ovariectomy does greatly decrease sexual behavior in the mature rhesus, and it disturbs monthly rhythms of various heterosexual interactions. The experiments show that under these test conditions estrogens secreted by the ovaries are necessary both to female sexual behavior and to male perception of female attractiveness. Artificial administration of estradiol (a form of estrogen) re-induces sexual behavior in the ovariectomized female, administration of progesterone decreases receptivity and attractiveness, and administration of testosterone may increase proceptivity without a concomitant increase in attractiveness to males.

158

The ovariectomy experiments have been many and varied and have followed a long-term, developing research program over the years (see Herbert 1970; Michael & Bonsall 1979 for reviews). Probably their outstanding accomplishments, as Rowell (1972) noted, have been the separation of the two concepts of female receptivity and female attractiveness, and the resultant documentation of a vaginal pheromone. A pheromone is a chemical substance produced by an animal which acts as a signal in olfactory communication. By careful manipulation of hormonal treatments and observations of subsequent behavior in the females and in their male partners, these experiments have shown that female motivation to mate is influenced by circulating levels of estrogen and testosterone, while the male's motivation to mate with her is influenced by a vaginal pheromone whose presence is dependent on estrogen.

From this outline of the principal results of these studies, an important question which follows is: what can we conclude from these findings? The question can be approached from both the physiological and the behavioral aspects of the research. Firstly, what does ovariectomy and hormone replacement treatment tell us, or allow us to conclude about the normal physiological cycle of the rhesus female and other primate females? Rowell has given the best expression of the answer:

> The ovary is a complex organ and can hardly be considered adequately replaced by a known quantity of estradiol. The physiological changes which take place during menstrual cycles, pregnancy and lactation have been related to dynamic endocrine processes; some of the most important events occur when the output of one hormone is changing at a particular rate and in a particular direction in relation to similar changes in one or more other hormones. The whole system is characterized by a remarkable absence of steady states. It seems likely that behavioral events are also influenced by these dynamic processes, and where in this do we place the castrated female with steady replacement therapy of one or two hormones in constant proportion? Studies of such animals do, however, produce consistent and reproducible answers; for the behavioral zoologist they can provide valuable pointers, but they should not be accepted unreservedly as illustrating events in normal cycles without confirmation (1972: 71).

In other words, these experiments may indicate some of the physiological processes accompanying sexual behavior, and even suggest the way in which some elements interact to produce sexual behavior, but as yet, they are far from describing or explaining the whole process. As Rowell points out, these experiments have not yet tested the dynamic nature of physiological change, indeed present research techniques and designs have to assume steady physiological states. If, as I have suggested, these physiological states are so dynamic that even the manipulations involved in experimentation change them, then, as Rowell notes, we must look elsewhere for confirmation of their findings.

Socially-Living Rhesus Monkeys

Similarly, what do the behavioral findings of these experiments permit us to conclude about normal rhesus monkey behavior? Both Herbert and Michael stress individual variability and sexual preferences for certain partners as overriding factors in the sexual behavior of their subjects; indeed, they try to control for this by using the same limited number of individuals and the same pairs throughout a set of experiments. Of course, this technique may then limit the generalizability of their research, as it could be argued that they are studying idiosyncratic rather than normative patterns. However the alternative of using very large samples of animals in these experiments would be impractical. Variability in response, even in the most rigorously controlled testing situation, has probably been as troubling to this research as has the previously mentioned dependence of primate behavior on the social environment. For example, some rhesus monkeys continue to mate long after ovariectomy without hormone replacement treatment; the absolute frequency of mating varies both within and between pairs; monthly rhythms in interactions do not occur in all pairs and may change in one pair over time; and the introduction of an additional female to the paired-situation changes the behavior patterns so much that heterosexual interactions are not predictable from knowledge of hormonal status.

Given that rhesus monkeys are socially-living animals and that there will always be additional females and males as an integral part of heterosexual interactions, the implication is that these laboratory studies cannot tell us a great deal about behavior in wild or free-ranging monkeys, or even captive social groups. For example, although two experimental studies have shown that even *intact* (not ovariectomized or hormone treated) laboratory rhesus under the highly-controlled, paired test situation will exhibit a mid-cycle peak in sexual behavior which presumably correlates with estrogen peaks and ovulation (Michael & Herbert 1963; Phoenix et al 1968); this cyclic pattern appears to "virtually disappear" (Rowell 1972) when the monkeys are maintained in social groups. Thus Rowell (1963) tested untreated rhesus in small captive social groups and found that they would mate throughout the cycle. In other words, females continued to be receptive and to attract males under different hormonal states, and actually mated more often in the luteal phase of their cycles when progesterone levels were higher.

In free-ranging conditions rhesus macaques exhibit a definite breeding season, and mating outside of these months is rare. However the relationship of the monthly menstrual cycle to sexual behavior is very difficult to determine (mature rhesus do not exhibit sexual swellings and menstruation is difficult to observe), and, probably as a result, the reports are mixed. Carpenter (1942b) and Kaufmann (1965) suggested that sexual activity was fairly well restricted to the middle part of the cycle when ovulation occurs, but as Rowell has pointed out, their data actually show considerable variability, with estrus periods lasting from 1 to 95 days. More recently, Loy (1970) has suggested two peaks in rhesus sexual behavior — one at mid-cycle and one peri-menstrually. These rather mixed findings may well be due to not only the difficulties of determining menstrual cycles in free-ranging monkeys, but also to variability in the animals themselves.

It has also been found that rhesus monkeys (as well as other species) in social groups will continue to mate in the first months of pregnancy (Conoway & Koford 1964; Kaufman 1965; Loy 1970) when circulating levels of progesterone are high. This, and Rowell's finding of high frequencies of mating in the luteal phases of menstrual cycles, would appear to contradict the experimental finding that progesterone *decreases* receptivity and attractiveness (see discussion in Michael & Bonsall 1979). Hinde concluded that: "under natural conditions, social factors become more important than hormonal ones" (1974: 302). Clearly, these data imply that under social conditions, behavior in the rhesus monkey is not easily predicted solely from a knowledge of hormonal status.

Other Nonhuman Primate Species

How well do the findings from the rhesus ovariectomy experiments generalize to other nonhuman primate species? One problem in answering this question is that these experiments have not been replicated in many other species. Saayman (1972, 1973) found marked individual differences in the responses of ovariectomized baboons treated with known doses of estrogen. Anderson and Mason (1977) ovariectomized squirrel monkeys and found that hormone replacement therapy did effect heterosexual interest and attraction, but unlike the rhesus studies, copulation was not stimulated between males and ovariectomized females through the use of estrogen replacement treatment. Slob et al (1978a, 1978b) using the ovariectomy and carefully-controlled paired-test technique described earlier, found that stumptail macaques remained as attractive and as receptive to males, and as sexually active after ovariectomy as before, without replacement hormone treatment. There was a partial decrement in the frequency of copulations, which they think was due to the increasing difficulty of gaining intromission, because estrogen helps to maintain the elasticity and lubrication of the vagina. Furthermore, intact stumptail females showed little or no variation in receptivity and attractiveness throughout the regular cycle, and administration of progesterone did not decrease sexual activities. Even adrenalectomy, removal of the adrenals, the other gland besides the gonads which may secrete sex steroids, did not significantly decrease sexual behavior. Since stumptails are in the same genus (*Macaca*) as rhesus monkeys and are therefore considered to be quite closely related to them, it can be seen again that one must be very cautious of generalizations or extrapolations concerning the effects of ovariectomy, and estrogens and progesterone on all primate species.

Some studies of other species have not used ovariectomy experiments, but instead have monitored behavior in untreated monkeys during normal menstrual cycles. For example, Rowell (1970, 1971) and Bramblett et al (1975) tested for frequencies of copulation correlated with physiological data concerning the reproductive state of intact vervet and Sykes' monkeys, maintained in captive social groups. Rowell characterized the results of her studies as a "sum of extremely varied patterns" (1972: 80). Although they did not have direct physiological data, Hanby et al (1971), Wolfe (1979) and Gouzoules and Fedigan (unpublished data) were not able to find evidence of discrete estrous periods which might be correlated with specific stages in the menstrual cycle in provisioned colonies of Japanese macaques. These monkeys, like the free-ranging rhesus, may be in estrus for periods anywhere from one to ninety-five days. In laboratory paired-test experiments, pigtail macaque females did not change their sexual behavior patterns during the ovarian cycle (Eaton and Resko, 1974). On the other hand, a definite mid-cycle estrus associated with a sex-skin swelling and high estrogen levels *has* been reported for social groups of baboons (Rowell 1967), mangabeys (Chalmers 1968), talapoins (Herbert 1970) and chimpanzees (McGinnis, cited in Hinde 1974). Once again, the great diversity in the response of individuals, groups and species to hormonal influences is the outstanding impression one receives from reading these studies.

Human Females

In spite of this variability, can the findings on ovariectomized rhesus monkeys give us any clues about the behavior of human females? First of all, the point needs to be made that the popularity of the rhesus monkey in laboratory experiments is not due to its singular aptness as the best model for human biology and behavior, as occasional reports of its similarity to humans might lead one to believe (e.g. Michael 1975). Rather it is so often used because, until recently, it was the easiest and cheapest Old World anthropoid to obtain and maintain in the laboratory, and over the years scientists have built up the best library of information on the biology of this species. Medical research has shown that the rhesus monkey can make a reasonably good model in the study of some, but not all, human diseases (the chimpanzee makes a better model but it is too rare and usually too expensive to use). Whether the rhesus makes an appropriate model for the study of other aspects of human biology, and especially for the study of human behavior is a different and highly debatable question.

We know that, as in rhesus monkey females, estrogens promote lubricant secretions of the vagina in human females; however they do not produce consistent behavioral effects on the latter. Androgen-producing tumors of the ovary or administration of exogenous testosterone can apparently increase female libido (c.f. "proceptivity" and "receptivity"), while progesterone can decrease it, but different studies do not offer consistent results (Rossi 1973 O'Connor et al 1974; Swyer 1968; Barfield 1976).

Michael et al (1974a), in a study of 47 women found evidence of a vaginal secretion similar in composition to a sex pheromone of the rhesus monkey, but there has been little or no testing of its behavioral function, if any (however see suggested function in Udrey et al 1973). Udrey and Morris (1968, 1970) reported signs of a mid-cycle peak in copulation and female orgasm rates in two different studies of American women. The first study in particular has been widely referenced as "convincing evidence of some underlying biological mechanism" (1968: 595). One must question such generalizations drawn from this first paper, which reported data obtained incidentally to a different study, with no controls, and questionable techniques for gathering data. (Southern black women of low socioeconomic status were paid to report frequencies of intercourse and orgasm, and the authors were not certain that the women understood the meaning of orgasm). A mid-cycle peak was *not* found in a later study by Udrey & Morris (1977), which demonstrated a maximun level of intercourse occurring soon after the end of menstruation. Incidentally, this last study also demonstrated that very different results may be obtained from different methods of measuring menstrual cycles in human females.

According to Stoller (1968), ovariectomy of the human female does not produce any consistent change in behavior patterns, including copulation patterns, and ovarian hormones do not play a significant role in the human female's motivation for sexual behavior. The variability in reports concerning human females, the apparent differences between the behavioral effects of estrogen on the human female, compared with the laboratory rhesus female, as well as the obvious plasticity of behavior and importance of social learning in both the rhesus monkey and the human, all make it rather unlikely that the ovariectomy experiments can tell us anything universal and concrete about the effects of gonadal hormones on behavior in socially-living female primates. Other than, that is, the general observation that aspects of primate behavior are likely to be related to hormones in multiple, complex and probably circuitous ways.

MALE CASTRATION, TESTOSTERONE AND BEHAVIOR

The effects of removing primate male gonads have not been nearly so extensively studied as the effects of ovariectomies, and the results of male castration seem, if possible, even more fragmentary and inconclusive. Infant male rhesus monkeys castrated at birth or weaning and maintained in laboratory conditions, exhibit no significant behavioral differences from intact males until puberty (Goy 1968). The castration of adult rhesus has variable results, depending at least partially on the amount of previous sexual experience. In general it can be said that post-pubertal castration reduces, but does not eliminate, sexual behaviors; however, its effect on aggression and dominance rank is unclear. Phoenix (1974a, 1978a), Resko and Phoenix (1972), Michael and Wilson (1974) and Wilson et al (1972) have reported on laboratory studies of castration of adult male rhesus. Phoenix and his co-workers concluded that no significant correlation exists between testosterone levels and levels of sexual performance (ejaculation frequency, latency, intromission frequency), either before or after castration, but that a sudden sharp drop in testosterone, such as follows castration, does bring about a decline in sexual behavior. Replacement testosterone treatment brings sexual behavior back up to its pre-castration levels. Surprisingly, castrated males then continue to exhibit normal levels of sexual activities up to one year *after* exogenous hormone treatment ceases. Obviously testosterone does not have a simple causal relationship with male sexual behavior. Phoenix particularly emphasized the high degree of variability between the responses of his subject males to castration, and the importance of non-hormonal variables. For example, three of his ten castrates displayed all sexual behaviors, including ejaculation, one year after castration even without replacement testosterone treatment.

Wilson and Vessey (1968) described the long-term results of castrating ten free-ranging rhesus males from the colonies of Cayo Santiago and La Parguera. Again, one is struck more by the variability than the consistency in the effects of castration. However, the authors did conclude that all the castrated males showed an irregular decrease in sexual and aggressive behavior and an intriguing tendency to associate with each other and form alliances against non-castrates. Recently, in the first of what will be a series of articles on behavioral effects of castration in 17 rhesus males and females, Loy et al (1978) reported on play differences between castrated and control animals maintained in small corrals and small social groups. The nine experimental males involved were all castrated pre-pubertally, and it was found that they continued to play significantly more often than the control males during the four years of the study. Normally, play levels decrease as rhesus males mature. The results of this study could simply mean that levels of play are inversely correlated with levels of testosterone, but Loy has suggested that a more complete explanation is that pre-pubertal castration produces a retardation of several aspects of male growth, development and maturation, including behavioral patterns. Loy also stated that the castrates were all very aggressive monkeys arranged in a clear dominance hierarchy, although the possibility remains that male-male relationships among the castrates may have been friendlier than those among the controls.

As well as the relationship between male sexual behavior and androgens, an association between aggression and testosterone has long been postulated. Rose et al (1971) found low, but significant, correlations between plasma testosterone levels and dominance rank and aggressive levels in a group of 34 male rhesus. Rather than assuming that the amount of circulating testosterone was determining dominance rank and frequency of aggression in the monkeys, these researchers went on to demonstrate that behavioral and environmental events were themselves influencing testosterone levels. For example, males subjected to sudden and decisive defeat by other males exhibited sharply fallen levels of testosterone afterwards, while males introduced into a group of receptive females, in which they could rise in dominance and also copulate, showed two to three-fold increases in testosterone levels. Rose, Bernstein and Gordon have explored extensively the ramifications of the finding that testosterone levels are influenced by multiple environmental, social, and species variables, performing numerous experiments on a variety of species (Rose et al 1972, 1974, 1975, 1978a,b; Bernstein et al 1974a,b, 1977, 1978, 1979; Gordon et al 1976, 1978, 1979). The following factors have been shown to influence circulating levels of testosterone: ontogenetic status, circadian rhythms, relative access to females, seasonality, alterations in social rank, successful and unsuccessful agonistic encounters.

In a related study, Eaton and Resko (1974) were unable to find linear correlations between testosterone levels and rank and aggressiveness in a naturally-formed group of Japanese monkeys. They suggested that the correlations found in the original Rose et al (1971) study were due to the artifically-formed, all-male group in which dominance was determined by repeated fighting rather than social tradition and long-term relations. Of course species differences or methodological differences could also be implicated. The linear correlation in the Rose et al (1971) study was only demonstrated for the upper 25% of the ranked males. Similarly, Kling & Dunne (1976) found correlations between rank and testosterone levels only in the top 33% of their group of stumptail macaques. Since the original study, Bernstein et al (1974) and Gordon et al (1976) reported a failure to find correlations between rank and hormonal levels in several on-going, mixed-sex groups of pigtailed, crabeating, and stumptailed macaques. They also reported (Bernstein et al 1979) on an agonadal male with the testosterone levels of a juvenile, who became top-ranking in a group and performed the alpha male role. It is agreed by all these different researchers that the studies demonstrate the lack of a stable or fixed amount of testosterone in each male which determines his aggressiveness and his social rank. Rather, the hormonal levels are at least in part a function of social stimuli, and may fluctuate in concert with them. In turn the levels of circulating testosterone may facilitate the appropriate pattern of assertiveness in the individual, given his current social standing.

Studies of the human male have generally shown no correlations between androgen levels and aggression levels (Barfield 1976; Doering et al 1974). However Kling (1975) does suggest that: "Although not yet adequately substantiated, there appears to be increasing evidence for positive correlation between plasma testosterone levels and agonism in man and other primate species" (1975: 309). He reviews four studies of testosterone levels and agonism in human males, three of them on groups of prison inmates and psychiatric patients. Significant correlations were found in three of the four studies, but only for some of the subjects with certain qualifications, so further substantiation of this evidence is needed. Kedenburg (1979) argued that correlations reported between testosterone and aggressive levels in institutionalized human males are the result of the confounding variable of "stressful environment." We have seen that even in the rhesus monkey the relationship between testosterone, aggression, and dominance is complex (Rose et al 1974; Eaton and Resko 1974), and one would not expect it to be less so in humans. Neither have studies of human males demonstrated consistent correlations between impotency and levels of testosterone, or homosexuality and prenatal or adult levels of sex hormones (Kolodny et al 1971; Doerr et al 1973; Brodie et al 1974). In general, while the castration studies do indicate that the general presence of testosterone may be important to the maintenance of sexual behavior, there is no evidence to indicate that *levels* of testosterone have significant effects on aspects of virility and the "quality of masculinity" in the male population at large, in spite of the popular misunderstanding of this hormone.

If a human male is castrated prior to puberty, that is, before secondary sexual characteristics have been developed and before sexual intercourse has been experienced, then puberty and normal adult sexual functioning will not occur, although Money and Ehrhardt (1972) note that pre-puberal castration does not eliminate all libidinal behavior. Castration after puberty has variable effects and although there is much documentation of these effects on the human male body, there is very little information on behavioral effects (Kling 1975; Meyer-Bahlburg 1974). Kling's review of voluntarily castrated human males convicted of a variety of sex crimes (see Stürup, 1972; and Bremer, 1959), indicates that habitual sexual offenses and all sexual behavior are greatly reduced after castration, while aggression is not. Given the important psychosocial complex of fears and beliefs surrounding male castration documented by social scientists in Euroamerican societies, and the ambiguities of "volunteering" while a prisoner or mental patient, it seems myopic, to say the least, to approach this topic considering only the variables of hormonal levels and resultant sexual performance or aggressive behavior. For a human, male or female, being castrated is not reducible to simply having an endocrine structure removed, and one cannot study castration as though it were comparable to having one's tonsils taken out.

In sum, the data on primate male castration show inconsistent results, so variable that one must again conclude that hormonal influences are interactive in complex ways (masked, augmented, counteracted) with learning and experience. Laboratory experiments show that in a general sense testosterone plays a role in maintaining normal levels of male sexual behavior, and that in some cases testosterone levels are correlated with male aggressiveness, but a simple causal relationship has not been demonstrated. Indeed, experiments with socially-living male monkeys, such as those by Rose, Bernstein and Gordon, have begun to elucidate the multiple feedback loops in the relations between testosterone, behavior and environment.

PRENATAL HORMONE ALTERATIONS

Studies of "deprived" infant rhesus monkeys, such as those by Harlow (1965) and Mason (1960) described in detail in Chapter 13, found that these monkeys begin to exhibit some sexually differentiated patterns of behavior quite early in life. Specifically, infant males threaten more often, engage in more rough-and-tumble and chasing play, and mount more often than do infant females. This is particularly interesting because infant male rhesus in the first two and a half years of life do not produce measurable amounts of testosterone, and so it seems very unlikely that these behavioral differences can be attributed to the effects of circulating testosterone in the systems of the young males (Resko 1967; Goy 1968). Further, Goy (1966) castrated rhesus males and females at birth and

165

at weaning, and found that they still exhibit the differentiated frequencies of threat, play and mounting, in the absence of gonads and gonadal hormones. So, endocrinologists have been led to wonder why these sexual differences in behavior occur even in neutered monkeys.

Embryologists have been aware for some time now that androgens produced by the genetic male fetus have a decisive effect on the prenatal development of the male reproductive structures themselves. What appears to happen is that the mother's own estrogens and progesterone will dominate fetal development unless the fetus itself produces androgens. Various endocrinologists have suggested that prenatal androgens may also affect the developing central nervous system, and thus behavior, so that the basis for "psychological" as well as physical sexual differentiation may be established prior to birth, and infants are then born with a male or female behavioral "bias."

A large number of experiments, especially on rodent species, have been conducted to test this hypothesis, and hormones have been added and/or gonads extracted, in almost all of the possible combinations. What is of note here is that significant species differences have been found (for example, some species appear to be affected by *pre*natal androgens and some by *neo*natal androgens), but in general prenatal androgens appear to have behavioral effects on long-gestation species such as the guinea-pig and, as we shall see, the rhesus monkey.

Following the experimental research on rodents, testosterone was injected into pregnant rhesus females for variable periods of from 25 to 50 days (Goy 1968, 1970; Young et al 1964; Goy and Resko 1972; Phoenix 1974b; Phoenix et al 1967, 1968; Young et al 1964). It was found that while the resulting genetically-male infants do not exhibit any behavioral or physical difference from untreated male rhesus infants, a fetus which is a chromosomal female, similarly subjected to abnormally high levels of prenatal androgen, develops into what is called a virilized, or androgenized, or "pseudohermaphroditic" female, possessing both male and female reproductive structures. Eight such pseudohermaphroditic females have been born and tested. They have a penis and a scrotum (empty), so that their external appearance is male; however they also have ovaries, and once they reach puberty, which is delayed, at least some of them exhibit menstrual cycles.

The behavior of these pseudohermaphrodites, as well as their reproductive physiology, appears to be influenced by the prenatal androgens, as demonstrated in the following tests. The experimental infants as well as the control infants were left with their mothers for the first 90-110 days of life; then all were permanently separated from their mothers and kept in individual cages, except for five test periods each week. During the latter, four to six infants were placed in a group for 20 to 30 minutes and their interactions observed. It was found that in play, threat and mounting frequencies, the androgenized females were more active than normal females, and that their test scores tended to be intermediate in averaged frequencies between those of untreated males and untreated females.

Interpretation of Results

The researchers appear to be aware of the importance of experience and other environmental factors on sexual differentiation of behavior, because the very behaviors they are testing — playing and mounting — do not occur in any normal fashion in isolate-reared rhesus monkeys and remain abnormal in frequency even in socially-reared laboratory rhesus (Phoenix 1978b). What they do conclude from these experiments is that prenatal androgens "organize" the central nervous system, such that there is a "predisposition to acquire specific patterns of conduct that are normally characteristic of the genetic male" (Goy 1968: 160), and that "the contributions made by experience act upon a substrate that is already biased in either a masculine or feminine direction" (Goy 1970: 28). They do not mean that an infant is born with female or male behaviors springing full-blown from its central nervous system (as instincts were once thought to do). Rather, in the normal process, males are more inclined to acquire certain behavioral characteristics than others, and genetic females have a different set pattern of predispositions, or a readiness to learn and/or express certain behaviors.

The interpretations of the researchers then, are usually circumspect and qualified, as they attempt to recognize and take account of the interactive nature of the hormonal and experiential variables. I will suggest at the end of this chapter that much questionable argumentation, supposedly based on this research, stems not from the research results themselves, but from a common misconception of the dichotomy or mutual exclusivity of female and male patterns. However, even the fairly cautious conclusions of the original investigators have been the subject of much debate among experts, and this also will be briefly described. First, let us consider, as we did with the castration studies, the generalizability of the experimental research data.

Socially-Living Rhesus Monkeys

Very few studies focusing directly upon the development of behavioral sex differences in the rhesus monkey have been conducted, but Rose et al (1974) reviewed two studies of play in socially-living rhesus which did *not* find significant sex differences in play patterns of young monkeys (Kaufmann 1966; Hinde & Spencer-Booth 1967a). Rose suggested that sex differences in play of the type found by Harlow (1965) in his deprived rhesus may not be directly applicable to animals raised in social groups. Of course, Goy and Phoenix's experimental rhesus were raised in deprivation conditions similar to some of Harlow's, and Goy has said only that he is able to corroborate Harlow's findings on sex differences in behavior "under the specific conditions of rearing used" (Goy & Goldfoot, 1973: 179). It is therefore important not to assume that there are well-established and/or universal sex differences in the behavior of young rhesus monkeys under all environmental or even all experimental conditions.

Further, female and male patterns of play, mounts, and threats are overlapping rather than discrete patterns, and are measured as mean frequencies. Thus the data do not show, for example, that males engage in rough-and-tumble play and females do not; rather that males do so on an average of 9 times in ten test periods, and females do so on an average of 5 times in ten test periods (Goy 1968). The frequencies differ but the distributions are overlapping. Any given individual female can, and sometimes does, exhibit higher frequencies of play, mounts and threats, than any individual male. These variable, overlapping patterns may also partially explain why different studies of play sometimes produce different results, especially given the small number of monkeys used in these studies, and the relatively low frequencies of the behaviors. This research is of course looking for, focusing, and reporting upon *differences*, but it is worth noting that the few sex differences in behavior which are found may be much less striking than the similarities which are not reported, but which occur in many behaviors of young rhesus of both sexes.

In socialization studies of sex differences in young rhesus monkeys, it has been shown that mothers respond differentially to the genitalia of their male and female offspring, and that they treat female and male infants differently from birth. This behavior probably contributes to developing behavioral differences in the infants (see Chapter 12). Since the pseudohermaphroditic females have the external appearance of males, the possibility that they were treated like males by their mothers in the 90-110 days that they were allowed to remain together, is one question that has been raised. At least one of the investigators, Phoenix, has agreed that maternal treatment might be influencing pseudohermaphroditic rhesus behavior. He also points out that two androgenized females were reared from birth in a nursery without mothers, and still exhibited increased male patterns of behavior (see discussion in Friedman et al 1974: 79). Nevertheless, given that these pseudohermaphroditic females do possess scrotum and penis, capable of erection and, in later life, ejaculation, one is led to wonder what effect these male genitalia have on their own behavior, upon that of their peers, and, especially with regard to behaviors such as mounting and thrusting, upon the reinforcement of these behaviors by self and peers. Money and Ehrhardt (1972) and Quadango et al (1977) also have suggested that the presence of a phallus makes it difficult to ascribe increased mounting frequencies to prenatal androgens alone. Goy, Phoenix, and others have further emphasized the ability of pseudohermaphroditic females to develop the "double-foot clasp mount" typical of mature male rhesus. However, since some *untreated* free-ranging female macaques exhibit this same "male" pattern, and since not all of the pseudohermaphroditic females displayed foot-clasp mounts (Wallen et al 1977), the researchers recognize that this pattern cannot be attributed solely to the masculinizing effects of abnormally high levels of prenatal androgens either.

Other Nonhuman Primates

The prenatal hormone experiments have not been replicated on other nonhuman primate species, although similar results have accidentally occurred in human females (see below). Because of the individual and species variability, described earlier, in such physiological and behavioral events as length and appearance of estrus, timing of puberty, effects of castration, etc., it is not necessary to assume that prenatal androgens administered to fetal females will have the same effects, especially behavioral effects, on all females of all species. For the moment this must remain a theoretical argument, for if studies of the development of sex differences in young rhesus monkeys themselves are rare, such studies in non-macaque species are nearly non-existent (see Chapter 12).

Nevertheless, if we were to assume that prenatal androgens always have similar effects on the central nervous system such that it is biased into a "masculine"* pattern or predisposition, a most significant question presents itself. What exactly is this "masculine" bias, and is it universally manifested, or at least common to all nonhuman primates?

---

* "Masculine" and "feminine" are terms which are sometimes used in this animal experimental literature, but they refer really to gender roles and gender identity, which are sociological and psychological *concepts* of maleness and femaleness, and should be used in reference to nonhuman primates with great caution, if at all.

Would the masculine bias of a spider monkey's central nervous system, or a ring-tailed lemur's, or a talapoin's, be the same as that of a rhesus monkey? This does not seem very likely, as the behavior patterns of male spider monkeys, ring-tailed lemurs, talapoins and rhesus are all quite different one from another. (However see Wolfheim 1978, discussed in Chapter 12). If androgens do organize the central nervous system such that there is a predisposition to acquire male behavior, it seems there must be some shared key features to all these male behaviors, or else androgens must have species-specific effects such that male behavior is acquired, whatever its normative pattern may be in a given species. The latter would be a remarkable mechanism, but perhaps not impossible to envision because of genetic differences between species. On the other hand, the former requires the assumption of universal male behavior, or at least a universal basis to male behavior. Such a basis has been suggested, or alluded to, by Goy (1966), Money and Ehrhardt (1972), and others, as a tendency to higher energy expenditure. Since this argument has been applied to humans also it will be returned to later.

Androgenized Human Females

For many people the most significant aspect of prenatal hormone research in rhesus monkeys is its apparent similarity to a body of research using clinical cases of abnormal exposure to prenatal androgens in human genetic females (Money 1968; Money and Ehrhardt 1972; Baker and Ehrhardt 1974). Human females may be prenatally masculinized in two ways which closely parallel the experimental studies of rhesus monkeys. The human syndromes result from giving progestin, a synthetic pregnancy-saving hormone (no longer in use) to the pregnant mother; or from an adrenal dysfunction in which the female fetus itself produces abnormally high levels of androgens. In either case, the females are born with masculinized external genitalia and female internal reproductive structures, and must undergo long-term surgical and hormonal corrective treatment.

In conjunction with the corrective treatment of such females, Money and Ehrhardt and their colleagues have conducted interviews with 37 androgenized females and their mothers in order to determine if the behavior of these females differs from that of normal females, especially their unaffected sisters. The principle conclusion has been that abnormal levels of prenatal androgens modify certain traits of temperament and personality in these females, such that they are called "tomboys." Money and Ehrhardt list the following traits as characteristic of "tomboyism": (1) preference for vigorous outdoor activities; (2) self-assertiveness in competition for position in a dominance hierarchy (they stress that this is not aggressiveness, which they believe is not a trait of "boyishness" either); (3) the spurning of self-adornment in favor of utility in clothing, hairstyle, etc.; (4) negligible interest in maternalism; (5) priority of career and achievement over marriage; (6) responsiveness to visual and narrative erotic images such that the opposite-sexed figure is objectified as a sexual partner (1972: 10).

They go on to stress that the temperamental modifications in androgenized females are independent of gender identity formation, that is, these females feel themselves unquestionably to be females. Indeed, all of Money's research on problems of sexual development and gender identity has consistently underlined the fact that what an individual feels herself or himself to be, is far more significant than what the genitals look like, or what the hormones and chromosomes suggest. In addition, these researchers point out that sex-differentiated behavior is not necessarily the exclusive property of either sex. Any specific temperamental trait can be incorporated into either a female or a male gender identity, if not as the norm, then as a culturally-acceptable variant of the norm. That is probably why they chose or accepted the word "tomboy," which is a term referring to a socially-accepted version of a young female's behavior in Euroamerican cultures, which is, nonetheless, "like" the behavior of a boy.

It is this "tomboy" behavior which strikes many researchers as being very similar to the behavior of androgenized rhesus females, but it is also this concept of tomboy which is the weakest link in the argument for strongly sex-differentiated behavior in both androgenized girls and monkeys. Not only is "tomboy" a culturally-biased (ethnocentric) concept, but it reflects an out-moded stereotyping of behavior patterns and an androcentric bias. One that we could have hoped biological and behavioral scientists would avoid using, whether they initially apply the label or simply pick it up from their subjects and reinforce it. The elements of so-called "tomboyism," such as being vigorous, achievement-oriented, and self-assertive human beings, are common and normative for many young girls, are highly subject to socialization factors, and really do not constitute convincing evidence of behavioral masculinization due to prenatal hormones. As Williams (1977) noted, if androgenized females identified with the male sex, or displayed erotic interest in other girls (which they do not), then the effects of prenatal androgens on behavioral sex differentiation would be more impressive.

Even if we were to say that these researchers are simply reflecting and working within the gender stereotypes held by most people in our culture and by the families involved in the study, there remains the equally ethnocentric implication of universal normative male temperament in the tomboy list. Yet anthropologists have shown that there is no universal human female or human male temperament or personality factor (e.g., Martin and Voorhies 1975; Mead 1949, 1963), and there is no universal sexual differentiation in the ascription of social roles or social tasks. For example, it would be meaningless to suggest that a woman in a horticultural society, where women do much of the backbreaking, physically active work, was behaving like a "tomboy." It would be equally nonsensical to suggest that a man of the Dugum Dani people of New Guinea, where men spend large amounts of their time in self-adornment, was behaving like a "sissy."

The ethnocentrism is so obvious one could go on with examples like this for each of the supposed tomboy traits. Money and Ehrhardt appear to be well aware of cultural variability; they include a chapter on cultural variations in sexual behavior in their book. Although it seems unlikely that they would suggest that abnormal levels of prenatal androgens produce tomboy behaviors of the same sort in every culture, the alternative, as we have seen in the discussion of rhesus monkeys, is that prenatal androgens somehow bring about a "male organization" of the central nervous system such that males and androgenized females acquire "masculine" traits, in *whatever* manner these are interpreted and expressed in different cultures.

Surely the most parsimonious explanation would be that prenatal hormones affect behavior indirectly by producing morphological traits (male external genitalia) and perhaps some physiological patterns, the possession of which affect the nature of the socialization process. One of the persistent problems with all phases of this research concerns the uncontrolled and uncontrollable variable of socialization effects. The androgenized females involved are patients, undergoing long-term medical corrective treatment. They and their parents are definitely aware of the nature of their disorder as it has been explained to them by the medical experts. What effect does this have on their views of how androgenized females should behave? Further, the data are collected through the interview method rather than direct observation of behavior, and thus relates to how these people *think* they behave, which may not be the same as how they actually do behave.

It seems that many endocrinologists and physiologists working both with monkeys and with humans, in spite of accepting the importance of environmental factors, continue to assume there are fundamental and universal male patterns of behavior (Money and Ehrhardt would say male traits of temperament or personality), and by inference, female traits of personality which are biologically determined. Whether one views female and male traits as totally discrete patterns (see below), or as being on a continuum as Money and Ehrhardt suggest, the assumption is still that there are demonstrated, consistent polarities in behavior.

The most extensive review of the literature to date on human behavioral sex differences (Maccoby and Jacklin 1978) concluded that there are few significant or reliable differences, and we can see now that these differences (aggressiveness, visual-spatial skills, mathematical ability, verbal reasoning) do not correspond to any of Money and Ehrhardt's "tomboy" traits.

ANDROGENS AND THE BRAIN

The prenatal androgen studies speak repeatedly of the "organizing" effects of androgens on the central nervous system, such that, with the pervasive computer metaphor for what happens in the brain, one envisions an entire network of circuits being wired up throughout the system to produce a male framework for behavior. Or, perhaps one imagines the development of an elaborate structural "center" to regulate male behavior throughout the system. In fact there is no direct evidence from studies of the mammalian nervous system for either of these models.

In one of the few entertaining as well as informative articles on hormones and the sexual differentiation of behavior, Beach (1971) uses a mythical animal, the "ramstergig," which is unambiguously female or male in behavior, as a foil to highlight vaguely-defined concepts such as "organization of the CNS," and unwarranted extrapolations, such as from morphological to behavioral development, which are common in this area of research. He argues that "organization," which was borrowed from structural embryological processes, is an inappropriate concept as far as the neurology of behavior is concerned, and, as such, is a "non-explanation" of the neurological aspects of behavioral development. Further, Beach proposes that hormones present early in development may produce functional rather than structural changes in neural activity, by altering the sensitivity of brain circuits to the gonadal hormones which circulate later in adulthood.

What we do have the best direct neurological evidence for, is that prenatal androgens (or neonatal, depending on the species), have some effect on the hypothalamus, that phylogenetically old structure at the base of the brain which regulates internal body functions such as hormone production, in coordination with neural information about the environment. There are documented neuroanatomical differences in one part of the hypothalamus of newborn female and male rats, as detected by electromicroscopy, which are attributed to the effects of neonatal androgens (Beach 1978; Raisman and Field 1971, 1973; see also Nottebohm & Arnold 1976 on sexual dimorphism in the brains of songbirds). The hypothalamus regulates gonadal hormone release such that in the adult male, levels of these hormones are relatively constant, rather than cyclic as in the female. It has been suggested that prenatal androgens also effect the hypothalamus in other ways, for example, in males there is a tendency to higher activity rates or greater energy expenditure. This is thought by some researchers to explain the finding that infant human males in the first days of life are more "restless" (Moss 1974) and that androgenized females show tendencies to higher frequencies of energy-expending activities (Money & Ehrhardt 1972), and that females and males differ in *rates* of behavior rather than *kinds* of behavior. This might even suggest the underlying, "universal" basis to the male-differentiated patterns of behavior I mentioned earlier. Even if such a pattern were found (and it has not been adequately documented), it still falls far short of what is implied by a "male organization of the central nervous system." In fact, other scientists have questioned the relevance of activity levels and metabolic differences to the *central* nervous sytem at all. Quadango et al (1977) pointed out that in the human case, androgenized females are maintained on cortisone therapy, and such steroids are known to affect energy-expenditure levels. Vande Wiele noted that a single injection of prenatal testosterone permanently alters an animal's metabolism of steroids and concluded: "These data suggest that effects of sex steroids attributed to differential *central* nervous system sensitivities presumed to reflect dimorphic brain differentiation, might really reflect *peripheral* metabolic differences" (discussion in Friedman et al 1974: 78, emphasis mine).

In sum we know very little about what happens in the mammalian brain as a result of prenatal androgens, and what does happen appears to be limited to one small area of the hypothalamus, although these limited neurological modifications may give rise to divers physiological patterns which are, in turn, related to a variety of behavioral patterns. To date, we must consider suggestions of extensive sex differences in the neuroanatomical structure of the central nervous system in newborn primates as hypothetical.

MALE *OR* FEMALE BEHAVIOR: A FALSE DICHOTOMY

When a person brings an infant home from the hospital, or "sexes" a new litter of kittens, each individual is discretely labeled female or male on the basis of genital appearance, and it is thought to be as simple as that. Only when an unfortunate "sex error" (Money 1968) occurs in an infant, or a scandal arises over sex testing of female athletes, do we generally recognize the complexity of biological sexual identity and sociological and psychological gender identity. In fact, Hampson & Hampson (1961) have identified *seven* variables of sex: (1) chromosomal or genetic sex; (2) gonadal sex (ovaries or testes); (3) sex of internal reproductive structures (uterus, vas deferens, etc.); (4) sex of external reproductive structures (penis, clitoris, scrotum, vagina); (5) hormonal sex; (6) sex of assignment and rearing (the sex ascribed to an individual by society); and (7) psychologic sex (a person's own view). A vast body of literature exists documenting the effects on people when these seven variables of sex fail to correlate.

Beach (1976, 1978) has suggested that the obvious discreteness of genital sex (an individual cannot have *both* a penis and a clitoris, or both a scrotum and a vagina), has led us falsely to assume that all other aspects of sex, including behavior, are similarly dichotomous. The separation and opposition of female and male biology and behavior is really not supported by the currently available evidence. For example, the so-called "female" and "male" horomones, androgens and estrogens, are produced by the endocrine system of both sexes. Genetic males normally produce more androgens and genetic females more estrogens, but Resko (1974) for example, has found overlapping values of testosterone levels in female and male rhesus monkey fetuses. As Money and Ehrhardt note, hormonal sex differences are a matter of ratios, not absolutes. Further, estrogen is not the antithesis of androgen, either in chemical structure or behavioral consequences. In fact estrogen, testosterone and progesterone are closely related chemically and have variant forms with multi-potential effects depending on the circumstances.

In respect to this, Money and Ehrhardt point out that the female pattern of reproductive and neural differentiation occurs in the virtual absence of gonadal hormones, and one significant point not stressed often enough is that if gonadal hormones are artificially administered to a genetic female fetus, it does not matter whether they are androgens (testosterone) or estrogens (estradiol). Either will disturb the developmental processes in the direction of abolishing the female type of differentiation. (Some forms of estrogen, artificially administered, may also bring about miscarriage and hence are not used in animal experiments.) In fact, progestin is a synthetically-produced variant of estrogen and yet, as we have seen, it has a "masculinizing" effect on female fetuses, such that they are born with male-like external genitalia. While prenatal hormone alteration studies do not prove that androgens cause "male" structures of biology and behavior, they do demonstrate that the delicate balance of the developing neuroendocrine system is easily disturbed.

Moreover, sexual exclusivity is not even true of some aspects of morphological sex differences. As an initially undifferentiated fetus develops into either a female or a male, there are two sorts of processes which occur. In the first process, one rudimentary structure may develop into either a female or a male structure, as with the genital tubercle which becomes either the penis or the clitoris. In the other process, *both* female and male rudimentary structures are present, and during development, one set will grow and differentiate while the other atrophies with only small remnants remaining. Both of these processes have been used as models of how the central nervous system may also develop along female or male lines. Goy & Goldfoot (1975) for example, used the second process as a model to explain what happens in the brain of pseudohermaphroditic rhesus monkeys. Beach (1976a, 1978) and Diamond (1976) make it clear that neither of these analogies from the differentiation of morphological structures to the differentiation of behavior is appropriate, because both "female" and "male" behavioral systems can coexist in one individual. The development of mechanisms to mediate male behavior does not necessitate the suppression of mechanisms related to female behavior. Thus Beach has said it is inappropriate to conceive of behavior on a single male-to-female continuum, because an individual is potentially both female and male in behavior and psychology, instead of merely one or the other. Money and Ehrhardt note that sexually differentiated behavior is bisexual in potential, though predominantly female or predominantly male in manifestation.

Prenatal androgens do affect the development of the hypothalamus, which regulates metabolic patterns and endocrine secretion in adults, and other physiological patterns which in turn influence rates and patterns of behavior. This is a far cry from the notion, sometimes expressed, that androgens *cause* specific male behaviors or even male "organization" of the central nervous system. Kessler & McKenna (1978) aptly noted that, for some people, prenatal androgens appeared to be the "*deus ex machina*" of the 1970's. The evidence we have so far is from data that are still fragmentary and sometimes contradictory, and research designs that are often unsatisfactory because of small samples, short test periods and abnormal animals. Nonetheless the impression that emerges from the more careful considerations of this research is that the individual primate is not born psychosexually female or male. Rather it is born with varying potentials for acquiring given behaviors. These potentials are at least partially related to the prenatal environment. Although these potentialities involve quantitative, not qualitative, differences, from the moment of birth social forces begin to mold these possibilities into one of two diverging patterns of behavior, "female" behavior and "male" behavior. In humans this is often a self-fulfilling process which is both based upon and reinforcing of the belief that maleness and femaleness are discontinuous, mutually-exclusive, and opposing categories. The socialization processes, whereby differences in potential are shaped and given social meaning, are the subject of the following chapters.

Chapter TWELVE

Ontogeny and Socialization

Barbary Macaque

Macaca sylvanus

The only macaques outside of Asia, Barbary macaques are found in N. Africa and Gibraltar. They live in multi-female, multi-male groups with extensive male (right) care-taking. They are moderately dimorphic.

There exists a sizable literature concerning ontogeny and socialization, much of which is beyond the reach of this book. So, rather than attempting what would necessarily be a superficial summary of this body of information, the present chapter will concentrate on those studies which relate the ontogeny of sex differences to the effects of socializing agents such as parents, alloparents (parents' helpers) and peers. For reviews and extensive discussions of primate ontogeny and primate socialization, see Poirier (1972a), Chevalier-Skolnikoff & Poirier (1977), McKenna (1979a), and Hinde (1974). What I propose to discuss here is the development of sexual differentiation, especially from the point of view of socializing agents and social learning.

In so structuring the chapter, I hope not to lose sight of the fact, often mentioned but not always fully emphasized, that the infant or adolescent, upon which socialization studies usually focus, is an *active* partner in socialization processes. Perhaps no one has documented this more fully for infants than Robert Hinde, who views the infant-mother relationship as a miniature social system, in which each influences the behavior of the other. Ideally, it might have been preferable to emphasize this by organizing the chapter along developmental lines with subheadings such as: "neonatal sex differences," "sex differences in infants and social learning," "sex differences in juveniles," etc. However, the primate data for such a description are simply not available in an appropriate form, and with a few exceptions (e.g., Bramblett 1978; Baldwin & Baldwin 1977; Mori 1979b; Rosenblum & Coe 1977) the many studies of primate developmental stages are in reference to *the* primate, rather than the infant female or male primate. Because it is entirely possible that the similarities between male and female behavioral maturation in one species are more important than the differences, sex differentiated schemes of development may not always be feasible or even meaningful.

DEFINITIONS OF SOCIALIZATION

Although we cannot review the entire topic of primate socialization, a few comments concerning the meaning and the nature of the concept are in order. There are nearly as many definitions of socialization as there are researchers in this area. The following descriptive definitions (as opposed to operational definitions) from researchers who have devoted some thought to this concept, are presented as representative and useful starting points for the discussion which follows in this chapter.

> Socialization can be viewed as that process which links an on-going society to the new individual. Through socialization a group passes its social traditions and ways of life to succeeding generations . . . The term socialization will be used here to refer to the sum total of an individual's past social experience, which, in turn, may be expected to shape its social behavior (Poirier 1972a: 8).

> Socialization is the total modification of behavior in an individual due to its interaction with other members of society, including its parents (Wilson 1975: 595).

... Socialization is an enormous, ubiquitous concept which resists a unitary definition. But most researchers generally agree that it refers to the lifetime process by which behavior is learned through observation and social participation. Socialization refers not to particular events, but rather to an interplay of processes guiding individuals as they move from one ontogenetic stage to another ... (McKenna 1979a: 252).

These descriptions, in spite of their rather all-encompassing nature, do highlight several important aspects of primate socialization. Firstly, it is a process, rather than a static phenomenon, and secondly, it is a conceptual abstraction or classification of behaviors, rather than a direct transcription of observed behaviors. An ethologist would not watch a monkey mother push her infant away from her nipple, or pull her infant back from a dangerous situation, and record one unit of socialization. Rather, socialization is our general model for how a young primate, dependent on social learning for survival, develops appropriate behaviors and becomes channeled into the life-way of the species.

As with role analysis, I should note that the study of socialization does not explain in any direct or conscious sense what causes or triggers an individual's behavior. Nonhuman primates do not punish, restrain, protect, or otherwise influence an infant's behavior because they are directly motivated to "socialize" the infant, and indeed one could argue that not all of human socialization is so motivated either. Instead, socialization is an analytic tool borrowed from the social sciences and used to identify and classify certain regularities in social interactions, especially in adult-young interactions, and to explain how these might operate in social learning, with particular reference to the behavioral development of the young.

SEX DIFFERENCES AND OTHER DIFFERENCES

If we wish to determine how the sex of a young primate acts as a variable in the socialization process, there are a number of confounding variables of which we must take account and, if possible, for which we must control. This is particularly true when very small samples are used, as they are in most of the studies described in this chapter, and where the impact of individual variability on results can be quite marked. For example, if we find that one particular mother and male-infant rhesus pair interact differently from another mother and her female infant, there are a number of variables other than, or as well as the sex of the infant which may be at work. As Poirier (1977) has noted the number of infants a female has previously raised is known to affect the nature of mother-infant interactions (see Mitchell et al, 1966; Mitchell & Stevens 1969; Altmann 1978; for example). This is due to factors ranging from significant if ill-defined aspects of the mother's capabilities, such as her "experience" and "confidence," all the way to apparently trivial factors, such as the length of her nipples. Many descriptions and analyses of "mother rejection" are based on measures of mothers removing infants from their nipples, and Ransom & Rowell (1972) among others have discussed the possible repercussions of the fact that multiparous mothers (females who have borne more than one infant) have elongated nipples, which are greatly stretched and probably desensitized by past suckling and are easy for infants to reach and hold. Primiparous mothers (females who have their first infant) on the other hand, must repeatedly raise and assist their infants to their nipples, and show signs of discomfort as their nipples are stretched by the suckling and the weight of the infant. They remove their infants from the nipple more often than multiparous mothers and set them down, which can in turn result in greater fussing on the part of the infant and a less harmonious mother-infant relationship. Nipple length is just one variable in maternal characteristics, and one which may seem humorous or at first glance trivial, but as I have tried to show, can be significant in its impact on differences in maternal care measures. Although laboratory studies of rhesus macaques often indicate that males play more than females, two studies, comparing multiparous to primiparous mothering, found a reversal in the offspring of primiparous mothers — in this case young females played more than young males (see Mitchell et al 1966). Parity (number of previous infants) and maternal experience are variables which may thus confound sex differences, and it is not clear in studies such as those by Jensen et al (1968) discussed later, if account was taken of how many previous infants the mothers had raised.

Rank of the mother is also known to affect the nature of an infant's general socialization experiences (Gouzoules 1975, 1980; Altmann 1978), and there are a number of other variables concerning both the mother and the infant (eg. maturational age of the infant and genetic differences between infants which have nothing to do with sex) which may influence differences in interactions. Complexity of the physical environment (Jensen et al 1973) and composition of the social group also are known to have effects on how the infant primate learns and interacts with its social world (Suomi 1976). Then too, there are repeated statements in the literature that female primates mature faster than males (for example Rosenblum & Coe 1977) and thus infants nominally of the same chronological age may not, in fact, be at the same ontogenic stage.

Finally, there is the problem that many of the behavioral measures used to determine the nature of the caretaker relations (approaching, leaving, cuddling, retrieving, rejecting, punishing, etc.), differ significantly in frequency and pattern of occurrence between even closely related species. For example, Chalmers (1972) demonstrated that in the genus *Cercopithecus*, mothers in more arboreal species are more restraining of their infants, male or female, than mothers in more terrestrial species. Sussman (1977) suggested a similar difference between the more arboreal brown lemur (*Lemur fulvus*), and the more terrestrial ring-tailed lemur (*Lemur catta*). Kaufman and Rosenblum (1967, 1969; Kaufman 1975; Rosenblum 1971a,b) in an excellent set of longitudinal laboratory colony studies demonstrated numerous differences in the socialization patterns of pigtail macaques versus bonnet macaques. Thus sex differences in socialization patterns which may be established for one species do not necessarily hold true for other, even closely related species.

CONTINUITY OF SEX DIFFERENCES IN BEHAVIOR

One common assumption in studies of primate ontogeny and socialization is that sex differences in the behavior of young individuals will correspond directly to, and are precursors of, the adult sex differences in that species. And some evidence, at least from the well-studied Old World terrestrial species, has appeared to support this idea. Thus higher levels of rough and tumble play and "aggression" and overall locomotor activities as found or suggested for juvenile male rhesus macaques, bonnet macaques, squirrel monkeys, chimpanzees and humans (Hansen 1962; Simonds 1974; Baldwin 1969; Nadler & Braggio 1974; Brindley et al 1973) are considered to presage the greater aggressiveness and dominance of the adult male in these species (Hansen 1962; Nadler & Braggio 1974; Suomi 1977; Gray 1971). Of course, not everyone would agree that adult male rhesus macaques, for example, are more aggressive than adult females of the species (see discussion in Chapter 6), but the notion that trends in sex differences begin in infancy and continue along gradually differentiating lines is a well-accepted one.

Poirier (1977) suggested that early social experiences "condition" females and males and prepare them for their adult social roles. As a specific example of this Mitchell et al (1967) pointed out that exposure to punishment is correlated with later hostile behavior in monkeys, and that if young males receive more rough treatment and aggression from their mothers, this may reinforce their own tendencies toward aggressive, rough activities as juveniles and adults. For humans, a number of studies found correlations between levels of maternal punishment and measures of aggression in maturing children, although some of the data concerning female and male reactions to punishment are conflicting (Becker 1964; Gordon & Smith 1965). Further, many studies of allomothering or "aunting" in young female primates suggest that this early exposure to infant-care prepares females for their roles as mothers; and many studies of play have proposed that play is practice for adult behavior, so that sex differences in play are often expected to correspond to sex differences in adult roles.

In the first analysis then, it would appear perfectly reasonable that sex differences or similarities in the young of a species should correspond to those in the adults, and that this hypothesis would be well worth testing in a variety of primate species with different sorts of adult sex roles — monogamous species, for example, in which the behavior patterns of males and females are not as sexually differentiated as in macaques and baboons, or species in which the adult male is not considered to be more aggressive than, or dominant over, the adult female. One such study was conducted by Wolfheim (1978) with talapoins, an Old World monkey species in which adult males are reported to be less active, less aggressive, and lower ranking than adult females. However contrary to what we would have predicted from the above hypothesis she found that juvenile sex differences did *not* correlate with adult sex roles. Juvenile male talapoins were more active, less affiliative, more assertive and more playful than juvenile females. It was not until social and sexual maturity that males switched over to adult male patterns of interaction. Wolfheim concluded that juvenile sex differences might be similar in all primates, perhaps due to prenatal endocrine factors, but that adult sex differences cannot be accurately predicted from those of the juveniles. Now we must await the testing of her hypothesis that juvenile sex differences will prove to be similar in all primate species.

Two other studies which speak somewhat less directly to this question, but are nonetheless relevant, are Burton's (1972) study of socialization in Barbary macaques, and Rosenblum & Coe's (1977) laboratory study of squirrel monkeys. Burton found that Barbary macaque females lead a comparatively solitary existence in their juvenile and subadult years, with few affinitive interactions such as grooming, but become very sociable and integrated into their group as adults. Once again this does not conform to a picture of universal juvenile sex differences in primates, since young females are generally considered

to be the more sociable sex, especially as measured by grooming, while males are portrayed as the more active and aggressive sex (Mitchell & Tokunaga 1976; Tokunaga &
Mitchell 1977). However it must be said that in some respects Barbary macaques as
adults, do not fit our standard view of primate sex roles either (see following section
on Male Care).

Rosenblum & Coe (1977) investigated the effects of adults on preadult social
structure in captive groups of squirrel monkeys. This species is unusual for anthropoids
in that adult males and adult females tend to remain spatially separated and to interact
infrequently, except during the mating season. They do live in the wild as year-round
groups, but very segregated groups, and females are more active in both affiliative and
agonistic interactions and form the "core" of the social group. An interesting question
then is whether sex segregation would be apparent in young squirrel monkeys. Through
manipulation of group composition, the researchers found that, left to themselves,
neither juvenile nor sub-adult squirrel monkeys revealed adult forms of social and spatial
organization. However, sub-adults were strongly attracted to adults should they be
present, and were differentially accepted into the female and male segregated groups by
the same-sexed adults. Again, as with talapoins and Barbary macaque females, sex role
patterns in the young squirrel monkeys do not appear to correspond to those in the
adults; rather, at about the time of sexual maturity individuals exhibit changes in behavior which may appear to be rather sudden, and even reversals of past patterns of behavior. Recently, Raleigh et al (1979) suggested that in vervets also, juvenile sex differences are not necessarily predictive of adult sex differences.

In conclusion, the relation of sex differences in the young to sex differences in
adults is obviously one requiring further investigation. For many of the familiar Old
World species, such as baboons and macaques, it has appeared almost self-evident that
behavioral as well as morphological sex differences will gradually unfold and increase as
part of the ontogenetic and socialization processes. However, for at least the three species
described, it is suggested that adult behavioral patterns of sexual differentiation do not
represent end products of linear trends beginning in the young.

SEX DIFFERENCES IN THE BEHAVIOR OF YOUNG PRIMATES

For the majority of primate species we simply do not have evidence about the age
or the manner in which behavioral sex differences appear (one exception is Japanese macaque ontogeny, see Gouzoules 1980b), or even evidence as to whether they appear much
before adulthood. This is especially true of behavioral secondary sexual differences or
non-sexual differences as opposed to specifically sexual behaviors such as thrusting. From
the data, much of which will be described in this chapter, it appears there are some
established behavioral sex differences in infants and juveniles of at least seven primate
species: rhesus, bonnet, pigtail, and Japanese macaques, talapoins, vervets, and squirrel
monkeys.

For some other species, common baboons and chimpanzees for example, reports
in the literature are conflicting. Thus, while Ransom & Rowell (1972) reported consistent
differences between female and male baboon infants as early as the second month of life,
(in their interactions with mothers and peers) Nash (1978b) using one of the same study

groups, noted a lack of sex-related differences in her study of mother—infant relations in these wild baboons. Young & Bramblett (1977) conducted a detailed study of captive baboon infants raised with peers only, or in groups of mothers and peers. In neither environment was there support for their hypotheses of greater frequencies of locomotion, play, aggression, exploration, and sexual behavior in male infants than in females. They did not find behavioral sex differences in baboons in the first three months of life and a later study of baboon infants raised in "harem" groups also failed to find sex differences (Young and Hankins 1979). As parts of these studies replicate the deprivation research on rhesus monkeys, they will be discussed again in Chapter 13.

Turning to chimpanzees, Nicholson (1977), in a study of behavioral development in wild and captive chimpanzees, found no evidence of sex differences in either infant development or maternal response. Interestingly, neither Goodall (1967) nor Clark (1977) mention sex differences as being an important variable in their separate studies of development in free-ranging chimpanzees at Gombe Stream, although they both discuss at length significant variables, such as individual differences in both mothers and infants. However, Nicholson does acknowledge that subtle sex differences not detected in her study might be amplified in certain captive environments; Sackett (1972, 1974) demonstrated that stressful, captive-rearing environments have a greater impact on male behavior. (Sackett's study is further discussed in the next chapter.) In light of this suggestion, it is of note that Nadler & Braggio (1974) found that nursery-reared chimpanzees (raised with peers but no mothers or other adults) showed sex differences such that males were more socially "assertive" and active than their female counterparts. The authors concluded by arguing that greater male assertiveness represents a common phenomenon in all higher primates, including chimpanzees and humans. Their generalization would seem rather premature, if for no other reason than the lack of confirmed behavioral sex differences in the young of feral chimpanzees as opposed to those raised in laboratory settings.

FEMALE PARENTAL CARE

Mothers

The primary, most intense, and longest-lasting social bond for the primate is usually the mother-offspring relationship. Infant primates are born relatively helpless or "altricial" compared to other infant animals, and their survival is completely dependent upon the care given them by others of their social group, primarily by their mothers in most primate species. Although the concept of "imprinting" upon the mother is considered too simplistic and mechanistic a model for the complex processes of learning in young primates, nonetheless many socialization studies emphasize that primate infants learn first and foremost from their mothers.

It is through watching and interacting with the mother that an infant learns what to eat and what to fear, who to dominate and who to submit to, who its friends or kin are, where to travel, drink sleep, and so forth. Infant primates with few exceptions model their behavior most closely upon that of their mothers, and at least in the early stages of development, mother and infant may seem "as one" in many social situations. For example, in Japanese macaques, if a mother sees one of her offspring threatened, although she may be some distance away, she will grimace in fear or counterattack exactly as if she herself were being threatened. That others perceive this extraordinary closeness of bond borne out by the attention and submission directed to the young of a high-ranking female, even when the mother is not directly present, and by agonistic encounters in which threats or chases are first directed at a female, and then if she escapes, at any of her offspring who happen to be in the vicinity, whether or not they are participants in the encounter. You will also recall from Chapter 9 on kinship, the many social and personal similarities between mothers and offspring in Japanese monkeys, and probably in other multi-male, multi-female societies.

A primate mother is much more than a warm body with nipples and a milk supply. She is also the infant's introduction into the complex social world in which it must survive, and her behavior sets the stage upon which her offspring's social life-drama will unfold. Mother-infant relationships in primates are not only complex and intense, but also of great variety. This variability may depend upon how many infants the mother has had, or the temperament and personality of the mother, or even whether the mother herself had good mothering. Dissimilarities in mother-infant interactions may also be due to differences in the infants, and some research has focussed upon differences in macaque mother-offspring relations according to whether the infant is a female or a male.

Jensen et al (1967, 1968) studied sex differences in the development of mother-infant relations in ten pairs of laboratory pigtail macaques. Their subjects lived as pairs in four-foot square cages and their social and environmentally-directed behaviors were measured. The researchers hypothesized that male infants would be more active, climb on their mothers more, receive more punishment from their mothers, and achieve earlier independence than female infants. While their findings indicate that the relationships between these monkey mothers and their sons do differ from those between mothers and daughters, and that male offspring do become independent earlier, the roles that mothers and offspring play in this pattern were not as they anticipated. Specifically, Jensen et al found that female and male infant pigtails do not differ in activity rates in the first 15 weeks of life. However, mothers with male offspring are more punitive in the first 6 weeks of life, rejecting, punishing and leaving (if one can speak of leaving in a four-foot square cage!) their male infants significantly more often than female infants. Later in life, females receive similar treatment, so that maternal rejection is not more severe in total for males, but begins earlier. The authors conclude that largely as a result of this differential maternal handling, male infants spend more time spatially separated from their mothers' bodies, and *in this separated position* (labeled as the most "independent" position the males show higher activity rates and become more manipulative of the environment. If the cage is completely bare, in what the authors term a "privation" environment, rather than an "enriched" environment with objects for climbing and handling, male infants also bite their mother significantly more often than female infants.

The findings from these experiments were largely duplicated for rhesus macaques by Mitchell (1968a), using a larger sample of 32 mother-infant pairs. He found that mothers of males withdrew from, rejected, and played with their offspring more often, while mothers of females restrained their infants from leaving, held them in body contact, and groomed them more often. Mitchell's results differed from Jensen's somewhat. He observed that mothers of males not only punish their infants earlier, they also punish them more in total than do mothers of females. Mitchell's work also supports the contention that monkey mothers play an active role in promoting differential amounts of independence in the two sexes; males are "pushed" into other activities and relationships, while females are kept close and protected. What is more, in another set of experiments, Sackett (1974) found that even "motherless mothers"* are more abusive to male than female infants.

Researchers in this area generally offer two possible reasons for differential maternal handling. The first is that mothers react to the biological or physical differences they perceive in their offspring by sight, touch and possibly smell; the second is that mothers react to behavioral differences in their newborns. "Genital inspections" have been reported for a wide range of primate species, during which the mother and other group members carefully and closely examine the genitals of newborns, and perhaps in some way "classify," although not necessarily consciously, the infant as female or male. Some species, such as vervets, carry this interest in genitals to such an extreme that nursing infants spend large parts of their early days tilted upside down, while "inspectors" lift their hindquarters or tails and raise their bodies like a cellar-door from their mothers' ventrum for a closer look.

The explanation that mothers react differentially to the external genitalia of their infants does not require female and male infants to behave in any way differently from one another. However, the second possibility is that female and male infants do behave differently from the moment of birth, and mothers simply respond to these differences in neonatal behavior. This is the initial hypothesis with which Jensen et al began their research. Although they were unable to substantiate the expected difference of overall activity rates in infants, it is possible that differences in female and male neonatal behaviors are more subtle and have not yet been detected. For example, mothers may reject male infants earlier because they have more vigorous muscle movements, or "fidget" more often. White & Hinde (1975) did find that male infants initiate more ventral contacts with the mother than do female infants. This entire problem is very tangled, because different researchers report conflicting findings as to whether or not female and male neonates *do* differ, in what ways they differ, why they differ (is maternal handling already at work or are the differences inherent?), and finally, it is not at all obvious why mothers should be more rejecting toward active infants, if this is, indeed, the crucial finding. Perhaps the environmental circumstances are confounding variables here because almost all these studies take place in extremely small laboratory cages, where isolation, deprivation, and confinement may make the mothers hypersusceptible to slight behavioral differences in their offspring, and lack of outside stimulation may force the infants to direct all their activities toward the mother. If this is so, it is not clear how relevant the findings are to maternal handling in free-ranging groups. One wonders how human mothers and their infants would interact if they had to live together in a closet, and what relevance we would consider such interactions to have to normal mother-infant relationships.

The explanation that differential behavior in the infants leads to differential behavior in the mother is probably the more widely-accepted explanation for variability in maternal handling, perhaps partly because the idea of monkey mothers simply responding to different behavioral stimuli from their infants more closely approximates our conception of animal capacities, than does the notion of monkeys somehow "sexing" their

*Females raised in isolation without mothers, who are artificially inseminated and give birth to their own offspring; see next chapter for a description of their behavior.

infants by examining the genitals. In point of fact, neither explanation requires that monkeys have a concept of female and male. The question is, do they respond to behavioral or physical cues from the infant, or both? Another reason for the popularity of the mother-responds-to-infant-behavior model, is the expectation that genetic and/or prenatal factors should lead to sex differences in the behavior of infants at birth, irrespective of later handling. The problem in testing such an expectation is that innate differences in infant behavior as distinct from socialized differences cannot easily be delineated, for as soon as the infant is born, different experiences may begin for female and male.

Two approaches may be used to attempt the distinction between innate and socialized factors in the development of sex differences: deprivation studies in which primate infants are deprived of normal social and physical stimulation; and neonatal studies. Deprivation studies will be the subject of the next chapter, but their findings can be summarized as demonstrating that some social exposure and interaction is necessary for behavioral sex differences to develop. Total isolation of primate infants in the first year of life is reported to "destroy" the normal female-male differences in agonistic, sexual, dominance, and nurturant behavior (Mitchell 1968a, 1968b; Chamove et al 1967). Of course, total social isolation also destroys almost all normal facets of primate behavior, and I shall argue later that the completely abnormal monkey and mentally disturbed monkey which results, is one of the impediments to generalizing from extreme deprivation research. Sex differences, specifically more play, threats, and mounting in males, appear in some, but not all (see Chapter 13) studies of "peer-reared" monkeys, that is, monkeys who are exposed to other young monkeys, but not to mothers or other adults (Harlow 1965; Chamove et al 1967). These results have sometimes been used as evidence that certain sex differences are innate because they appear in the absence of normal socializing caretakers (see Harlow 1962) but this is not the only possible interpretation of the findings. On the contrary, it can be said that some social experience during development has proven to be necessary to the appearance of behavioral sex differences and thus the traits, like so many other traits, probably have innate components, but require certain environmental contingencies for their particular expression. It was argued in the last chapter that primates are probably born with differing potentialities to acquire behaviors, which potentialities are dependent upon the prenatal environment and the genetic make-up of the individual. These potentialities are in turn differentially realized, according to the influences of environmental variables after birth. Peers, while obviously not adult caretakers, are documented socializing agents in many "normal" situations, and can apparently provide an appropriate environmental context or adequate environmental stimulation for the development of some behavioral sex differences. Interestingly, macaque monkeys raised by their mothers only, exhibit greater sex differences than those raised with peers only (Chamove et al 1967).

Human infants are, for obvious reasons, not subjected to deprivation experiments in order to determine what their behavior would be like had they no mothers or other social companions and caretakers. Thus the second approach to the same question, is to attempt to study infants before their caretakers have had a chance to have very much effect upon them. This is not to say that neonatal studies are usually conducted in an effort to *separate* innate and learned factors. Quite the contrary, researchers usually examine early mother-infant *interactions;* however neonatal studies may be used, and have been used, to speak to this question. Perhaps the best known of these studies are those by Moss (1967, 1974). He conducted four studies of infants and their mothers during the first three months of life for a total sample of 128 mother-infant pairs. The sex difference which he found replicated across the studies was that male infants "fuss" more, or exhibit behavior indicating they are more irritable (defined as unsettled, "whiney" behavior, but not crying). Mothers of males had more interactions with their infants in apparent efforts to quiet them, that is they held, attended, stimulated and looked at their infants more than mothers of females. That this was in response to the infant's behavior is indicated by the finding that when the "amount of infant fussing" was controlled for as a variable, the maternal differences in treatment of the sexes disappeared. In contrast, mothers (and fathers) of female infants spent more time evoking positive social behaviors (smiling, vocalizing) from their infants. Moss has stressed that the infant sex differences he found were minimal, of marginal statistical significance, and many were not replicated in different studies. It is his suggestion that "greater irritability" is not an intrinsic characteristic of infant human males, but reflects the documented fact that males are more prone to physical distress at this period of life, and are less frequently at an optimal state for assimilating environmental events.

Moss does not mention the point brought out by Korner (1974) that almost all North American male infants (which Moss' subjects were) are circumcised in the first week of life, an experience which has at least a temporary effect on infant irritability, and for which it does not appear that he controlled. Richards et al (1976) were unable to confirm Moss' neonatal sex differences in behavior when they compared female and uncircumcised male infants. In her review, Korner summed up the neonatal studies as demonstrating that newborn human females have greater tactile and sweet-taste sensitivity, while males have greater muscular vigor and startle more easily, but do not show higher activity rates than females. She also stressed that newborns are immediately subject to differential handling, as studies have shown that mothers still in hospitals talk to and smile at their baby girls more often during feedings, than do mothers of boys. It would not be surprising to learn that hospital personnel also handle the blue-wrapped bundles differently from the pink-wrapped ones.

Returning to nonhuman primate research, although Jensen et al (1973) believe that mothers play the most important role in instigating differential independence in their infants, a number of studies, including one by Jensen himself, have shown that maternal handling may not be quite so all important if other group members are present (Rosenblum 1971; Hinde & Spencer-Booth 1967a,b; Jensen et al 1973). Looking at the mother-infant relationship quite explicitly from the infant's point of view, Rosenblum conducted studies of attachment to the mother and avoidance of strangers in infant pigtail and bonnet macaques (Alpert & Rosenblum 1974; Rosenblum & Alpert 1977). His extended general hypothesis on which the studies are based, is that female infants are attached to their mothers and more afraid of strangers; during maturation females tend to remain closer to their mothers and female-focal units. Male infants on the other hand, show less fear of strangers and attachment to mothers, and a greater readiness to establish new social relations; therefore during maturation males gradually leave their mothers, peripheralize from the female-focal units, and finally transfer social groups.

In order to test this hypothesis, Rosenblum separated mothers and infants and placed them in a test apparatus made up of glass-partitioned chambers such that the infant had a choice of remaining close to his mother's chamber, or a stranger female's chamber. One-way glass ensured that infants could see into the chambers, but mothers and strangers could not see the infant. The time which infants spent close to mothers, strangers and empty chambers was compared, and three test groups of subjects were used: group-reared bonnets, mother-only reared bonnets ("single-dyads"); and group-reared pigtails. Of the three test groups, only one showed any significant sex differences: group-reared bonnet females showed a significantly greater avoidance of strangers. These same bonnet females showed an earlier and greater preference for the mother than did male bonnet infants, although *both* sexes in fact showed a significant preference for mother throughout most of the first year of life.

Sex differences in "dependence" is a hotly debated issue in human studies, as some studies have shown females to be more dependent than males. Nevertheless, one of the problems with extrapolation from monkey research is that "dependence on the mother," "attachment to the mother," "preference for the mother" and "ability to discriminate the mother," often are not clearly distinguished. Rather, they are all inferred from the same proximity measure of an infant approaching and remaining close to its mother (Kaufman & Rosenblum 1969; Rosenblum & Alpert 1977; Alpert & Rosenblum 1974; Rosenblum 1974). Although Rosenblum's hypothesis of "very marked differences" in female and male infant attachment to mothers was only substantiated in group-reared bonnets, it is noteworthy that mother-only reared bonnets and group-reared pigtails showed a common *lack* of significant sex differences in attachment to mothers and in avoidance of strangers. Pigtail macaque mothers, unlike bonnets, raise their infants very exclusively, almost as though they *were* in isolation with them, and Rosenblum suggests that experience with other animals in addition to the mother during rearing is a prerequisite to the development of sex differentiated attachment to the mother.

Thus, human and laboratory-reared macaque infants have been reported in at least some of the research to exhibit sex differences in behavior in the first weeks of life. However Gouzoules (1980b) found that sex differentiated behavior did not become significant in semi-free-ranging Japanese macaques until four to six months, and even then not as an isolated variable. Further, we have not been able to prove that these differences will occur in the absence of influential environmental, especially social environmental, factors. Without denying that the propensities which partially give rise to these differences might be intrinsic, it is nonetheless true that powerful social forces, particularly in the form of differential maternal handling, may be at work from the moment of birth. It is doubtful whether neonatal or deprivation studies can be used to prove that some sex differences are innate and others are learned, and unclear why we should want to do so, given the accepted interactions of biological propensities and environmental circumstances at all levels in the development of behavior. It is certainly important to avoid becoming trapped in the circular and simplistic question of which came first, or which causes the other, differential behavior in the infant or differential treatment by the caretaker. Although here we have necessarily focussed upon certain studies emphasizing either the mother's influence or the infant's influence, in fact, most research in this area involves studies of intricate feedback mechanisms and mutual influences in the course of mother-infant interactions.

Allomothers or "Aunts"

Almost all feral primate mothers have had some previous experience with infants, and although they may react to and care for their young in slightly different ways, mothers in the wild generally clean their newborns of birth fluids, inspect them carefully and gently, cradle them in their laps and hold them to their chests so that the nipple is quickly found. Thus primate mothers react positively to their infants and quickly provide them a warm, secure, and nourishing environment.

The reaction of others living in the social group to these new members is much less uniform. In some species of nonhuman primates, other group members are highly interested in infants and will perform caretaking behaviors such as carrying, protecting, retrieving, and cuddling infants. There is great variability between species in terms of the amount of interest shown in infants and how it is expressed, who is interested, and whether or not the mothers allow others to approach and handle their infants. Nonetheless, with the exception of orangutans and some prosimian species, mothers and infants do not live as isolated pairs in the wild, and studies have shown that group-living, and social companions other than the mother, usually have a significant impact on mother-infant relations.

For example, in some New World monogamous species, males are reported to "assist" births by pulling on the emerging infant, cleaning the newborn, carrying and caring for it entirely, except for returning it to the mother for nursing (Langford 1963; although he may have overstated the male's role somewhat, see Chapter 16). In other species, such as the pigtail macaque, the mother is very restrictive and does not allow anyone else to hold or even to touch her infant for a long time after birth. Offering yet another contrast, langur mothers readily allow their infants to be carried off, even in the first hours and days of life, but only other females of the group are interested in doing so. Vervets may cluster excitedly around a mother, trying to get close enough to touch and sniff and peer at the newborn, and sometimes a juvenile even "makes off" with a new infant, running bipedally with the baby clasped awkwardly (perhaps upside down!) in its arms.

While vervet interest in infants is quite obvious, Japanese macaques are past masters at studied indifference. One first has the impression no one has even noticed the tiny black infant now clinging to the ventrum of a female, and certainly no one gathers around the new mother. However, she receives lingering looks as she passes by, and some monkeys loiter uncommonly long and often in her vicinity. In a few days close relatives will approach the mother circumspectly and groom the fur of her back or side, interrupting their apparent attention to the mother with rapid sniffs and close visual inspections of the tightly-clinging, suckling infant. The first times the infant ventures from the secure cradle of its mother's arms and lap, monkeys of the area may "lipsmack" and "warble" (terms for particular affinitive facial expression and vocalization) and peer closely with their faces lowered to the ground at the infant's own level. But they watch the mother's reaction carefully, and adult males may even leap up with a fear grimace, and rush off when approached by a very young, stumbling infant. Thus, although an infant's closest and most stable social bond is with its mother, it soon enters the wider social world of the group.

Other female group members who are interested in infants have been referred to as "aunts," from the British custom of teaching children to so address women friends of the family, often maiden women or women without children of their own, who form a special relationship with someone else's children. However, as the term "aunt" may lead those unfamiliar with this connotation to assume a biological relationship between the other female and the infant, Wilson (1975) has proposed the term alloparent ("other" parent) or specifically for females, allomother.

187

Age and Parity of the Allomother

The interest of other females in infants was first studied in detail in rhesus monkeys and vervets (Spencer-Booth 1968; Hinde & Spencer-Booth 1967; Rowell, Hinde & Spencer-Booth 1964; Lancaster 1972). In these species, as well as in a few others which have since been studied, it is the juvenile and subadult females who are primarily involved with infants. Perhaps because of this, there is a lingering impression in the literature that "aunting" is basically an age-related stage or phase through which young female primates pass.

Without denying that this may be true for at least some primate species, several researchers (Lancaster 1972; Hrdy 1976; Breuggeman 1973) have pointed out that this general interest in infants may be as much related to nulliparity (not having an infant) and the novelty of interacting with infants, as to a particular age-determined stage. Thus in some species, *older* nulliparous females are reported to show the most interest in infants (see reviews of allomothering by Hrdy 1976, and McKenna 1976). It may well be impossible to make a general statement about age and parity characteristics of primate allomothers, because in yet other species, such as ring-tailed lemurs (Jolly 1966) and bonnet macaques (Kaufman & Rosenblum 1969; Simonds 1974), it is adult females with infants of their own who are likely to show interest in, and to be allowed to care for, other infants. They may sit in nursery groups of mothers and young and actually "trade" infants or allow them to be passed from mother to mother, or they may adopt another infant when they already have one of their own.

Sex Related Factors

It is equally difficult to offer a general answer to the question of whether an interest in infants is sex-related or not. The overall impression one receives from the literature is that female primates are far more interested in infants, but quantitative comparisons within a species or group are rare. In two such quantitative studies of rhesus monkeys, it was found that preadolescent females direct significantly more positive social behavior and less hostility toward infants than do pre-adolescent males (Chamove et al 1967; Spencer-Booth 1968). So in this species females seem more likely to exhibit alloparental care. In the Chamove study, these sex differences in behavior toward infants were absent in isolate-reared monkeys, moderately present in peer-raised monkeys, and most marked in mother-raised monkeys, indicating that socializing agents are important in the development of these preferences. Other researchers (Poirier 1977; Baldwin & Baldwin 1977; Ransom & Rowell 1972) also have suggested that differential socialization experiences may lead young females into greater contact with infants, and a model reflecting this idea will be presented at the end of the chapter. That females are more interested in alloparenting than males has also been suggested for other primate species, such as common baboons, langurs, and vervets. However, as Hrdy notes, studies involving direct comparisons have only been undertaken for a few species, and they do not indicate that differences between the sexes will exist to the same extent or even in the same direction for all primate species. Examples of extensive male interest in infants will be discussed in the next section. There do not appear to have been studies of whether allomothering is more commonly directed toward female or male infants, which would indicate differential handling by "aunts," such as was described earlier for macaque mothers.

Ehrhardt (1973; Ehrhardt & Baker 1974) and Lancaster (1972), in an earlier paper on allomothering in vervets, suggested the possibility that prenatal hormone differences may be related to differential "maternalism" in primates. However, Lancaster in her later papers (1973, 1975, 1976) emphasizes the lack of inherent psychological differences between the sexes, and the ability of male primates to play caretaking roles. It is difficult to evaluate the suggestion relating prenatal hormones to maternalism with the data currently available. Androgenized female rats do *not* exhibit reduced maternalism (Quadango et al 1977), and both female and male rats, after a period of exposure to litters not their own, will begin to care for infants. An evaluation of maternalism does not seem to have been part of the research program with eight androgenized rhesus monkeys carried out by Goy et al (see Chapter 11).They present no data concerning interest in infants; their pseudohermaphroditic females were ovariectomized at puberty and treated with further male hormones.

Ehrhardt also reported that prenatally androgenized girls were *less* maternal than other girls, as measured by interest in infant-care and fantasies of pregnancy and motherhood. Thomas (1977) has countered that the masculinized girls in this study were prepubertal, averaging 7-10 years of age, and that younger girls show more tomboyish activities, and fewer interests in their adult reproductive role, than do post-pubertal girls who made up the "control" group in Money and Ehrhardt's 1968 study. Quadango et al (1977) have also disagreed with Ehrhardt's conclusion that lowered maternalism is due to prenatal hormones causing brain changes. They argue instead that androgenized females, as children and adolescents, are aware of their reproductive abnormalities and unsure of their role as potential mothers due to their biological problems.

To sum up, it has not yet been documented that there is a general trend in primates for females to be more interested than males in alloparenting, but many primatologists have the impression that they are. Perhaps future studies, involving direct comparisons in a variety of species with different social organizations will be able to confirm this impression. Further, it is not yet clear what mechanisms, social or ontogenetic or both, are involved in the development of interests in infant-care by alloparents.

Kinship of the Allomother

I stated in Chapter 9 that in relatively few primate studies are kinship relations known by the researchers. When they are known, it has sometimes been found that allomothers are most likely to be older siblings of the infants for whom they care, for example in rhesus monkeys, baboons, chimpanzees and Japanese macaques (Spencer-Booth 1968; Rowell, Hinde & Spencer-Booth 1964; Goodall 1967). However, in other species such as bonnet macaques (Simonds 1974) and vervets (Lancaster 1972) allomothers are just as likely to be unrelated to the mother and infant.

There has been some disputation in the literature as to the function of allomothering. When "aunts" are clearly related to the infants for whom they care, it can be argued that this is a form of kin selection (see Chapter 18) since the allomothers are promoting the survival of individuals who share some of their genes. In addition, other benefits have been suggested for all three parties involved in allomothering: (1) "aunts," particularly if nulliparous, gain important practice in infant care; (2) mothers benefit from a "babysitting" service which temporarily relieves them of the almost continuous burden of a young primate infant; (3) the infants enlarge their social contacts and have extra caretakers, who may even adopt them if they are orphaned.

However, Quiatt (1979), Hrdy (1976), Hinde (1974) and Krige & Lucas (1974) all have noted that aunting, carried to an extreme, may also have detrimental consequences for the mother and infant. A nursing infant may be "kidnapped," and if the mother is unable to retrieve it, the infant will weaken and die and the mother herself may become ill. Often allomothers are inexperienced and may sometimes handle the infant roughly or inappropriately. Thus mothers in many primate species are reported to be very discriminating in their choice or acceptance of alloparents.

As I noted earlier, infants are not just passive recipients of "socializing experiences," and young primates may struggle, scream and/or refuse to cling to someone who is attempting to carry them away, or conversely, infants may dash for and cling to individuals who are surprised into the role of alloparent when the infant becomes frightened in the absence of the mother (e.g. Mori 1979a). Lancaster (1972) referred to allomothering as "play-mothering," and I have observed in Japanese macaques what might be referred to as "playing-baby." Yearling infants may actively approach juveniles, sometimes their older female or male siblings, initiate ventral contact as though they were nursing, even nuzzling for the nipple, hunch their bodies down as though they were smaller, and give characteristic infant weaning reactions if the chosen caretaker leaves or refuses to cradle them. Thus in alloparenting, both infants and mothers may play a determining role in the interactions, along with the other caretaker. If we find that female primates do more alloparenting than males, it could be due to a differential preference on the part of the mothers and the infants for female caretakers, as well as to differential interest on the part of potential caretakers. It has sometimes been suggested that there is an inverse correlation between maternal restrictiveness, especially toward males, and the amount of male care exhibited. I shall try to illustrate this in the following section.

MALE PARENTAL CARE

In the order Primates, the interactions of males with infants range from hostility in which infants are sometimes killed, as in some langur groups, through apparent indifference to infants, as in patas and Sykes' monkeys, to nearly complete care of the infants, as in marmosets. There are differences not only between species, between groups of the same species, and between males of the same group, in terms of the parental care exhibited and the characteristics of the infants who receive it. This, of course, makes generalizations about primate male parental care very difficult. However, Hinde (1974) suggested that most primate species fall between the extremes mentioned above, and that generally primate males protect infants from predators and show some interest in, and tolerance of them, although they rarely exhibit anything approaching maternal behavior. Until recently, male care had not been very much studied or discussed in primatology. In an extensive review paper of adult male-infant interactions, Redican (1976) offered several possible reasons: (1) close, nurturant relationships between males and infants are rare in primates and just not as important to infant survival as maternal care; (2) the primate species most accessible and most studied — rhesus macaques, baboons and chimpanzees — also happen to show infrequent male care of infants; (3) the effect of males upon the young of a group may be far more subtle and difficult to measure than maternal influences. For example, a researcher may only become aware of an adult male's importance once he is experimentally removed from the group (see Poirier 1977).

190

When interactions between primate males and infants were first described, they were referred to as "paternal behaviors" (Itani 1959; Alexander 1970; Mitchell & Brandt 1972). However in multi-male, multi-female primate societies females mate with several males and thus true biological fathers are unknown to the researchers, and, in so far as we have evidence, unknown to the animals themselves. Therefore "male care" has been proposed as a term which is free of genealogical connotations. In polygynous (uni-male, multi-female) and monogamous (mated-pair) societies, it is usually assumed that the single adult male present in the group is the biological father. Since several researchers have suggested that males may behave differently when they are clearly related to infants, (Redican 1976; Wilson 1975; but see counterargument by Grafen 1980); I will divide this discussion into two sections, care by male kin, and care by males of uncertain or unknown kinship to the infants.

Male Kin (Fathers and Siblings)

I have already described how in monogamous species fathers commonly care for infants, taking neonates soon after birth, carrying, protecting, and otherwise providing infant care we normally tend to think of as "maternal" only. Longitudinal studies of monogamous species are only now being undertaken, but the impression has been that offspring approaching maturity are aggressively peripheralized by the parent of the same sex as the offspring. Interesting contrary findings are reviewed by Redican (1976) from studies which show that in marmosets older male siblings are well-tolerated within the groups and will care for infants. In fact they require this caretaking experience if they are to show adequate paternal care as adults.

In polygynous primate societies, the father is also likely to be the only adult male present in the group. In the case of the polygynous-living geladas, hamadryas, patas, langurs and some Cercopithecus species, there is no evidence that the adult male of the group exhibits special nurturant behaviors toward these infants likely to be his own, although such males will usually protect these infants, in the same way as most group-living adult primate males protect most infants. Rather, in all of these species, adult males are reported to be relatively "indifferent" to the infants, at least to those of their own groups. However, there have been some interesting studies of interactions between males and infants of *neighboring* groups in polygynous species.

In hamadryas for example, subadult and young adult males attract, adopt and/or kidnap yearlings and juveniles from the reproductive units especially young females, which they then carry and protect. Kummer (1967) suggested that these young females transfer their attachment from their mothers to the males which have adopted them. When the females reach sexual maturity, they will mate with these males and form a reproductive unit with them. Bernstein (1975) has also suggested for geladas that "bachelor" males exhibit caretaking behaviors toward young, but "harem" males do not; therefore Redican concluded that male care in these polygynous societies is directed toward the formation of new reproductive units. However Mori (1979a,b) has argued that bachelor male geladas direct care almost exclusively to *male* infants, which could not lead to the formation of new reproductive units, and thus the gelada and the hamadryas examples of male care must be examined separately. In complete contrast to these examples, langur males in some areas have been reported to kill infants of neighboring polygynous groups during "invasions."

Male siblings, either an older brother of the infant or its mother's brother, are reported to be important socializing agents in Japanese monkeys, rhesus macaques, common baboons and chimpanzees (Yamada 1963; Southwick et al 1965; Koford 1965; Rowell 1972; Goodall 1967). These represent almost all of the species for which we have good kinship information. Older brothers are known to show interest in newborn relatives, and later to draw them into play, to protect, retrieve, and carry them, especially in the absence of the mother; and in macaques they are even said to form a "social bridge," aiding younger brothers to transfer social groups.

Male Care of Uncertain Kinship

It is in multi-female, multi-male primate societies that adult (breeding) males are reported to show only general protective responses and a certain degree of tolerance toward all the infants of the group. However, as Redican has described, there are certain prominent exceptions. Itani (1959) and Alexander (1970) were the first to draw attention to the variable pattern of male care in Japanese monkeys. Some males in some troops will care for infants, even to the extent of adopting orphaned young if they can survive without milk. In both the Oregon troop and the Arashiyama West troop there are recorded cases of a male adopting and raising an orphaned female infant and maintaining a close relationship with her into adulthood. In both cases the female temporarily moved away from the adopted male parent while in estrus and mated elsewhere, but returned to her close relationship with him before and after the birth of her infants. Japanese macaque males tend to avoid mating with the females who form special year-round relationships with them (see Chapters 10 and 17). Thus in this case male adoption of female infants does not lead to mating as it is thought to do in hamadryas adoption or as it is reported to in the common baboon described below. Male care in Japanese macaques is reported to be both a "cultural" phenomenon, varying in occurrence from group to group, and a seasonal phenomenon, being observed more often during the birth season in the spring.

Other macaques show a variety of patterns of male care. Bonnet, pigtail, and rhesus macaques all tend to ignore infants, but in all three species experimental removal or incapacitation of the mother has been shown to result in much higher levels of male care (Redican 1976). Barbary macaques offer a great contrast to other species of the genus, *Macaca*, as several studies have reported the adult males to be highly important socializing agents, and in some cases, more significant for infant development than the mother (Burton 1972; Burton & Bick 1972; MacRoberts 1970; Taub 1980a,b; Deag & Crook 1971). Males literally take over infants from their mothers and protect, punish, and care for them. Barbary macaque males were also reported by Deag and Crook (1971) to exploit infants in a manner known as "agonistic buffering," in which infants are used as buffers in potentially agonistic encounters with other males. Their interpretation was that a subordinate male carries an infant when approaching a higher ranking male in order to reduce the likelihood of being attacked. However, Taub (1980b) has recently tested the agonistic buffering hypothesis by analyzing interactions between two males and an infant. He found that males choose to participate in these "triadic" interactions, not because of potential agonism or their respective dominance ranks, but because they both share a special care-taking relationship with the same infant. Taub proposed that the term "agonistic buffering" be dropped, at least for Barbary macaques.

In a study of common baboons, Ransom & Rowell (1972) suggested that adult males may use the carrying of infants to "enhance their effectiveness" in agonistic encounters. The close proximity of very young infants, especially those still retaining their natal coat, which in many primate species is of strikingly different coloration from the adult coat, is sometimes thought to inhibit attacks from other males. In addition, these baboons were reported to form close relationships with particular infants of the group, depending on pre-existing ties between the male and the mother of the infant. These close relationships with infants may lead to future consort choices. This study and earlier work by DeVore (1963b) indicates that adult and subadult male baboons are important to infant socialization, often "babysitting" infants, alone or in groups, carrying them and supporting them in agonistic interactions.

Thus while the general picture of the non-nurturant primate male is based on studies of multi-female, multi-male species, such as baboons and macaques, even here there are many reported examples of males as infant caretakers. There are also multi-female, multi-male primate societies for which we do not yet have adequate data. Redican (1976) formulated the hypothesis that adult male parental care correlates positively with more relaxed, permissive maternal infant interactions. He suggested that the data from different free-ranging groups of macaques and baboons support this hypothesis and that results of experimental studies of removing the mother, also lend support to the hypothesis.

In a similar vein, Redican and Mitchell (Mitchell 1977; Mitchell et al 1974) conducted studies on interactions between adult male rhesus and infant rhesus in the laboratory. They found that males and infants maintained as pairs in small lab cages would form very close attachments to each other, even though rhesus males in the wild are seldom seen to interact with infants. Rhesus male care in the experiment differed from female care in that adult males made less ventral contact with their infants, played with them more often, protected them from outside threats by moving between the infant and the danger, instead of retrieving the infant and withdrawing as mothers do; and by maintaining their attachment to the infant over time instead of "weaning" them as mothers do. While these studies seem to lend support to the intriguing possibility that male primates have far more potential for infant care than they usually exhibit, and that mothers in some species may actually "inhibit" males from parental behaviors, the laboratory studies suffer from the same lack of generalizability as similar research in artificial settings, of mating and dominance.

Primates are intensely social animals, and denied the needed interactions with the standard companions they would have in normal rearing are known to become very attached to kittens (Harlow et al 1963), dogs (Mason & Kenney 1974), and to cloth-covered dummies. That rhesus males, previously maintained in isolation, will interact positively with an infant, does not in and of itself prove these males are behaving parentally. However the researchers involved do suggest that the males exhibit actual care behaviors (not just tolerance of clinging for example), that they form intense attachments to their infants, and that infants raised by males alone are socially and physically healthy.

There are conflicting reports in the research literature as to whether males are more likely to care for male than for female infants. Itani (1959) stated that male parental care was irrespective of sex, although his sample did show a preponderance of two-year old females receiving adult male care. Mitchell and Brandt's review (1972) used Kummer's study of hamadryas adoptions and Itani's data to suggest that female primate infants were more likely to receive male care. On the other hand, in rhesus macaques, bonnet macaques, geladas, and common baboons it has been found that adult males are more interested in the genitals of infant males and are more likely to interact positively and care for male than female infants (Breuggeman 1973; Rowell 1968; Simonds 1974; Mori 1979a,b). Male care is a heterogeneous category, probably comprising parts of very different social mechanisms in different primate societies. For example, in hamadryas baboons male care is believed to be related to the adoption of future mates, so female infants are preferred; while in squirrel monkey society which is sex-segregated, male infants are accepted more positively by adult males; or in societies where adult male interactions mainly take the form of rough-and-tumble play, as in rhesus macaques, young males may be preferred.

PEERS

Most primate infants appear to be attracted to other infants, and if peers are available, young prosimians, monkeys and apes will soon enter into play groups, often under the watchful eye of adult caretakers. Primate species which live in small monogamous groups usually do not offer the young a chance for peer contact, and Ellefson (1974) concluded that this is why the young gibbons of his study did not play. However, even in small groups, infants are often seen to play with the other non-adults present, usually siblings. And in many primate species large numbers of peers are available, and are important in socialization, to the extent that Bramblett (1976) proposed that primate birth seasons may be an adaptation for the production of peer groups (all the infants are born around the same time of year), rather than as an adaptation to annual climatic changes.

Harlow (1959, 1962) suggested at one time from his deprivation studies that peers were even more important for socialization than mothers, because peer-only raised infants grew up to be "normal" in behavior, while mother-only raised infants did not. Later, he retreated somewhat from this conclusion (Harlow and Harlow 1966), and said that both mothers and peers are important. Although Harlow originally overstated his case, it does appear that peer interactions are significant socialization experiences in many primate species (e.g. see Poirier & Smith 1974). It is often through peer interaction that a young primate enters the wider social world of the group, first encounters non-relatives, and learns something about its position in the society.

The most important component of peer interaction is usually play. Until recently play was a much underrated and understudied topic (see Loizos 1967), and although research on primate play is proliferating today (see for example, books on play edited by Smith 1978, and Bruner et al 1976), much of the essence of this behavioral complex, including a widely-accepted definition, continues to elude us. While various theoretical approaches to the topic of play are now being developed, and the list of definitions and suggested functions continues to grow (30 proposed functions of play are given in Baldwin & Baldwin 1977), perhaps for no other behavioral phenomenon does there seem such a discrepancy between what the observer sees, and what can be categorized and recorded as data. This is all the more curious because everyone remarks on the wide-spread recognition of, and agreement upon, the occurrence of animal play. Untrained observers may mistake agonism, or even sexual behavior, but they almost always identify play when they see it. Play seems easy to recognize, but difficult to define and measure.

There may be certain almost subliminal signals in all primate play, known as "meta-communication" signals, which allow us to recognize behaviors as playful, and there are definite standardized motor patterns, such as "chasing" and "wrestling." But beyond these common features, play is characterized by its lack of predictability from one behavior to the next, a characteristic which is both its fascination and its challenge to the ethologist. Few tasks have ever absorbed and frustrated me as much as the study of monkeys at play. What do primates do when they play? At a simple motor level, they cuff, hop ("gambol"), pull, pinch, grab, grapple, tickle, mouth, embrace, wrestle, roll, chase, perform acrobatics, and interact in ways analogous to the human games of tug-of-war, tag, follow-the-leader, king-of-the-mountain, and hide-and-seek. More general descriptive categories of play have also been used, such as social play, solitary play, locomotor play, aggressive play, rough and tumble play (social behaviors involving body contact, especially wrestling), and approach-withdrawal play (rapid alternation of chasing and fleeing). Although play behaviors are flexible and largely unpredictable in sequence, we sometimes recognize fragments of adult behavior patterns such as agonistic or maternal behaviors, albeit out of context, which indicate that play patterns are not completely random, and some may be quite species-typical.

Young geladas, a species which lives in precipitous, treeless terrain, will scramble and push each other up and down vertical rock faces with the agility and fearlessness of mountain goats. Immature vervets, who may be found in grassy open woodland areas, will spar and cuff on the ground with stiff-limbed bounces so effortlessly high that they appear to be on pogo-sticks. Young patas catapult their bodies sideways off springy bushes, an action very similar to the adult male patas diversionary display. Japanese monkeys, who dwell on the islands of Japan, are one of the expert swimming species of the primate world, jumping, diving and swimming under water, and playing familiar water games such as dunking other swimmers and standing on their heads. Spider monkeys, acrobatic experts that they are, hang from branches by their prehensile tails and use all four limbs to grapple with their neighbors, until they become so intertwined that only their tails, curled around a branch all in a row, can be used to count them. In the lowlands of Guatemala, one's attention is first drawn to this kind of play by a curious and infectious many-voiced "chuckling" which sounds out through the trees. High overhead the tangle of black bodies heaves and shifts like one multi-limbed creature, and occasionally one monkey lets go, or loses its grip and plummets from the mass onto a lower branch, only to hurdle back up to rejoin the group.

Play patterns differ from species to species, from group to group, and some primatologists have said, from males to females. For some species of baboons, macaques, langurs and squirrel monkeys, there have been studies either showing or suggesting that young males tend to play more often, more roughly, and longer into their adolesence (e.g. Kummer 1968; Dolhinow & Bishop 1970; Owens 1975; Gouzoules 1980b; Harlow & Harlow 1965). A number of studies, such as my own on vervets (1972b) found sex differences in play, but failed to control for age of female and male subjects. A longitudinal study by Bramblett (1978) in which age was a controlled variable, has shown that male vervets do play more often and longer into subadulthood, but that as adults, female vervets play more. Raleigh et al (1979) found more play in males to be one of the few sex differences in juvenile vervets. A few studies, such as Kaufmann (1966), Hinde & Spencer-Booth (1967a) and Baxter (1979) found no significant sex differences in the frequency or roughness of play in rhesus and spider monkeys.

Because Harlow (1965) found that laboratory rhesus females initiated less play, engaged in less rough and tumble play, and were often chased in approach-withdrawal play, he referred to female play patterns as "passive" as opposed to the active play of males, and claimed for many years that this difference presages later passive/active differences in the adult roles of females and males. To judge from other studies of this species, "passive" is a questionable label to apply to the behavior of rhesus females, whether they be youngsters or adults; and is, to my mind, a tendentious interpretation of data which show that females also engage in active play, albeit at lower frequencies and/or levels than males. This same falsely dichotomous active/passive labeling has sometimes been applied to human play. While some studies of play in preschool children found no significant differences in overall rates of play (McGrew 1972; Hammer & Missakian 1978), others, such as that by Blurton-Jones (1967) found more running, chasing, fleeing and laughing in nursery school boys. He felt that these patterns were equivalent to Harlow's rough and tumble play in rhesus monkeys, but in fact they correspond to the definition of approach-withdrawal play. McGrew (1972) also noted a few sex differences in nursery behavior (e.g. girls bite more, boys throw objects more often), but he concluded that sex differences are very difficult to evaluate. The entire area of sex differences in the play of young human children is problematic.

Goldberg & Lewis (1969) conducted a widely-quoted study of sex differences in the play of year-old infants. They found several differences — boys were more independent of their mothers, more exploratory, active, noisy and vigorous in their play. Boys reacted to frustration (a barrier separating them from their mothers and their toys) by attacking the barrier actively, while girls cried helplessly. Later Maccoby & Jacklin (1973; Jacklin et al 1973) attempted to replicate the various elements of this study, but they were unable to reproduce these findings. In their extensive review of human sex differences, Maccoby & Jacklin write:

> The two sexes are highly similar in their willingness to explore a novel environment, when they are both given freedom to do so. Both are highly responsive to social situations of all kinds and although some individuals tend to withdraw from social interaction and simply watch from the sidelines, such persons are no more likely to be female than male (1978: 354).

The authors continue that girls may exhibit a tendency to play more quietly, but their play is also very active, organized and planned. While boys tend to be more aggressive, girls are not more submissive and do not yield or withdraw in the face of aggression any more than do males. "In sum, the term 'passive' does not describe the most common female personality attributes" (1978: 355).

Clearly, the entire question of sex differences in play needs systematic, quantitative examination in a number of primate species, before we can adequately assess the validity of a widespread assumption in primatology, that males play more roughly and more energetically, as well as more often, than do females. And should they prove to do so, one can offer a less androcentric interpretation of this difference than "passivity" of females, such as that proposed by the Baldwins in the ontogenetic model described below.

I noted at the beginning of this chapter that there have been few sex differentiated schemes of behavioral development in primates. The Baldwins (1977) offer one such scheme in their description of sex differences in the learning and ontogeny of play, which Poirier (1977) reiterates, and which by implication could be used as a model to describe the development of many sex differentiated patterns in young primates.

The Baldwins argued that young male primates of several species (7 species are cited) spend more time in exploration and play than do females, and that if we chart frequencies according to age, we find that females tend to withdraw from, or cease to engage in, rough and tumble play earlier in life and prefer other types of activities, such as object manipulation and "play-mothering," which have the effect of integrating them with infants and other females. The Baldwins' first premise is that for each individual at every period of its development there are optimal levels of sensory stimulation, and different levels of stimulation are involved in given social and physical activities. Too much stimulation (overarousal) is frightening and causes anxiety and withdrawal, but too little stimulation (underarousal) has negative results ranging from grogginess and lethargy to behavioral abnormalities such as those observed in deprivation-reared primates. Thus individuals seek to engage in activities with just the optimum amount of "stimulus novelty;" amounts vary between individuals, as well as during the lifetime of one individual.

A number of factors are suggested to contribute to the differences between optimal stimulation patterns for female and male primates. Firstly, the prenatal environment may result in neurophysiological differences such that physically vigorous activities tend to be somewhat more reinforcing to males. Secondly, male primates in the seven species under discussion (squirrel monkeys, langurs, vervets, bonnet and rhesus macaques, hamadryas and common baboons) tend to be physically larger and stronger than females, and additional size, weight and strength are thought to result in higher levels of energetic play and exploration, and higher levels of optimal stimulation. The Baldwins believe these size and strength differences alone may be sufficient to account for some differences in frequency and duration of play.

As growing males move into rowdier play patterns then, the females are drawn to other activities, characterized by lower arousal levels, such as manipulatory exploration, grooming and play-mothering. Although not noted by the Baldwins, another factor that could be relevant here is the finding that rhesus and pigtail macaque mothers maintain much closer social and physical bonds with their female offspring, which may influence the possible sources of stimulation for young females. According to Jensen's work, males are pushed into earlier and greater "independence," or distance from their mothers, and once in this position, engage in more vigorous physical activities. Finally, the Baldwins point out that female primates of many species reach socio-sexual maturity at an earlier chronological age than males, and becoming pregnant and giving birth to an infant may hasten the termination or alteration of female play patterns.

SUMMARY

The studies reviewed in this chapter focussed on the manner in which social companions respond to the sex of a young group member, as well as how their behavior, in turn, affects the developing female and male patterns of behavior. It should be remembered that sex differences in behavioral development can be difficult to sort from other sources of variability, and it well may be that these other variables are of equal or more importance than sex differences in the processes of socialization. The very asking of the question, "Are there sex differences?" may place a misleading emphasis on differences rather than similarities. However, since adult females and males in many primate species do appear to behave differently, it is reasonable to ask if and/or how social learning contributes to these differences.

It is often assumed that adult sex roles are the result of steadily diverging trends which began in the young, although the manner in which such diverging trends might occur has not been well studied. Some research has found that primate mothers handle their male and female infants differently from birth, forcing earlier independence in males. However, it is not clear whether the mothers are responding to differences in the behaviors or in the physical characteristics of their infants. Other members of the social group often show interest in infants, interacting and/or caring for them, but there is great variability in alloparenting from species to species. Although there has been a widespread belief among primatologists that young females are more interested in this kind of infant caretaking than males, this hypothesis has only been tested in a very few species. In some of these cases, females without offspring of their own do show a great interest in infant care.

Male care of infants has been little studied or discussed, but there exist primate species in which the adult male takes almost complete care of the infants. Many primate males protect infants in general, but do not form close individual relations with particular infants, although experiments show they are capable of this. It has also been suggested that some primate mothers may restrict interactions between adult males and their infants and thereby inhibit male care patterns. Whether or not alloparents prefer female or male infants seems to depend on the species social organization, although data are lacking for many species, and the sex of the infant may well be irrelevant in some cases. Peer group interaction, especially play, is thought to be an important agent of socialization in many primate species, and young male primates are believed to play more often, and possibly more roughly than young females, although this too is an idea in need of more careful documentation.

The model of sexual differentiation described at the end of the chapter suggests that at least four factors may contribute to observed differences in the behavior of young female and male primates: prenatal hormones, variability in size and strength, differential maternal handling, and differential maturation rates. It seems that this model could be used to explain the development of sex differences in the better studied species of nonhuman primates, especially baboons and macaques, but there are prominent exceptions to this pattern, and it is not yet clear how general it will prove to be in the primate world.

Chapter **THIRTEEN**

Deprivation

Rhesus macaque

Macaca mulatta

Found in the towns and the countryside of South Asia, these monkeys are used extensively in medical research. The female (right) grimaces and flees from the open-mouth threat of the male.

In approaching the question of how experience affects behavioral development, the deprivation study is a powerful methodological resource; so powerful, in fact, that its effects can sometimes be likened to the blow of a sledgehammer, when a few gentle taps might have been more instructive.

Mason 1968a: 80.

The deprivation experiment has been a traditional ethological technique for studying innate and learned behavior in animals. The goal of such research is to rear animals in the absence of one or more environmental factors which have been hypothesized as necessary to the normal development of behavior in that species. If the deprived animal exhibits the behavior in the absence of learning experiences, then the behavior may be considered to be relatively independent of learning, or closely innately circumscribed. If, on the other hand, the behavior fails to appear in a deprivation-reared individual, learning or environmental factors may be considered to be essential to the expression of the behavior. For example, if one hypothesizes that the characteristic song of the white-throated sparrow is learned or otherwise stimulated by hearing the parents or other conspecifics sing, the bird can be reared in isolation, or with birds of another species, to determine if it will still sing like a white-throated sparrow. Since social learning is considered to be essential to primate behavior, most primate deprivation studies have focussed upon the effects of rearing monkeys and apes in the absence of normal social experiences. However the effects of sensory and/or physical deprivation are also recognized to be important in development, so that for example, a monkey reared alone in a small, bare metal cage suffers from both social and sensory (physical) deprivation.

Ideally one would begin a deprivation study with a thorough knowledge of the animals' normal behavior (that is, of the behavior shown by free-ranging or feral animals), and a knowledge of the socialization processes common in the study species, and then observe the results of carefully subtracting one environmental factor considered to be essential to normal development. This has almost never been accomplished. Because studies of social learning usually take place in the laboratory, there has been no clear separation of deprivation studies, learning studies, and socialization studies. In captive settings infants are often reared, and adults are often maintained, with a limited variety of social companions, in a group of abnormal composition (for example no adult males, or only one adult male, in a multi-female, multi-male species such as the rhesus macaque), and in a physically impoverished environment, with limited space and variety. Thus many of the socialization studies described in the last chapter are, in fact, deprivation studies; and primatologists have seldom begun their experimentation, either in deprivation research or in hormonal alteration studies, with a knowledge of the standard developmental parameters of the behaviors which they are experimentally altering.

An enormous number and variety of primate deprivation studies have now been conducted, the vast majority of them with rhesus monkeys; although chimpanzees, pigtail macaques, bonnet macaques and squirrel monkeys also have been used. This chapter will briefly describe the major deprivation rearing conditions under which the rhesus macaque has been studied, summarize the primary findings of those studies, and then turn specifically to sex differences reported from deprivation research.

The most famous of the deprivation studies of primates are those of Harlow and his students, which began in the 1950's at the Wisconsin Regional Primate Research Center. Although I have suggested that knowledge of behavior in natural settings should be a starting point for deprivation research, Harlow readily avowed that contrary to this ideal of beginning with knowledge of the normal developmental processes and subtracting or altering important variables, his laboratory began at the other extreme, by accidentally rearing monkeys in the absence of those things essential for normal behavior. Gradually,

200

over the years they added more and more of the needed opportunities for social experiences as researchers became more aware of these needs, and also as they refined their experimental techniques (Harlow et al 1971).

Rhesus monkeys have been raised in a great variety of deprivation environments, but for simplicity I will list here a slightly modified version of the rearing conditions described by Sackett (1974):

1. total social isolation — infants are removed from their mothers at birth and maintained in completely enclosed metal cages with no physical, visual or auditory contact with other animals;
2. partial isolation — infants are removed from mothers and maintained singly in bare, wire-mesh cages from which they can see, hear and smell, but not touch, other infants.
3. surrogate-reared — infants are removed from mothers and live individually in a partially or totally enclosed cage containing a cylindrical wire-device, known as a "surrogate mother," which provides food and/or contact comfort;
4. peer-only reared — infants are removed from mothers and placed in wire cages containing groups of two to five infants;
5. mother-only reared — mother and infant are maintained in a wire-mesh cage as a pair;
6. mother and peer reared — four mother-and-infant pairs live in wire cages surrounding a central playpen area. Usually the testing procedure involves allowing the infants a limited amount of access to the play pen and to one another by raising small doors between the cages and the playpen for a short test period during the day. The doors are small so that only the infants and not the mothers can go through into the playpen area.

What happens to infant monkeys growing up under these conditions? To some extent it depends on the length of time spent in the impoverished environment, the severity of the conditions and on the primate species studied (Sacket et al 1976). The list above is roughly in order of decreasing severity of deprivation, although researchers disagree about the relative placement of rearing conditions 4 and 5. Stated simply, deprived monkeys grow up into severely disturbed adults, incapable of normal patterns of social interaction. Labels such as autism, schizophrenia, psychosis, and depression have all been used to describe them, although thus far it does not seem that the disturbances in these monkeys correspond to any one human mental disorder (McKinney 1974). Deprived infants cling to anything available, such as the "surrogate mothers" often provided; and one of Harlow's early and major findings was that they prefer terry cloth-covered surrogates to milk-dispensing bare wire devices (1959). They also cling tenaciously to other infants if available; and in isolation, to their own bodies, sucking their fingers, toes, or penises. When they reach physical maturation, motherless monkeys are incapable of normal sexual behavior. If such females are artifically inseminated or otherwise coerced into pregnancy (for example, in the past Harlow's team used a device they referred to as a rape rack [Harlow et al 1971]), these monkey mothers are described as being brutal, rejecting, or indifferent toward their infants, and may kill them if the infants are not removed. Almost all primate mothers protest vigorously when attempts are made to separate them from their infants, but motherless mothers do not attempt to hold on to their infants, and on the contrary, may redirect agonism *toward* the infant when frightened by the experimenter.

Since, to varying degrees, all deprivation reared rhesus monkeys exhibit consistent abnormalities, Mason (1968a) described a general "deprivation syndrome" consisting of stereotyped repetitive patterns (swaying, rocking, self-clutching, etc.); motivational disturbances (excessive fearfulness and aggressiveness); sexual inadequacies, and in the case of females, maternal inadequacies; and deficiencies in social communication. The deprivation studies in all their variety leave no doubt that social learning is absolutely necessary for the normal expression of primate behavior. Indeed, it is considered a most important finding that aspects of behavior as fundamental to survival and reproductive success as copulation, and a mother's willingness to help and allow her infant to suckle, are dependent upon social learning for their proper expression.

There are however, two major problems which plague this type of research. The first is the assumption, held by more researchers in the past than in the present, that deprivation studies will allow us to distinguish innate behaviors from learned behaviors; an inadequate and falsely-dichotomized manner of approaching behavior, as I suggested earlier and to which I shall return at the end of this chapter. The second problem is that these experimental monkeys have not simply been deprived of the opportunity for a certain type of learning experience; they have been reared in an extremely stressful environment, such that social learning is subtracted and stress is added. As a result, they develop such gross abnormalities in behavior that it is very difficult to sort out the manner in which specific environmental elements are relevant to the expression of certain behaviors. For example, if an isolate-reared male rhesus is unable to copulate with a female, it can be because he is terrified of her and/or the test situation; because he has never seen a female nor seen mating take place; because he does not know how to communicate his intentions to her; because his own motivations are abnormal; or because his standard response to sexual arousal is to become absorbed in sucking his penis, chewing his wrist, or performing other disturbance behaviors. Possibly all these factors and more are at work in this one demonstrated inadequacy of an isolate-reared monkey. Motherless mothers may injure their infants because they are hostile toward them, or simply because they have not the slightest notion as to what an infant is, never having seen one — as might be indicated by isolate-reared females who chew off the umbilical cord of newborns, as do feral mothers, and then calmly and without agonistic signals, proceed to chew on the infant itself (Ruppenthal et al 1976). To follow the conclusions of some of the researchers, and not to put too fine a point on it, all isolation-reared monkeys and many deprivation-reared monkeys are grossly disturbed or mentally ill, and this seriously complicates research which attempts to delineate specific variables in the development of social behavior. As Jolly (1972) wrote of primate deprivation experiments, "bad rearing" may not only shock us sentimentally, but may also destroy the conclusions scientifically. Or to pursue Mason's metaphor quoted at the head of this chapter, much deprivation research attempts to crack a most delicate problem with a sledge hammer, thus obliterating the kernel of behavioral development rather than disclosing it.

This is less true of "separation" research such as that carried out by Hinde's team in Cambridge (e.g., Hinde & Spencer-Booth 1970, 1971; Spencer-Booth & Hinde 1971; L.M. McGinnis 1979) and by Kaufman & Rosenblum (1969b) and see also the review of separation research in Mineka & Suomi (1978). In these research programs infants have been reared in stable, complex social groups, which is very important, and the deprivation or separation variable has been carefully controlled. In the Cambridge research, either the infant or the mother was removed for very brief periods from the group, and the behavioral changes during the separation and after the reunion were followed. Hinde reported the noteworthy finding that rhesus mother-infant relations are less disturbed if infants are removed from the group than if mothers are removed, and that mother-infant separation for only *six days* in the first year of life has effects on the behavior of the infant up to *two years* later, although such effects are also dependent on the nature of the pre-separation bond with the mother. Further, Kaufman and Rosenblum have been able to show that the effects of mother-infant separation on the infant are dependent on the type of society in which the infant lives, and the nature of the social bonds which the infant forms with other social companions.

It is this type of research which begins to demonstrate just how complex and delicate the processes of social learning really are, and it is this type of research which may have some relevance to human developmental research, especially studies of bonding and enforced separations (hospital stays, etc.) in children. Bowlby (1973) described a separation syndrome in children involving three stages after the removal of the mother or principle caretaker: (1) protest; (2) depression (despair); and (3) detachment or denial upon reunion with the caretaker. Experiments in macaque separation such as those described above have documented protest and depression in infant monkeys of some species, but generally have failed to find the third stage of detachment. However, Mitchell (1969) reported that a few young male infants have been known to react temporarily to reunions by withdrawing, and langur infants usually remained with their adopted caregivers when mothers were returned (Dolhinow 1980). Usually when mothers and infant macaques are reunited they vigorously seek ventral contact; and according to Hinde's research, such mother-infant pairs undergo a lengthy period of readjustment in their relationship even after a separation of only a few days. Suomi & Harlow (1977) argued that separation research such as that reported above, and some studies they conducted, are different in nature and results from isolation research, because in the latter monkeys are never allowed to develop social bonds and therefore suffer from intense behavioral anomalies. On the other hand, separated monkeys suffer from disruption of social bonds and affective ("emotional") or motivational disorders associated with increases in fear and aggression, which disappear with renewed social contact. However, it seems to me that isolation, deprivation, and separation can also be viewed as imposing different degrees of social and sensory stress and not everyone agrees that motivational disorders can be sorted from learning disorders in these experimental monkeys (see, for example, comments by Rosenblatt and Tinbergen in Beach 1974: 262-263).

SEX DIFFERENCES AND DEPRIVATION

Play behavior in the playroom is typically initiated by males, seldom by females. However, let us not belittle the female, for they also serve who only stand and wait. Contact play is far more frequent among the males than the females and is almost invariably initiated by the males. Playpen data . . . show that real rough and tumble play is strictly for the boys. I am convinced that these data have almost total generality to man . . . These secondary sex behavior differences probably exist throughout the primate order, and moreover, they are innately determined biological differences regardless of any cultural overlap (Harlow 1962: 5).

Some twenty years ago Harlow began raising infant rhesus monkeys in two test situations, both called "limited peer access rearing," and both of which are versions of rearing conditions 3 and 6 described earlier. In the first situation, infants were removed from their mothers at birth, and raised on surrogate dummy mothers, with brief (one half to one hour) periods of daily peer interaction with other similarly-reared infants in a toy-filled playroom. This is referred to as the surrogate-reared, play*room* case. In the second test situation, infants were reared with their mothers as pairs living in cages surrounding a central playpen, which they could enter (but their mothers could not) for one or two hours daily. This is the mother-reared, play*pen* case. There were other variations on these two test situations, but for the moment we will stay with these two as described in Harlow (1962: 3). In both cases, infants played in groups of four — two males and two females — and Harlow felt that the behavioral results, at least in the infantile heterosexual stage, were so similar he freely mixed findings from both test situations in his descriptions (e.g., 1962: 3,4; 1969: 344).

In the quotation at the beginning of this section, Harlow drew conclusions from both limited peer access rearing situations. Specifically, he found that from two and a half months of age on, male rhesus infants exhibited more threats, earlier and more frequent sexual behavior, more play initiations, and more contact or rough and tumble play. Female infants exhibited more of what Harlow called "passive responses," described as "immobility with buttocks oriented toward the male and head averted" (1962: 4). This posture sounds very much like what primatologists call a "present" and which in feral animals occurs in females and males in greeting, sexual, infant-care, and agonistic contexts (see discussion in Chapter 7). Female infants, when approached, were also found to exhibit more withdrawal and "rigidity," a posture in which the body is stiffened and fixed. Harlow's conclusions from these data, some of which he reiterated and redescribed in subsequent papers (1962, 1965, 1969, 1975; Harlow & Harlow 1962; Harlow et al 1972; Harlow & Lauersdorf 1974; Suomi & Harlow 1976) are that these experimental monkeys demonstrate the normal and intrinsic development of sex differentiated patterns of behavior in primates, which will occur even in the absence of regular socialization or social learning factors, and which therefore are innately determined. Moreover, he believed, adult sex role behavior develops from these prototypical "basic response" patterns in the infants; young males are "forceful and active," while young females are "passive and withdrawn," and these are the roles they will exhibit as adults.

There are several assumptions and ideas explicit and implicit in this model of sex differences, I would like to examine in turn. First is the assumption that "limited peer access rearing" controls for social learning and therefore allows us to conclude that behavioral patterns which occur in monkeys so reared are "innately determined." It has already been noted that total isolation rearing of rhesus monkeys destroys the standard female-male differences in behavior, and thus some social experience is considered to be necessary to the appearance of sex differences in agonism, play, and sexual behavior of the kind reported by Harlow. Peer interaction *is* a social experience in which learning occurs, and indeed Harlow has argued frequently that peer-only reared monkeys (rearing condition 4 from above) are more normal in their behavior than mother-only reared monkeys (condition 5), so that peers are considered by him to be the single most important social factor in the development of normal rhesus behavior. What he means then, in his description of innately determined sex differences, is not that peers are excluded as socializing agents, but rather that they cannot form models and do not handle or otherwise influence female and male infants differentially, as adults in social groups might do. However, Harlow's proclivity to freely mix data from the two test situations, the surrogate-reared, playroom infants, and the mother-reared playpen infants, is a problem for the reader at this point. For example, in one paper he stated that the playroom data on sex differences in rough and tumble play were not graphed, but the playpen data were just the same, and therefore he presented that graph (1969: 344). However the playpen infants in question (from a study by Hansen 1962) were reared by real mothers, who may well have handled their female and male infants differentially and stimulated differences in their behavior, of the kind reported in Mitchell's and Jensen's studies of maternal handling and sex differences (see Chapter 12). Although Harlow definitely does say in other places that sex differences such as he observed will occur without maternal influences, it is not easy for the reader to determine from the data presented that surrogate-reared as opposed to mother-reared infants exhibit the described differences in play, agonistic and sexual responses.

At any rate, it should be said that limited peer access monkeys, whether they are reared with surrogates or real mothers *have* been exposed to some social experiences during peer interactions, making it difficult to claim that any sex differences observed are purely "innately determined." However, this is not to argue, as Harlow did, that these limited peer access reared monkeys are "normal" in their behavioral development and so can be used as models for primate development. Indeed, a number of other researchers argue against this claim (for example, Hinde 1974; Goy & Goldfoot 1973; Goldfoot 1977; Rose et al 1974). Studies such as those by Goy, Goldfoot and Wallen, show that infant rhesus monkeys reared with only limited access to peers differ in several important respects from group-reared infants; namely the deprived infants are tenser animals, exhibiting more threats and more fear-reactions and more intense agonism. They also engage in more play, particularly rough play, they groom less, and their adult copulatory abilities are clearly deficient. Only 40% of limited peer access reared males achieve intromission, and only 25% display ejaculation, while close to 100% of feral rhesus display these behaviors. Further, Phoenix (1978b) recently published data demonstrating that almost all laboratory-reared rhesus monkeys, even those reared in groups of mothers and peers to ensure they were not socially deprived, exhibit abnormally low levels of sexual activity in comparison to feral males. That it is doubtful if the behavior of laboratory-reared rhesus can ever be used to establish "norms" for primate behavior is evident in Phoenix's conclusion: "any rearing condition that involves a social situation different from the condition found in the wild could be considered suspect (i.e., socially inappropriate)" (1978: 191). This is a contentious and sometimes irritating issue, as made clear by the following exchange, which took place between two endocrinologists discussing the use of laboratory rhesus as models for human sexuality:

Dr. Rose commented that, in his opinion, monkeys should be studied in a natural setting: caged monkeys are crazy. Dr. Goy replied that 'crazy' is perhaps an exaggeration, and the term 'legally insane' is more accurate (from "Group Discussion" in Goy & Goldfoot 1975: 418-419).

Further abnormalities in the behavior of limited peer access reared monkeys have been reported in the literature. Harlow concluded from his study of these monkeys that the double-foot clasp mount of the male rhesus only develops at puberty, but other researchers believe his experimental monkeys to be retarded in their sexual development, since this sort of mounting appears in the first year of life in group-reared male rhesus (Wallen et al 1977). Even the "rigidity" response described by Harlow as very prominent in females, and necessary to the development of adult heterosexual behavior and the "passive" female role, is quite likely abnormal, for it rarely occurs in group-reared infants, has not been described for feral rhesus, and is considered by Wallen to be a pathological fear response which occurs when an infant is very frightened or bullied, and has no safe haven to turn to (it "freezes"). Hinde believed the "rigidity" response should not be linked with the female sexual posture, because the two are quite different in a motor sense, while DeVore suggested that it is a response to extreme fear (in Beach 1974: 262).

I noted earlier in this section that the "passive response," deemed by Harlow to be predominant in, and necessary to, the development of the female sex role, is not different in motor description from the behavior known as "presenting," which occurs in both males and females of group-reared rhesus, and clearly is an "active" behavior, functioning as a form of solicitation in several contexts. Whether any behavior performed by an animal can be properly called "passive" is doubtful, and in this case, where the labels "active" and "passive" are used to categorize repertoires of unrelated and sometimes pathological behaviors in an apparent attempt to support cultural sexual stereotypes, what is most clearly revealed is the bias of the researcher, rather than the behaviors being researched.

Finally, Wallen et al (1977) countered Harlow's argument that sex differentiated patterns of play (active, forceful males, withdrawn, passive females) are necessary to later proficiency in female and male adult sexual behavior. They argued that many limited peer access reared males are highly playful, but deficient in sexual behavior as adults, and that *too much* play may be as indicative of a deficient social environment as too little play. More evidence of this comes from separate studies, such as Phoenix (1978b) who also found that sexually deficient laboratory-reared monkey males were highly playful, and Anderson & Mason (1974) who concluded that deprivation-reared males played *more* than socially reared males. It is one of the standard axioms in the play literature that play is necessary to, and indicative of, healthy, normal young primates (see Baldwin & Baldwin 1973, for a counterargument), but we cannot assume that because animals play, they are therefore normal, and we should not allow playful behavior to obscure deficiencies in the development of limited peer access monkeys.

Thus far then, I have argued that Harlow's deprivation studies relevant to the development of sex differences do not eliminate, and in the case of mixing playroom and playpen data, do not control for, socialization factors; and that monkeys reared with limited access to peers are not "normal" or at least similar in behavior, to group-reared monkeys. Criticisms also have been offered of his fundamental concepts of "rigidity" and "passivity," and his model of sex differentiated play patterns developing into adult sexual patterns. One wonders if Harlow's conclusions concerning sex differences observed in his test monkeys, and his generalizations to all primates might have differed, had he obtained, or had access to, data on the parameters of behavioral ontogeny in free-ranging group-living rhesus. Perhaps one should state this as euphemistically as Bramblett's comment on deprivation studies: " . . . assessments of social development in caged animals generally suffer from an insufficient awareness of the abnormality engendered by the cage environment" (1976: 40).

STRESS AND SEX DIFFERENCES IN BEHAVIOR

Even if one does conclude, as I have, that the female-male differences in limited peer access monkeys are not necessarily indicative of "normal" sexual differentiation, one is still left with the question of why reported sex differences in experimentally deprived monkeys appear at all. First of all, two cautionary notes concerning the evidence for sex differences in deprivation reared monkeys should be sounded. Harlow argued for many years that his studies demonstrated innate sex differences, and one has the impression in reading his papers, written over the last twenty years, that data concerning sex differences are steadily accumulating in his laboratory, an impression reinforced by statements such as: "We would like to discuss differences between the male and female sexes that we have found to hold true after and over many years of research" (Harlow & Lauersdorf 1974: 348). Although such data may well be available in unpublished form, in the tables and figures actually presented in published papers, the data are frequently from one original study conducted prior to 1962. In some papers, Harlow does not give the exact source or sample size for each figure or table, but the same graph, usually with an N of 4 males and 4 females (sometimes 2 males and 2 females) can be found repeatedly for a given behavioral sex difference in papers ranging in publication from 1962 to 1975. Thus for example, identical data for sex differences in rough and tumble play are presented as graphs in Harlow, (1962 Figure 12, page 5; 1969, Figure 11, page 344; and 1975 , Figure 9, page 86); and data concerning female sexual posturing or the so-called "passive response" appear as Figure 16 (1965: 249), Figure 6 (Harlow et al 1972: 6), Figure 3 (Harlow and Lauersdorf 1974: 352), and Figure 11 (1975: 87) in several different papers. The repetition of the same data in many different papers is also characteristic of the other sex differences discussed by Harlow, such as mounting, threatening, grooming, etc. While there is certainly nothing wrong in republishing data, and indeed it is a common enough occurrence, this practice can be deceptive, especially if a general case for all primates is being made. Individual variability in behavior is one finding that has been established over and over again in primate research, and one would like to see data on sex differences for more than four rhesus males, and four rhesus females, reared in limited peer access conditions, before accepting any general statements about sex differences existing throughout the primate order.

The second caution regarding the evidence for secondary sex differences in deprivation-reared primates comes from the study by Young & Bramblett (1977) described in Chapter 12. These researchers raised twenty common baboon infants in limited peer access conditions very similar to those of Harlow's surrogate-reared, playroom, in that the baboon infants did not have real mothers, and their only social interactions were with a small group of peers in a play cage for one hour a day. Young & Bramblett were unable to demonstrate the expected sex differences ("expected" largely from Harlow's studies) for the baboons in the first three months of life — young male baboons were not more active or aggressive and did not engage in more sexual behavior or rough and tumble play. The timing of rhesus macaque and common baboon developmental stages is considered to be very similar (Rowell et al 1968), and the authors are uncertain as to why they failed to find the expected sex differences in baboon behavior. This study also indicates that we are far short of the point where generalizations concerning the primate order are justified.

This is not to suggest that sex differences in deprivation-reared monkeys are not "real," only that they may be rather different in nature and cause, and rather more variable between species than Harlow's writings indicate. Sackett (1972, 1974) also studied sex differences in deprivation-reared rhesus monkeys and offered a most interesting alternative hypothesis as to their cause: it is his thesis that rhesus males reared under social-sensory deprivation are more susceptible to the development of abnormal behaviors, and that female rhesus are somehow more buffered against the stresses of deprivation rearing. To demonstrate this he presented data from his own studies and an unpublished dissertation study by C.L. Pratt (1969). Rhesus monkeys were reared under three different deprivation conditions for nine months and then exposed to social interactions. The resulting sex differences in behavior may be summarized as follows:

1. total social isolation — females were more active and exploratory, males were more fearful. There were no sex differences in play, aggression and passive behaviors; play never occurred in these animals and aggression only rarely. When total isolates were compared to partial isolates and peer-only reared individuals of the same sex, the male total isolates showed more disturbance behaviors, such as self-clasping and mouthing, rocking, etc.
2. partial isolation — females were more playful, more aggressive and more exploratory; while males were more fearful, more self-aggressive, and showed higher levels of disturbance and passive behaviors.
3. peer-only reared — only minor sex differences were found in comparison to the two other groups; females were more exploratory while males were more disturbed and more aggressive. There were no significant differences in play or sexual behaviors.

Sackett concluded from these data that males are more adversely affected by the stresses of deprivation rearing, and that the kind of sex differences in this study *increased* with the degree of social deprivation experienced, as the conditions are again listed in decreasing order of severity, and peer rearing is considered the "most normal." He wrote: "Under socialized conditions of rearing, few important sex differences appear for any behavior not directly related to sexually dimorphic anatomical or physiological variables" (1974: 120). Differential response to stress is, in his opinion, the factor responsible for most of the sex differences found in deprivation studies. The more normally, or less stressfully, the animals are reared, the less these differences will be evident.

It may appear that Sackett's proposal contradicts Mitchell's (1968a) contention that total isolation rearing destroys what he believes to be the normal male-female dichotomies in behavior. However, it only appears to do so. The important difference is that

Sackett's comments on isolates do not refer to the kinds of sex differences in agonistic, sexual, and dominance behavior, usually found in group-reared rhesus, as do Mitchell's, but rather to abnormal, disturbance behaviors which appear under stress. This becomes clear, for example, as Sackett discusses the finding that partial-isolate males are more self-agressive. Rather than being a natural result of the fact that rhesus males are always more aggressive than females (indeed several researchers, including Sackett. have argued the converse), he considers it an indication that males are more disturbed in this situation. If males do respond differently than females to the same environmental stimuli occuring in deprivation conditions, that is certainly a sex difference, and one which probably has important innate components; however it does not mean that differences observed under deprivation conditions are indicative of sex differences in all environments. The problem is that stress may not magnify "normal" behavioral differences, but instead may affect the behavior of females and males differentially, and therefore distort differences, and may very well "create" differences not usually present. In other words, it is not correct to assume that under deprivation or stress conditions, males simply act like themselves, only more so. On the contrary, they may behave abnormally.

Although none of the three deprivation conditions described in this experiment correspond exactly to Harlow's limited peer access rearing test situations, it is important to note that Harlow's postulated innate sex differences, which should appear regardless of social learning or rearing conditions, do not appear even as trends. Total isolate reared rhesus showed no sex differences in play, aggressive or passive behaviors, partial-isolate females were more playful, aggressive and exploratory (the opposite of Harlow's "passive" sex role stereotype), and for peer-reared monkeys, Sackett emphasized the lack of sex differences where Harlow has emphasized their prominence and importance. I previously criticized Harlow's characterizing of his findings on sex differences as exemplifying active/passive sex roles. But beyond the question of how (and why) he chose to label his female and male patterns, Sackett's theory of differentially-stressed male rhesus would indicate that the stresses of deprivation rearing, in this case limited peer access rearing, should be considered a contibuting and possibly confounding variable in interpreting any resulting sex differences. The large amounts of threats and rough play in infant males, and the high frequencies of "rigidity" and withdrawal in infant females Harlow found in his test monkeys, may indeed reflect sex differences in response to a particular form of environmental stress; however we do not have evidence for Harlow's claim that they also represent general primate patterns of innately determined sex differences occurring in all environments and all primate societies.

LOVE AND SEX ROLES MADE SIMPLE

After resolving the nurture and nature of maternal, infant, peer, and heterosexual love, the only thing that remained was paternal love (Harlow et al 1971: 541).

The goals of environmental control are first to isolate contributing factors and then to put the pieces of the puzzle together for a clear picture of 'normal' socialization. There is a flaw in this, for the sum is more than its parts. Dyadic

209

interactions do not re-present [sic] the full social repertoire. In addition, environmental restrictions have pervasive influences on other aspects of behavior than the particular one under study (Fragazy & Mitchell 1974: 568).

This chapter is by no means a comprehensive survey of deprivation research in primatology, as I have concentrated upon those studies which have particular relevance to the development of sex differences in primate behavior. For a comprehensive survey of the field, the reader is referred to excellent reviews of deprivation research by Mitchell (1969) and Suomi (1976, 1977). Experiments in which nonhuman primates are reared in impoverished environments demonstrate that physical, social and cognitive skills as divergent as tool-use and self-recognition in chimpanzees (Menzel et al 1970; Gallup 1975), and the forming of social coalitions and the development of the double-foot clasp mount in macaques (Mason 1978; Goldfoot 1977) are dependent for their expression upon maturation in a complex social and physical environment.

All of these skills, and behavioral sex differences too, are obviously dependent upon the biological potentialities and capacities of the animals. Carefully controlled deprivation research may allow us to determine some contributing factors in the development of behavior, but it cannot separate innate from learned behaviors because, as we have seen, such a distinction and such categories are meaningless, a conclusion that consideration of the deprivation studies reported in this chapter helps to underline. We cannot simply subtract the environment or the experiential life of an animal and rear an infant in a "vacuum" or the absence of these qualities. A deprivation or isolation environment is, nonetheless, an environment; in fact, a very stressful one. If a behavior *does* appear in an isolate-reared infant, we still cannot rule out environmental influences or learning as an aid in its appearance, but we can, for example, rule out maternal handling as a necessary element, since mothers are not present. Conversely, if a behavior fails to appear in an isolate-reared infant, we can infer that aspects of the normal environment are important to the expression and development of the behavior, but we still cannot rule out innate factors as also influential within that normal environment.

Deprivation researchers today are aware of this, and their goal has not been to separate innate from learned, but to trace links between some aspects of the environment and the development of certain behaviors. The problem has been that an adequate knowledge of the behavioral and social parameters which are being altered in this type of experimental research, and an awareness of the complex web of causal relations involved, has tended to lag far behind, rather than to precede, many of the studies. This has sometimes resulted in the reductionist approach described by Fragazy and Mitchell in the quotation above, and in the "overkill" or sledgehammer research procedures commented upon by Mason.

Harlow himself has described with self-deprecating humor how he and his colleagues attempted to induce psychopathology through elaborate experimental techniques and mechanical devices (such as surrogate-mothers which catapult, violently shake, blast air currents, or thrust brass spikes into the clinging infant), while the infant monkeys growing up alone in their cages in the laboratory were developing into severely disturbed individuals without the benefit of any further experimental treatment. Speaking of their hard-gained understanding of the infant primate's needs, Harlow wrote: "Thus by this ingenious research we learned what had been totally obvious to everyone else, except psychologists, for centuries" (Harlow et al 1971: 540). Although this was no doubt said tongue-in-cheek and obviously does not do justice to psychologists, it is nonetheless revealing of a tendency in some experimental research to "leap before looking." Menzel (1968) has said that the conceptualization of a primate mother in terms of static properties such as texture (e.g. cloth versus wire) and presence versus absence of nipples, appears something short of complete. I would add that the conceptualization of primate male and female roles as simply passive versus active, based on studies of laboratory-reared rhesus; the perception of the primate mother-infant bond as that which is revealed in experiments where stressed infants cling tenaciously to mechanical devices designed to repel them; and the belief that the results of such research allow us to say we have resolved the nature of primate love and sex differences, represents a view of primate social life which is also somewhat less than adequate.

PART IV

SOCIAL ORGANIZATION
AND BEHAVIOR:
FEMALE AND MALE ROLES

Chapter FOURTEEN

Multi-Male, Multi-Female Societies

Japanese macaque
Macaca fuscata
These semi-terrestrial, temperate zone monkeys of Japan are shown in their heavy, winter coats. The larger male (left) grooms a female, who in turn grooms her infant.

INTRODUCTION

In this and the following two chapters I shall present descriptions of the life-ways of nine selected primate species, with an emphasis on the roles females and males play in both social and ecological aspects of group life. As these nine are intended to be neither a representative nor a random sample of the approximately 200 primate species recognized by taxonomists, it is important to know how and why I selected these few for description. Out of the very large number of primate species which exist in the world today, only a fraction have been the subjects of any primatological research, and an even smaller fraction have been well studied in both field and captive conditions. Since these chapters are an attempt to provide comparable information on many different aspects of group life, I chose only those species for which adequate data are available. In particular, I looked for species with information from a variety of in-depth and if possible, longitudinal studies for example, of Barbary, Japanese and rhesus macaques, common baboons and chimpanzees. I also attempted, wherever possible, to describe species with which I have some first-hand observational and research experience.

The necessity of making these choices results in the omission of many fascinating social systems from these three chapters. For example, some howler monkeys, or *Alouatta* species, may be found living in monogamous, polygynous *and* multi-male, multi-female groups, even in similar habitats (Bolin 1980). Hanuman langurs in northern India are typically found in multi-male, multi-female groups, but in southern India are more often seen in polygynous groups, probably due to variable environmental pressures. Prosimian species, which exhibit equally complex societies, are not represented in these chapters. Studies of spider monkeys and talapoins are just beginning to present us with tantalizing glimpses of unusual social systems. These and other types of primates will not be totally neglected, however, since short descriptions of social behavior in these species will be provided in other chapters.

For each of the nine species discussed, the description is drawn from the currently available information on the following aspects of group life, with particular regard to sex differences:

1. Physical description of the animal, including sexual dimorphism
   Distribution or range of the species
   Aspects of niche − diet, arboreality and so forth;
2. Synopsis of social organization − group size, composition, mating system, membership stability;
3. Subsistence factors − foraging, resource and predator defense, group movement patterns;
4. Ontogeny and socialization;
5. Social dynamics and patterns of interaction − relations within and between the sexes.

Since each description of a primate life-way is a synthesis of the reports of numerous researchers, a list of source studies for each species is provided at the end of the chapter.

MULTI-MALE, MULTI-FEMALE PRIMATE SOCIETIES

Multi-male, multi-female primate social systems, as noted in Chapter 4, have received much attention from scientists. Partly this emphasis has resulted from the biases of the early field studies, which concentrated on the easier-to-observe ground-dwelling, open country primate species, many of which live in multi-male, multi-female groups. However, this social system has also interested researchers because it is unusual in the animal world for a number of adult males to live in year-round groups with a number of females and their offspring.

Eisenberg et al (1972) have argued that truly multi-male, multi-female groups are not as common in primates as once believed. It is their opinion that what appears to be a multiple adult male composition of a group, may often be, in fact, one adult male and his sons, in various stages of development prior to leaving the group. The one fully adult male is probably the only mating male, and therefore in spite of first appearances, this is a polygynous mating system. Such a grouping pattern they have referred to as an age-graded male group. This interesting assertion is difficult to substantiate or disprove for at least two reasons. As I noted in the chapter on sexual dimorphism, sexual maturity and aspects of physical or body size development usually appear to *precede* full social maturity in male primates. Thus in a field situation, especially when a new group is first studied, it may be very difficult and somewhat arbitrary to decide which males are actually "adult" males. I would have to agree with Eisenberg et al that there exists a tendency to ascribe "adult" status to male primates with which one is unfamiliar and which may have not yet reached full social maturity. However, both females and males develop and change socially over all the years of their lives, and deciding on a "cut-off point" for adult status is difficult at best. The real test of the prevalence of age-graded male groups rather than multiple-adult-male groups can only come with a knowledge of kinship, since a father-son relation between males is postulated. Paternity is always hard to ascertain in primates, and knowledge of kin relations over more than one generation requires a five to ten year study for species of monkeys, and a ten to fifteen year study for apes. Such longitudinal research has been conducted on chimpanzees and Japanese macaques, two of the three species described in this chapter. In these two cases, the multiple adult males of the group are either unrelated, or in some cases, brothers. Thus, chimpanzees and Japanese macaques can be said to live in truly multi-male, multi-female groups, as Eisenberg et al recognized. Vervet monkeys, the third species described in this chapter, have not yet been studied long enough in field conditions to determine if what has been described as multiple adult males in groups, are actually one fully adult male and his sons. Eisenberg et al do suggest the vervet as one example of a species sometimes found living in age-graded male groups.

Whether one ultimately decides that vervet societies are best described as multi-male, or as age-graded male societies, it should become clear from the following descriptions that group composition and mating systems are aspects of social organization which can function in different ways, in articulation with other aspects of social life. This should serve to remind us once more that categories such as "multi-male, multi-female" societies, or "diurnal, arboreal omnivores" are our constructs, and that dissimilarities

217

within the category may be as important as those similarities which identify the category. Thus some readers may well find the differences between the three species described in each chapter to be more striking than their common traits.

THE LIFE-WAY OF JAPANESE MONKEYS *(Macaca fuscata)*

Description, Distribution, Diet

Japanese macaques are medium-sized, quadrupedal monkeys found only on the islands of Japan, where they range from 31° to 41° latitude. Males weigh an average of 14.6 kg. and females 12.3 kg. Adult males have larger canines, somewhat larger body size and more muscle and fur on the shoulders and hips than adult females. This is a moderate degree of sexual dimorphism that appears accentuated during the annual breeding season when some males develop a shoulder "cape" of fur giving the impression of even larger size. Both sexes have pink faces and pink genitalia, which redden to a near-scarlet color in the breeding season. Fur-color ranges from golden to brown and from green to gray, the differences not being related to sexual dimorphism, but to individual, matrilineal, seasonal, and age variability.

Although often thought of as a ground-dwelling species, Japanese monkeys are adept in the trees and are more appropriately described as partially-terrestrial, partially-arboreal. Their typical habitat is mountainous broad-leafed or montane forest and common items in their diet are nuts, buds, berries, shoots, leaves and bark (Itani 1956). They also eat insects, crustaceans and bird eggs, and can be classified as omnivorous in diet. Japanese macaques are one of the very few species of nonhuman primate able to tolerate a cold winter with deep snows; temperatures in their most northerly distribution falling to −5°C, with 2-3 meters of snowfall (Izawa & Nishida 1963; Izawa 1978). Also, given one troop's successful adaptation to ambient temperatures up to 37°C, and novel flora and fauna in their new home on the Texas-Mexican border, Japanese macaques have proven to be a very adaptable primate species.

Synopsis of Social Organization

Japanese macaque society is structured around sets of matrilineal kingroups which are closely associated with a small number of central adult males, usually unrelated. These kin groups are also loosely associated with a larger number of peripheral males of mixed ages, some being natal males in the process of leaving the group and others being unrelated males in the process of joining the group. Thus Japanese monkey groups usually are composed of many related, breeding-age females and several unrelated, breeding-age males, with a mean adult sex ratio of approximately 1 male to 4-5 females. In addition to bisexual groups, all-male groups often consisting of peers or siblings who have peripheralized together, and all-female groups consisting of matrilineages that have broken away from a larger group, and solitary males also are known to occur.

Breeding is highly seasonal in this multi-male, multi-female species and both females and males form several consortships for the purpose of mating. There is no evidence that paternity is recognized by these monkeys or that close bonds are formed with possible biological fathers. Rather, intense bonds are formed between mothers and offspring. Group membership is stable for females, who remain bonded for life with their matrilineal kin, but unstable for males, the strength of whose life-long tendency to roam we are only just beginning to fully appreciate (see Sugiyama 1976 and below).

Japanese primatologists have conceptualized the society of this species as a series of concentric circles, with central males, females and offspring at the core, peripheral females in an outer circle and peripheral males in the outermost circle. Although some Western scientists have suggested this structure is simply the result of provisioning artificial foods in a circle, this criticism may not do justice to the theoretical sophistication of the idea, in which concentric circles are simply a graphic illustration of an abstract concept in a social theory. My own studies have substantiated the notion that these social groups may have a central core and a periphery in a behavioral and interactive sense rather than a physical sense, and that central-peripheral tendencies in group members is a useful concept in the study of within- and between-group integration.

Subsistence Factors

Each Japanese macaque is an independent foraging unit. Mothers do not give their offspring food to eat, other than milk, although they are very tolerant of infants in their laps picking up scraps or eating from the same food source, and they are known to brush inedible or toxic foods from infants' hands. Infants for their part show a strong interest in what their mothers eat as early as two weeks of life, nuzzling the mother's mouth and tasting her food. They are known to survive without mothers from as early as three months, if "adopted" by another group member. By six months of age infants are well acquainted with the troop diet, although they may also continue to suckle until they are one or two years of age. Most Japanese macaques appear to monitor what the others around them are eating throughout their lives. Sex differences in the feeding habits of Japanese monkeys have not been documented, although the "protoculture experiments" (see Chapter 9) would indicate that adult males learn new food habits less readily than other group members. On the other hand adult central males may be a focus of attention such that their own feeding habits are easily learned by others. Sex differences in juveniles are less apparent, but young females are often reported to be innovators of new food-processing techniques and females do better in food-related problem-solving tests in the field (e.g. Tsumori 1967).

Leadership in group movement is difficult to determine in this species, but again adult central males often appear to be a focus of attention for at least some of the troop members. These males do *not* walk out in front of the foraging group, but instead various subsections of the troop maintain proximity around them. Thus the movement of the troop during foraging sometimes resembles the movement of an amoeba, to borrow a metaphor from Kummer (1971a). A subgroup consisting of females and young and a central male move out of the main body of the troop like a pseudopod, then the male often climbs to the top of a tree or other promontory and looks around while the others move around him eating, passing in front of him some distance, but seldom going far in advance or out of sight of him. Meanwhile other adult central males have climbed nearby trees and helped to form other "pseudopods" and the whole troop gradually catches up and flows around these males. They then descend the trees, feed for a while either in one spot or while walking, and the "pseudopod" may become apparent again and the pattern is repeated. Young males and females sometimes climb trees and look around during foraging also. It would be very difficult to determine how decisions concerning

219

group direction of movement are made. One could perhaps generalize; certain adult males, at least, act as a focus during foraging and these males pay a great deal of attention to what is happening around the troop as a whole, while nearby females and young monitor the location of these males and continue feeding. However it should be added that not all adult males perform these vigilance activities and some only sporadically, while some females never follow or focus upon such males.

From their positions of heightened visibility, such males are often the first to sight other groups of monkeys or potential predators and therefore the first to utter warning calls, and they are the only ones to perform long-distance tree displays. Japanese monkeys are not territorial, but males of neighboring groups quite often display to each other, swaying the tops of supple evergreens. Individual "stranger" males which approach a troop during mating season also display in this manner and troop males may respond with tree-shaking.

As I stated in Chapter 8, males, especially in the ground-dwelling Old World species of monkeys, are often said to perform a specialized group defense and protection role. To a certain extent this appears to be true in Japanese monkeys, although observations of actual defense against predators have been few (see Itani 1963). In a rare opportunity to see the Arashiyama West troop repeatedly defend itself against predators such as bobcats and coyotes, my colleagues and I noted that responses to predators were variable, depending on the nature of the predator and on the situation. Often the group as a whole would simply run in one direction after an alarm call was given or perhaps all climb neighboring trees. Thus, some "hasty" escape manoeuvre followed by strong cohesion was the major type of response. On some occasions adult males would react more slowly, or run a short distance and stop, with the net result being that they were between the predator and the rest of the troop. We occasionally received the impression that adult males in highly stimulating situations such as a troop alarm or being suddenly confronted with frightening, novel objects, would seem to "actually inhibit signs of emotional reaction" (Menzel 1966:137). On even rarer occasions alarm calls by other troop members appeared to be answered by central males coming to the source of the scare. Or if a watchful male in a tree gave the first alarm call, neighboring females and young might scramble into the tree with him, perhaps reinforcing the impression that they were relying on him in some sense.

On the other hand, some females and young were also watchful and occasionally the first to spot a predator and spread the alarm, and on the one occasion when a bobcat was seen to grab an infant monkey in its mouth, nearby monkeys of all ages and both sexes participated in chasing and mobbing the predator. From this, and descriptions in the Japanese literature I would suggest that adult males in this species may play a somewhat specialized defensive role which includes watchfulness or vigilance in particular, as well as a greater tendency to approach or remain close to a predator; however females and younger males are also quite capable of showing defensive behaviors. As described in an earlier chapter, all-adult-female groups, or troops led by females, have been recorded for this species. This rudimentary form of sexual division of roles during foraging is therefore not immutable, and while the males' activities are certainly useful to the females and young around them, the females should not be construed as ultimately dependent upon the males for resources, resource maintenance, or group defense.

Ontogeny and Socialization

It has not yet been determined whether Japanese macaque mothers handle their female and male infants differently; however this species does seem to be one where behavioral differences between the sexes begin to appear by six months of age (Gouzoules 1980). The literature often refers to the fact that young females remain close to their mothers and matrilineal kin, while young males, around the time of weaning, spend more time away from their mothers and come to show more and more interest in male peers and in the older males who frequent the periphery of the troop. Juvenile males of this species do engage in more locomotor play, a category characterized by the large muscle movements involved in acrobatic and rough and tumble play (see Chapter 12).

Somewhere between three and seven years of age most males peripheralize from their natal group, often making several temporary departures with playmates or sibliings before the final break. Since most male immigrants ("strangers") approaching groups are judged to be between seven and 10 years old, Sugiyama (1976) is of the opinion that many young males must spend several years living alone or in all-male parties. Such lone males are often sighted by Japanese villagers and are referred to as "hanarezaru." It is a difficult and slow process for males to enter new groups and yet they are often known to leave again three or four years after joining a new group. Sugiyama's conclusion is that a solitary, as well as a social life, is normal for male Japanese macaques. It should be noted that there are exceptions to this pattern and some males remain in their troop of origin (maintaining close bonds with their mothers and other kin), or troop of adoption, for many years. The mechanisms behind the solitary and social periods of a male's life, or solitary and social aspects of his temperament, are not well understood and offer great potential for future research.

Young Japanese macaque females, particularly those growing up in the center of the troop, are highly social animals, who both give and receive a large amount of affiliative, integrative behaviors, mainly interacting with their matrilineal kin but also within the larger social network of the troop. Thus by the time they reach sexual maturity these females have a very good knowledge of the other females of the group, with whom they will spend the rest of their lives. Juvenile peripheral females also interact frequently with matilineal kin, but in addition they may become familiar with peripheral males who are often in their vicinity and may be in the process of forming bonds with the mothers and aunts of the juvenile females. As adolescents, many females enter into a period of transition or instability when their mothers are more concerned with care of younger siblings and the subadults do not yet have offspring of their own. Once such a young female has her first infant, she often moves back into very close proximity to her mother, who may have borne an infant in the same birth season, and a very cohesive nursery group, often involving "babysitting" is formed.

Although mothers and offspring form very close bonds and it is the females who are largely interested in and responsible for the care of infants, some males at some times in their lives may adopt yearlings either temporarily or for life (see Chapter 12). Given that this is one of the more complete or extreme examples of male care documented in the nonhuman primate world, it is all the more surprising that the trait is individually, seasonally and "culturally" variable. The same males who live alone for years of their lives or who live in the group and appear indifferent to infants, have the potential to successfully rear infant monkeys. Short of adoption, many group-living males will occasionally carry, comfort, protect, punish, or play with young monkeys, and males during the social phases of their lives can be very gregarious.

Social Dynamics

Japanese monkeys have been the subject of some of the best known studies of dominance (see Chapter 7) and the conclusion of most researchers is that female and male dominance hierarchies should be considered separately. Certainly this species has a very complex system of power relations influenced by kinship bonds, alliances, coalitions, age, tenure in the troop, and so forth. In terms of dominance, as the Japanese primatologists usually measure it, by peanut or caramel tests (see Chapter 7), individual, adult central males generally rank over individual adult females. However while disputes *between* females (female-female) and disputes between males (male-male) are generally settled quickly and decisively, male-female disputes generally drag on and see-saw back and forth, especially as females attempt to enlist aid from other group members. When engaged in a confrontation with subadult and peripheral males in particular, a female will "back into" the vicinity of potential supporters and actively enlist aid while shrieking at the offending male. Should she receive support, she will rout the young male. Should she not, or should she run after him too enthusiastically and too far, he may turn suddenly and jump on her or chase her back to the invisible line that appears to mark the "group safety zone" for females. The young males are very cautious in such disputes, as their attacks or counterattacks often appear to arouse support for the female. While fully adult central males may attempt to dominate females with more impunity, even alpha males cease their attacks when confronted with a coalition of females or with an enraged mother defending her offspring. In fact, high ranking central males are less likely to enter into disputes with females than the more insecure, younger and/or peripheral males.

Japanese primatologists refer to most troops as having "leader males," these being the top-ranking central males who watch-out around the troop and break up long-lasting disputes within the troop (control interference, see Chapter 9). Most troops also have a top-ranking leader female, referred to in the literature as a "chief" or "nucleus" or "alpha" female. This role has been little studied, but some research that I have done indicates that such a female behaves differently from both adult central males and other adult central females. She does not generally appear to share the male's outward-from-the-troop orientation, but she is highly confident, intragroup oriented, interferes in on-going disputes, displays in trees, and courts estrous females like an adult male. She is a strong focus of attention for the other females of the group. She is also perhaps a more stable factor than the adult males who come and go from groups. Most of these alpha females appear to form a strong cooperative alliance with the alpha male, although groups such as the Minoo-B troop were led for years by an alpha female and her daughter without any adult central males. As I have already mentioned, the alpha female of the Arashïyama West troop remained in her position even when her allied alpha male left the group and was replaced by a male whose relationship to her was antagonistic. Alpha females usually rank just below alpha males and above all other members of the troop on food tests. Although young males spend most of their time with other young males, and females interact mainly with their female kin, and although adult males seldom form close bonds with other adult males, cross-sex friendship bonds and/or alliances are by no means rare.

Immigrant males appear to enter new troops by gradually forming affiliative bonds with the individuals, sometimes peripheral females, they encounter on the edge of the group. Reciprocal grooming, proximity, and mutual support appear to be important in the formation of these bonds and whole matrilineages may become bonded to one particular male. Sometimes new males attempt to enter a troop by force, perhaps attacking some of the group members, but they are driven away by coalitions of females and central males. As females are very hostile to strangers, successful entry into a new group seems to be largely the slow process of being around long enough and discretely enough until others become accustomed to one's presence, all the while gradually forming bonds with the females.

Another form of close and important cross-sex bond occurs between females and a related male (son, brother, mother's brother) who has remained in his natal group. Finally, but not least, males and females form temporary but intense bonds, known as consortships, during the mating season.

Synopsis

In this multi-male, multi-female society a female's major bonds are with her matrilineal kin, although she establishes and maintains bonds with all the members of her group. She bears and rears offspring and forms part of the on-going hub of Japanese macaque social life. Some females rely on certain males to support them in intragroup agonism and to watch out around the troop for danger. Some females also occupy positions of power or leadership in their groups. Male Japanese monkeys are more mobile than females and move not only from group to group, but also from social to solitary to social existences. During their social phases, males are bonded to females through kinship ties, through year-round friendship and support relationships, and/or through seasonal mating bonds.

THE LIFE-WAY OF VERVET MONKEYS (*Cercopithecus aethiops*)

Description, Distribution, Diet

Vervets, also commonly called green monkeys and grivets, are small monkeys in comparison to many other Old World species, with greenish-gold fur, black faces and long tails which are not prehensile, but which can be braced or wrapped for support around trees or around the mother's body by infants. Adult males weigh an average of 7 kg. and adult females average 5.6 kg. Both females and males have long, sharp canines and there are no marked color differences between the sexes except that in some parts of Africa the adult male has an egg-shell blue scrotum, red perianal region and white surrounding fur. In other parts of Africa, often in captivity, and on the islands of St. Kitts and Nevis, these bright hinder-end colors are lacking in males. Most researchers note that vervets are a relatively non-dimorphic species, especially compared to other semi-terrestrial, Old World monkeys.

Although they are the most terrestrial species within the genus *Cercopithecus*, and can live in drier areas than do other members of this genus, they are often found in forest and woodlands and can be classified as semi-arboreal, semi-terrestrial. This is a very difficult species about which to make any generalizations because of its extreme adaptability.

According to Struhsaker (1967c) vervets are the most widely distributed and abundant of all African monkeys next to baboons (the genus *Papio*). Vervets can be found from Senegal to Somalia, and from the southern edge of the Sahara to the tip of South Africa. The range of habitats they have been able to exploit is equally wide — the mangrove swamps of Senegal, the semi-arid savannahs of Kenya, the riverine forests of Zambia, the rain forests of Uganda, and urban parks in South Africa, being a sample of habitats where they have been studied. Vervets are opportunistic omnivores, eating a variety of plant foods as well as insects, bird eggs, crustaceans and lizards in some areas. A rather remarkable historical event concerns some vervets brought from West Africa as "boat pets" on the slave ships which sailed for the Caribbean in the 1600's. On the islands of St. Kitts, Nevis, and Barbados the monkeys which survived the journey either escaped or were released, and have successfully adapted to the local flora and fauna, reproduced and formed social groups highly comparable in organization to those in Africa, and all this despite continual efforts to eradicate them as a crop pest.

Social Organization

Along with their adaptation to a wide variety of habitats, vervets exhibit some variability in social behaviors (see Gartlan & Brain 1968). However they can be generally classified as living in multi-female, multi-male groups with a size range of about 7 to 53 members (average group approximately 20) and an adult sex ratio of 1 or 1.5 females to 1 male.

It is frequently said that mothers and infants are the focus of this society, and as the longitudinal data are published, it is becoming clear that matrilineal units are important structures in social life. Male mobility seems to differ from study to study; in some areas males are reported to transfer groups frequently, but in other areas groups are said to be "closed" or stable in membership. Many years of observations of Japanese macaque troops have shown that male emigration is one of the more difficult behavioral phenomena to document, and the question needs further research in vervets. Solitary or peripheral males are occasionally noted in Africa and all-male groups as well as one all-adult-female group have been reported from St. Kitts. Births are seasonal, but in contrast to some other seasonal primate species, mating is rapid, discrete and difficult to observe (see Chapter 10). Although males and females in these multi-male, multi-female groups do mate with different partners, consort bonds are not formed.

Subsistence factors

In many parts of their range vervets are found to sleep in trees in the forest at night and range out into more open areas to feed by day. Group movements are usually initiated by adults, and by females as often as males. In one study (Struhsaker 1978a) higher-ranking adults were said to be more prone to start group movements, while in another (Basckin & Krige 1973) mothers-with-infants were often the initiators. Adult vervet males are said to be very watchful or vigilant and often perform a "sentinel" role in which they sit very upright in an exposed location, closely monitoring the surrounding environment, and perhaps exposing their white chests. Interestingly there are no reports that such males are watched or act as foci of attention for the females and juveniles of the group. Perhaps this is related to the Galats' (1976) suggestion that the vervet male role is more diversionary (like the patas, see below) than defensive as open-country baboons and some macaque male roles are said to be.

When an alarm is sounded, disorganized flight or retreat is the general reaction and there is no grouping around the males. Predators such as pythons and crocodiles are occasionally mobbed by both females and males, although mothers with infants clinging are more prone to hang back (Lancaster 1971). Lancaster also reports that mothers may strip infants off their chests and sit them down next to a juvenile before running over to join in threatening a predator. It should be noted that in captive groups of vervets, adult males will attempt to defend members of the group against handling by researchers and frightened group members will cluster around adult males (Rowell 1971; and personal observation). How this relates to their lack of specialized predator defense in field studies is unclear but perhaps can be taken as another caution concerning the potential flexibility of sex roles.

Vervets are highly territorial in most areas of their range and both males and females take active part in territorial encounters, generally involving threats and displays rather than physical contact. There is some evidence that territorial behaviors may be more characteristic of males than of females. Firstly, only males show unassisted defense of a territory, and even in areas where males shift groups every few months, they defend the territories in which they reside, even against a group which was "their own" only a few weeks before!

Females, I should restate, live in one area and with their same matrilineal kin throughout their lives, and it is probable that what we call "territorial encounters" between two groups may be of a different nature for the adult females than for the adult males of these societies. In addition, it is never very clear how much of "sentinel" behavior, in this and other species such as Japanese macaques, is directed toward locating predators and how much toward locating other groups of conspecifics, particularly other males. Vervets are reportedly subject to much predation and are reknowned for giving different alarm calls depending on whether the predator is aerial (eagles, etc.), terrestrial (leopards, etc.), or a snake. Struhsaker (1967b) had reported a sex difference in vervet alarm calling such that only adult females and juveniles gave these discriminating vocalizations, while adult males gave one general purpose "threat-alarm-bark," whether a predator had been located, or simply another vervet group. This would indicate that males are indiscriminantly vigilant against both predators and other groups of vervets. However, more recent research by Seyfarth and Cheney (1980; Seyfarth et al 1980a,b) found that *both* adult males and females gave (and responded appropriately to) different alarm calls for different predators.

In sum, vervets may be an excellent example of the point made in Chapter 9, that in some primate species males may play an important role in predator protection and territorial defense but females in these same species can also protect themselves against predators and defend their group and its resources. Although this should be self-evident, too often theories of the evolution of primate social life are founded on the primacy of these male roles and the ultimate dependency of females upon these male functions.

Ontogeny and Socialization

Vervet females alone are concerned with the rearing of the young, as males are said to be not very interested in infants, although tolerant of infant contact. One is sharply reminded of Redican's theory (1976 and see Chapter 12) that the amount of male care is correlated with the amount of maternal tolerance; in this species males are chased and harassed by coalitions of females should they even be suspected of disturbing an infant.

Females are strongly and overtly attracted to infants, especially newborns, which act as a focus of attention. Genital inspections are common, although it is unknown if infants are handled differently by sex. By the time they reach their juvenile years, behavioral sex differences are apparent in juveniles, with young females spending much of their time with adult females and infants (this is the species which is most renowned for aunting), and young males spending much of their time engaged in vigorous locomotor activity, especially play (see Chapter 12).

Struhsaker (1971) has suggested that vervet young are more precocious and faster maturing than other species of the Old World monkey family, *Cercopithecidae*. Vervet infants are weaned earlier, groomed less and become independent earlier than other cercopithecines such as macaques and baboons. Although vervet sexual maturation (3.5 years in the female, 4.5 years in the male) is comparable to that of rhesus monkeys, full physical and social maturity occurs at an earlier chronological age in vervets than in macaques or in baboons (Struhsaker 1971) Thus he has hypothesized that vervets are a shorter-lived and a faster maturing species. For whatever reason, the period of intensive socialization appears to be somewhat shortened in vervets when compared to other Old World monkeys.

Adult males, who keep their distance from vervet infants, are more interested in the young by the time the latter reach their juvenile years, and in some study groups adult males will play with and groom juvenile males in particular. Adult males of the group are sometimes described as being socially aloof, particularly from each other as they "do not seek each other's company, nor do they antagonize and fight" (e.g. McGuire 1974). However, adult males are not as aloof from the females of the group as are patas males for example (see next chapter). Adult females of the group are very gregarious and form the hub of the social system.

Social Dynamics

Dominance relations, by whatever measure, do not appear to be as clear-cut or as important in vervet life as in some of the baboon and macaque species. Some researchers report themselves almost unable to determine dominance hierarchies because agonistic behavior is so rare (e.g. McGuire 1974; Rowell 1971), while others stress that male and female hierarchies may operate separately. For example, Lancaster (1971), in field feeding tests found that subordinates would wait their turn while dominants fed, but that representatives of female and male hierarchies fed side by side without regard to rank. However, according to other researchers, it is common for some of the adult females to rank over some or all of the adult males of the group. This means, as Rowell (1971) has noted, that the first ranking adult male of a group does not rank above all the females, in contrast to the "alpha-male role" prominent in macaques and baboons.

In daily life, adult and juvenile females are reported to engage in more intragroup agonistic encounters than do adult and subadult males. Mothers-with-infants may be more aggressive than other adult females and maternal defense of young does seem to be important in this species and is often the basis of the frequent female coalitions against males. Strusaker (1967b) suggested tentatively that males may occasionally appear to attempt to disrupt or terminate a fight by rushing toward the agonistic encounter giving a barking vocalization, which is reminiscent of "the control male role" in some macaques. But it is also said that adult females are far more likely to intervene in fights involving their juvenile kin. Studies of a captive group (Fairbanks et al 1978) suggested there may exist a female and a male "control" role, involving the suppression of aggression as well as receiving avoidance and appeasement. There was also evidence that females may compete to groom the alpha female.

The relationship between adult females and adult males is seldom discussed in the field literature, other than in references to the common "ganging-up" of females against males. However the studies from captive colonies indicate that adult males may be very sociable with, and become quite bonded to, some of the adult females, grooming and sitting with them often. Rowell (1971) argued that male-female interaction patterns in vervets are far more emancipated from gonadal hormonal control than has been described for baboons and macaques. There are no sharp changes in female-male relations during mating seasons and females and males do not groom, sit together, initiate interactions with each other, avoid each other or even copulate more often during the fertile phase of the female's monthly cycle.

THE LIFE-WAY OF CHIMPANZEES *(Pan troglodytes)*

Description, Distribution, Diet.

Chimpanzees are large, black-haired, knuckle-walking apes who spend about 70-80% of their time in the trees, sleeping and eating arboreally, but traveling from place to place along the ground. Many people are only familiar with the small, fluffy, round-faced infant chimpanzees seen so often in circuses and shows, and are somewhat surprised by the large and muscled body, craggy individualistic faces, and thin or grizzled hair of many adult chimpanzees. The mean weight of an adult chimpanzee males is 49 kg. and of the adult female is 43 kg. Males also have larger canines than females, and as I noted in Chapter 5, Hamburg and McCown (1979) have argued that the "anatomy of fighting" is strongly sex differentiated in the chimpanzee, so the "fighting strength" of the male is much greater than in the female. This would involve differences in the canine jaw muscle complex, neck muscle size, orbit size and many cranial dimensions.

The range of chimpanzees corresponds roughly to the distribution of tropical rainforest and surrounding savannah woodland from the west coast of Africa through to western Tanzania, or across what is known as the equatorial belt of Africa. Thus, chimpanzees can be found from longitude 15° west to 32° east and from latitude 12° north to 8° south, although the distribution of populations tends to be patchy, rather than even. Chimpanzees are found in forests (lowland, riverine or semi-deciduous) or in complex habitats of forest edge, woodland and savannah. There has been some disagreement among specialists, not yet settled, as to whether chimpanzees are basically, or originally, thick forest dwellers, or forest edge or savannah woodland dwellers. They may be found in all of these habitats today, and also often in mountainous or hilly terrain. Chimpanzees are omnivorous, eating fruits, seeds, nuts, leaves, flowers, honey, insects, eggs and meat, although the majority of their diet is vegetable. Chimpanzees sleep at night on platforms or nests built in the trees by bending or breaking and arranging leafy branches. Usually each individual builds a new nest every evening, except for dependent offspring, up to the age of about five years, who sleep in their mothers' nests.

Social Organization

The social organization of no other primate species which has been as well studied as the chimpanzee, has been so difficult to determine. At first it appeared to researchers that the only social grouping was at the level of small foraging parties, and even these were noted to be temporary and fluid rather than stable social groups. Next a team of Japanese primatologists studying chimpanzees described a larger grouping or higher level of social organization which was variously named the "unit-group" or the "regional population." Soon after, a similar level was described for the Gombe population of chimpanzees studied by Goodall and her team of researchers, and was named the "community."

The community (or the unit-group or the regional population) of chimpanzees is usually described as a bisexual social group consisting of about 15-80 individuals, with approximately equal numbers of adult females and males. Although the community is a generally stable social grouping, this was at first difficult to determine because of two factors: firstly the community never travels together cohesively like the social groups of most monkeys, and secondly, adolescent and adult females may transfer or be mobile between communities at some points in their lives. Chimpanzees generally forage and travel together in small temporary associations called parties, which are subsets of the larger community. Several different "types" of these smaller, temporary parties are described in the literature: all-male parties; nursery parties (adult females and offspring); bisexual parties of adults only; matrifocal parties (one adult female and dependent offspring alone); mixed parties, etc. Both females and males may occasionally travel alone. It was this complex pattern of associating in small parties of fluctuating membership which made the larger community difficult to recognize until after many years of research.

The nucleus of the community or unit-group is described as a cluster of strongly bonded adult males who are closely associated with a large range and with a number of mature adult females and dependent offspring, and loosely associated with a number of younger females. Older females with dependent offspring generally do not travel widely, instead they range as a matrifocal unit over a small core area within the community's larger range. Estrous and nulliparous females however travel widely, either in association with parties of community males, or they may even move temporarily or permanently into the range of neighboring communities. Females generally mate with several males, and outbreeding is probably one result of the mobility of females between social groups. The matrifocal unit is an important structuring principle in the sense that offspring are dependent on their mothers and half-siblings through the maternal line for many years. However, the transfer of females between communities means that larger matrilineal groups spanning several generations, such as occur in macaques, probably do not arise. Rather than female kin bonding most strongly, it appears that in chimpanzees, related males, such as "brothers" may be strongly bonded throughout life. (See Social Dynamics).

Thus, the commonly accepted picture of chimpanzee social organization is of large bisexual communities, which consist of smaller, fluctuating foraging parties, and between

228

which estrous females transfer. The males of the community have a large range while the non-estrous adult females and their young (matrifocal units) have smaller individual ranges. This pattern can be diagrammed after Wrangham (1979a) like this:

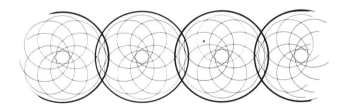

Larger circles are the community ranges and smaller
circles are the matrifocal unit core areas.

Recently, Wrangham suggested an alternative interpretation of chimpanzee social organization. It is his opinion that lactating females, or matrifocal units, disperse themselves geographically and evenly into small core areas which contain enough resources to support them and their offspring, in relation to other matrifocal units, but independently of adult males. A group of males collectively occupy a larger range which overlaps the core areas of several matrifocal units. This group of males attempts to defend their range against encroachment by neighboring males or neighboring groups of males. Other primate species, such as nocturnal prosimians and orangutans exhibit somewhat similar patterns, except that only one adult male, rather than a group of males, occupies a large range overlapping the smaller core areas of several matrifocal units. Wrangham diagrammed this suggested model of chimpanzee social organization as follows:

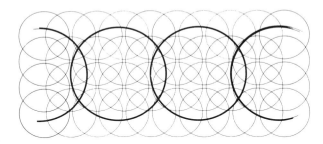

Larger circles are male ranges while smaller circles are female core areas.

Thus the major innovation of Wrangham's model is the suggested independence of female and male ranges and social bonds, as opposed to the interdependent bisexual community described by other researchers. Such a sexual separation of ranges is certainly not apparent when mixed-sex parties travel together as they often do. However many aspects of chimpanzee social life have proven to be not readily apparent. In addition, Wrangham argued that it follows that the "community" is a concept applicable only to males, while females are distributed individually and geographically in small matrifocal units. Pusey (1979) has

229

taken exception to the view that females do not belong to the community and has noted that there are not yet enough data to determine which of the two models of chimpanzee social organization is closest to the real situation. One thing that is clear from a reading of the literature on chimpanzees is that, with the exception of the mother-infant studies, the overwhelming majority of attention on the part of the researchers has been devoted to the study of male social behavior, and our overall impression of chimpanzee social life is based on that of the males (one exception is Wrangham 1979b). In addition, the very complexity and fluidity of social organization in this species means that even after nearly 20 years of continuous study, many fundamental questions still remain to be answered.

Subsistence Factors

In recent times chimpanzees have suffered predation only from humans, and possibly leopards, in most parts of their range. When a party of chimpanzees is frightened by humans, the adult males may seem temporarily more visible, but generally all the individuals in the group flee together. There is no evidence that adult males in particular exhibit a special tendency to defend themselves and/or other party members against predators.

During the first decade of chimpanzee studies, the species was described as non-territorial and the relations between communities as peaceful. However in the past ten years fierce encounters between males of different communities have been described at the Gombe study site, and males at the Japanese study site (Mahale Mountains) further south are also described as antagonistic to outsider males (see Chapter 6 for details).

An exception is the Budongo Forest study area, where chimpanzee communities are still described as associating peacefully and "stranger" males are said to be accepted into the group (Sugiyama 1973). Researchers at Gombe and in the Mahale Mountains regard the antagonistic interactions between males of different communities as territorial encounters. Since, as I reported earlier, chimpanzees travel and forage in small parties the membership of which fluctuates from day to day, there is no larger community group movement, or initator of group movement. Within the small parties there may appear to be temporary leaders of movement, but nearly anyone may play this role at one time or another and most of the time party movement does not appear to be "led." All-male parties and all-adult parties tend to travel more rapidly and more widely and may be the first to locate a new food source (e.g. fruit which has just come into season) to which they attract other parties with their excited food calls.

Chimpanzees sometimes share food and while the sharing of meat has received the most publicity, vegetable foods are more frequently shared, particularly when the food items are prized, scarce, or difficult to process (McGrew 1979). Mothers especially share with their offspring, and younger individuals usually receive from older ones, although "begging" and giving of food between adults of both sexes is also observed. Food-sharing has very occasionally been reported in other species of non-human primates, but chimpanzees are unique in the extent to which they have been seen to share. McGrew argued that there exist differences in diet between the sexes. Males do most of the hunting of mammals and although they may give some to females and other males, eat most of the meat themselves. Females, on the other hand, do most of the "insect gathering" or ant-dipping and termite-fishing and eat many more insects than do males. McGrew used both observational data and analysis of faeces to support his contended sex difference in diet.

It is in the foraging for insects and not the hunting of mammals that chimpanzees have been seen to use tools, and McGrew also argues that females are superior to males in manual dexterity and tool use. While insects are largely unrecognized in Euroamerican cultures for the highly concentrated source of amino acids, lipids, and essential vitamins which they contain and for the regularity with which they are eaten by many animals and most humans, the hunting of mammals by chimpanzees struck an immediate chord when it was first reported. Particularly fascinating were the descriptions claiming cooperative hunting, usually by males apparently acting in collaboration. More recently some researchers have suggested that chimpanzees may only give the appearance of cooperation during hunting (e.g. Busse 1978) but others continue to stress the significance of cooperative hunting and sharing of food. In this respect, Bygott (1979) notes that male chimpanzees have developed the greatest primate ability to live in peaceful proximity, and to take collective action against intruders and to cooperate in the hunting of mammalian prey of all male nonhuman primates. McGrew (1979), however, stresses the significance of female foraging activities which involve the use of tools and the common sharing of foods with offspring.

Ontogeny and Socialization

Perhaps the most striking aspect of chimpanzee ontogeny and socialization is the length of dependence on the mother. Pregnancy lasts eight months, nursing lasts five years and offspring up to eight years of age may die if their mothers die and leave them orphaned. Chimpanzee mothers then are the primary and almost sole caretakers of young. Adolescent siblings, female or male, may attempt to care for infants and juveniles, particularly if the latter are orphaned, but such care is always limited and does not often save the lives of young without mothers. Adult males may be very solicitous of young chimpanzees, carrying them, protecting them, and playing with them. However adult male chimpanzees are also very volatile animals (see below) and most chimpanzee mothers are very careful and very restrictive about leaving their young in the presence of adult males. Although other community members show overt curiosity about new infants, they are usually prevented from touching them in the first half-year of life.

Chimpanzee infants, then, form an intense and exclusive bond with their mothers, a bond which for males may last well into their own adulthood and until the death of the mother. Infants are not weaned until shortly before the birth of the next offspring, four or five years later; at this time they enter a juvenile phase in which they are still highly dependent on the mother, but no longer receiving milk. Although Goodall (1968b) suggested that there are no obvious differences in the behavioral development of male and female infants, she added that male infants tend to engage in somewhat more rough and tumble play, more aggressive display and are sexually more precocious. Chimpanzee females may begin to show small sex skin swellings, indicating estrus, at around nine years of age, but menarche does not normally occur until 11 years, and birth of the first infant is delayed until 13 or 14 years of age. Thus there appears to be a period of adolescent sterility in chimpanzee females. Around the time of sexual maturity females may begin to travel between communities, and bonds with the mother may be loosened, while new bonds are formed with consort males and finally with their own offspring.

Chimpanzee males may first produce sperm at nine years of age, but they are not considered to be "socially mature" for another four years, that is until they are 13 or 14 years old. According to Goodall (1975), this is a very difficult maturational phase for males, who are sexually, but not fully physically or socially developed. They begin to seek the company of adult males, but are very low ranking and must be very cautious, traveling and sitting on the periphery of adult male parties. At this time they also begin to attempt to assert dominance over adult females and often swagger and display in their presence. Close, and mutually supportive ties with the mother are maintained. Although the young male travels increasingly with all-male parties instead of his matrifocal party, he and his mother are still in the same community, unlike many adolescent females who move away and live at some distance from their mothers.

Social Dynamics

One of the most notable aspects of chimpanzee social life is that adult males are often described as being more strongly or closely bonded to one another than adult females are to each other. I have pointed out that more study has been devoted to males than to females, and one could argue that *stability* of male community membership is largely what is meant here by strong and close bonds. However in addition to being permanent group members, adult males of a community associate frequently and closely with one another, grooming and foraging together, sleeping in similar areas and simply being very gregarious. Females, on the other hand, have been portrayed as closely bonded to their mothers and then to their own dependent offspring, but seldom as associating closely with other adult females. One caveat to the foregoing is that only one case of adult female sisters is known from field studies, and these two sisters are described as very closely bonded (Goodall 1975).

Although strong bonding between adult males is quite unusual in the non-human primate world and should receive our careful attention for that reason, it would probably be more accurate to describe adult chimpanzee males as exhibiting a different rather than a stronger pattern of social bonding than females, as the former interact most often with peers and the latter with kin of various ages and sexes.

Aggression in chimpanzees has been described in some detail in Chapter 6 so here I will only reiterate that chimpanzees vary from being described as very aggressive to quite tolerant and peaceful in their intracommunity life. The charging display of the adult male is most impressive and receives a great deal of attention from other group members and researchers alike. Display may be initiated by males when parties reunite, when fighting over provisioned food, during torrential downfalls of rain, when individuals seem to be "frustrated," or sometimes for no apparent reason. One receives the impression that chimpanzee males can be quite mercurial, patting infants on the back one instant, dragging them across the ground as part of a charging display the next, and then comforting the shrieking mother of the infant almost before the display has ended (Goodall 1971b). Bauer (1979:397) describes charging displays as involving "rigorous and exaggerated locomotion with hair erect, occuring with a number of display elements, such as branch-waving, dragging, flailing, throwing and stamping." Nishida's description is very evocative:

It sometimes occurs that an adult chimpanzee suddenly goes wild without any apparent reason. He may suddenly begin to run about. He may slap at the ground, beat tree trunks with both hands, pick up and throw fallen leaves or sticks, roll stones, try to drag a large fallen tree, stamp on the ground or tree buttresses, and/or climb a tree very rapidly (1970:61).

According to Goodall (1975), Bauer (1979), and Bygott (1979), displays may function to help a male assert, reassert, and achieve dominance rank. Other community members generally stay out of the way of and appear frightened by a displaying male, especially a male in his prime.

Bygott (1979), and Riss & Goodall (1977) have done the most extensive work on dominance relations in Gombe chimpanzees and they report that certain males may use displays in a very manipulative manner, only displaying at opportune times, for example, when a rival is temporarily isolated, when a supportive brother is present, or when higher ranking males are absent. Generally speaking, adult males are dominant over adult females in the sense of priority to provisioned food and/or the direction of agonistic signals. However support relationships do exist. For example, sons defend their mothers, and older adult females do not tolerate displays from adolescent or young adult males and may chase or attack them. Adult males appear to account for most of the agonism seen in chimpanzee social life, as Bygott found that 90% of all agonistic encounters involved at least one adult male, that 39% of male attacks occurred at reunions of foraging parties and that 25% of attacks accompanied excitement during the consumption of popular foods, provisioned or hunted. Goodall (1975), does note that there are individual females who perform charging displays with the verve of the male, and there are males who show the anxiety in social situations usually more characteristic of females.

Bygott's study (1979) found that male dominance rank is *not* correlated with aggressiveness as measured by frequency of attack, nor with success in obtaining and eating meat, nor with copulation frequencies. Goodall (1975) adds that it is not correlated with the size and weight of the male either. Although Simpson (1973) earlier had suggested that the amount of grooming received correlated with dominance rank, Bygott pointed out several problems with the former study and states that age is a far more significant, and uncontrolled, variable in Simpson's grooming data. In fact, Bygott is of the opinion that age is a far more significant prediction of all interactions in males than dominance rank, and suggested that each male rises to his own particular peak in rank during his prime and then falls in rank during his post-prime.

Until recently, all reports on chimpanzee sexual behavior noted the "promiscuity" of mating and the lack of possessiveness concerning partners (e.g. Goodall 1968; Nishida 1970; Sugiyama 1972; Reynolds & Reynolds 1965; Kortland 1962). In fact line-ups of males waiting to mate with an estrous female without agonism were reported and certainly caught the public's imagination! However Goodall (1968) did note that although mature males were often tolerant of younger males copulating in their presence, low-ranking males were careful and discrete in their mating, and recently the reports from Gombe state that some males do attempt to isolate and monopolize estrous females by "forcing" the females to follow them away from other chimpanzees (Goodall et al 1979; P. R. McGinnis 1979).

The relative frequency of "consort relationships" is not clear as most of the data are inferred in retrospect from simultaneous absences from the provisioning area, rather than from direct observations. McGinnis's description of consorts would imply that they result almost solely from the male's ability to intimidate the female into following him although he found no correlation between consort initiation and dominance rank of males.

In some contrast, Tutin (1975, 1979a,b) stated that female cooperation is essential to consorts and that females do exercise choice of mating partners. She distinguished consorts from attempts by high-ranking males to monopolize females. In her opinion females choose males according to social and caretaking abilities rather than high dominance rank. However she notes that the highest ranking male has more success at disrupting or preventing copulations between an estrous female and other males. Therefore Riss and Goodall (1977) suggested that higher reproductive success is the "reward" for the energy expended in becoming the rop-ranking male.

Differing sharply with the Gombe researchers, who describe aggressive attacks as a rather common occurrence and dominance hierarchies as easily determined, Sugiyama (1972, 1973) stressed the rarity of aggressive encounters and the unimportance of dominance relations in male chimpanzees, and he draws a sharp distinction between provisioned and non-provisioned communities. He stated that chimpanzee society is based on tolerance, active cooperation and mutual assistance.

In reviewing all aspects of the behavioral literature on chimpanzees, one is struck by the strong diversity in the descriptions given by different researchers, and even by the same researchers at different times. Do they reflect differences between researchers? Between chimpanzees? Differences in the chimpanzees over time? Differences in the prevailing climate of opinion as to how our nearest primate relatives *should* behave? Or simply improvements in observation techniques?

While I cannot offer answers to these questions, two aspects of chimpanzee research do seem apparent to me. One is that the role of females in chimpanzee social life, apart from the mothering of dependent infants, has not yet been adequately studied or considered. And the second is that scientists and public alike have been too quick to extrapolate from chimpanzee data, still in the process of being collected, to generalizations about all of primate and especially human behavior. The instant and almost compulsive drawing of facile conclusions about human nature from the bits and pieces of studies of these animals has done little to further an objective understanding of chimpanzee behavior let alone our own.

REFERENCES

Japanese Monkeys (*Macaca fuscata*)
Alexander 1970; **Alexander and Bowers 1967**; Fedigan 1976; Gouzoules et al 1975; Hazama 1964; Imanishi 1957, 1960; Itani 1956, 1958, 1959, 1963; Itani et al 1963; Izawa 1978; Izawa and Nishida 1963; Kawai 1960, 1965a,b, 1967; Kawamura 1958, 1959; Koyama 1967, 1970; Menzel 1966; Nishida 1966; Norikoshi 1971; Norikoshi and Koyama 1975; Sugiyama 1976; Tokuda 1961; Tsumori1961; 1967; Yamada 1957, 1963, 1966, 1971.

Vervet Monkeys (*Cercopithecus aethiops*)
Basckin and Krige 1973; Bramblett 1973, 1975, 1976, 1978; Fairbanks 1980; Fairbanks et al 1978; Fedigan 1972b; Galat and Galat-Luong 1976; Gartlan 1968; Gartlan and Brain 1968; Lancaster 1971, 1976; McGuire 1974; Poirier 1972b; Rowell 1971, 1974b; Seyfarth & Cheney 1979; Seyfarth et al 1980a,b; Struhsaker 1967a,b,c,d, 1971.

Chimpanzees *(Pan troglodytes)*

Bauer 1975, 1976, 1979; Busse 1978; Bygott 1972, 1979; Goodall 1965, 1967, 1968a, b, 1971a, b,1973a, b, 1975, 1977, 1979; Goodall et al 1979; Halperin 1979; Hamburg and McCown 1979; Izawa 1970; Kano 1971, 1972; Kawabe 1966; Kortland 1962; P.R. McGinnis 1979; McGrew 1977, 1979; Nishida 1968, 1970, 1973, 1979; Nishida and Kawanaka 1972; Nishida et al 1979; Pusey 1979; Reynolds 1968, 1975; Reynolds and Reynolds 1965; Riss and Goodall 1977; Sabater-Pi 1979; Simpson 1973; Sugiyama 1968, 1969, 1972, 1973; Teleki 1973a,b; Teleki et al 1976; Tutin 1975a,b; Wrangham 1974, 1979a,b.

Chapter **FIFTEEN**

Polygynous Societies

Hamadryas baboon
Papio hamadryas
Egyptians worshipped these terrestrial monkeys from the arid regions of Northern Africa. A female, with sex-skin swelling, huddles behind the much larger, breeding male.

Before non-human primates were ever studied in the wild, they were all believed to live in polygynous societies dominated by single adult male "despots." Unfortunately, social groups of animals consisting of one adult male and two or more adult females have been dubbed, misleadingly by ethologists, "harems." Thus, the single adult male of the group often has been called either the "harem master" or the group despot. Criticisms of these terms, as reflecting misconceptions of the social structure of polygynous mammals, have been made by numerous workers. For example, from a review of the literature on sexual dimorphism in mammals, Ralls (1977) noted that in ungulates (hoofed mammals) and pinnipeds (seals and walruses) females may travel rapidly from area to area, with grouping patterns resulting more from the gregariousness of the females than the efforts of the territory-holding males to restrain them (see also Peterson 1968). Ralls concludes from this description and from the denoted dictionary definition of "harem" that the term is only properly applied to stable groups of one dominant male and several adult females, such as those found in hamadryas baboons. She suggests that long-term group stability and male control are the distinguishing criteria for use of the term harem. Even here, I would suggest that the connotations of the word, especially those of "enslaved females" and male autocracy, are always inappropriate to the dynamics of the roles which females and males play in the maintenance of primate societies, and that social groups of one adult male and several adult females would be better described in less anthropomorphic terms.

The first wave of primate field studies in the 1960's (see Chapter 4) led ethologists to believe that polygynous groups might be quite rare in primates, as only three species — the patas, the gelada and the hamadryas baboon — were found to exhibit this pattern. Thus, in one of the early reviews and syntheses of primate field data, Crook and Gartlan (1966) suggested that uni-male, multi-female groups in primates represent a special and relatively rare adaptation to harsh, semi-arid habitats with sparse food resources. But, field work in the late 1960's and the 1970's, on a greater variety of primate species, shows that polygynous social systems are not uncommon and occur in a variety of ecological settings. If one is seeking a pattern to the occurrence of this type of social system, it would seem, from the currently available information, to be most characteristic of arboreal primate species of certain taxonomic families or genera. Most of the species of the leaf-eating Old World monkey family Colobidae, including langurs and colobus monkeys, and all of the African, tree-dwelling Cercopithecus monkeys, live in uni-male, multi-female groups.

However, the ground-dwelling hamadryas baboons remain the first and the most widely known of the polygynous primate societies. Kummer (1968, 1971a,b) and his students (Kummer et al 1970, 1974) conducted many years of justifiably renowned observational and experimental studies on hamadryas baboons, both in Ethiopia and in captive settings. From Kummer's descriptions and experimental findings, the hamadryas baboon to many people, would seem to be the one primate species which might accurately be said to live in male-controlled "harems." Hamadryas males are described as kidnapping very young females from their natal groups and caring for them until they are old enough to become sexual mates. All females in the group are said to be herded together rigourously, and if necessary, aggressively, by the adult male. Experiments in transplanting hamadryas females into anubis baboon groups (multi-male, multi-female societies) and anubis females into hamadryas societies, demonstrated that hamadryas females do not remain

238

close to one particular anubis male since they are not being herded, but that anubis females can learn to be herded by hamadryas males. From this and other experiments and comparisons, Kummer concluded that hamadryas males are "responsible" for the form and maintenance of hamadryas society, since their herding of females appears to be what holds reproductive units together.

Complementary experiments in which anubis and hamadryas males are individually transferred into different social systems, particularly to determine if hamadryas males would continue to herd females outside the context of their own social system, were not to my knowledge conducted. It might be impossible to perform such experiments without severe fighting, and it might well be that "herding" is an inflexible trait in hamadryas males. However the lack of information on how males behave when removed from their own societies and transferred into a different type of system has always left me wondering about the impact on the individually transferred females of the rupture of their social bonds and their introduction to an unfamiliar environment. That is to say, these transplantation experiments may demonstrate elegantly one way in which bonds between a female and a male can be established and disrupted, but I am not certain how much they reveal about the underlying structure of on-going life in undisturbed reproductive units. Hausfater (1975) for example, suggested that the importance of females to hamadryas society has been underestimated due to a concentration on the behavior of the adult males.

Kummer had concluded that the several adult females in a reproductive unit of hamadryas baboons show little to no evidence of bonding to one another. Experiments in removing the single adult male from a group showed that neighboring males then "took over" the females, but whether one or several females together, is unclear. This is in contrast to the gelada, for which Kummer (1974) demonstrated that female-female bonds are essential to group formation and group maintenance (see also below).

It was a surprise therefore when a colleague, A.M. Coelho, showed me cohesive groups of hamadryas females without males in a captive setting. These females herded together in hamadryas-type social groupings in the absence of any adult males. Similarly, Stammbach (1978), a student of Kummer's, demonstrated experimentally that all-female groups of hamadryas, in the absence of adult males, will develop a social group similar in structure to the "natural" uni-male, multi-female groups of the species. The group is described as focussed upon a central, high-ranking female, rather than on an adult male, and female bonding and friendship are important. From this research, Stammbach concluded that the characteristic hamadryas social system is not merely a product of the male's herding behavior, but that females and their bonding tendencies also contribute substantially. Recently Kummer (Bachman & Kummer 1980) acknowledged that females should not be regarded as mere possessions that males conquer and herd, without regard to the females' preferences.

Nonetheless, hamadryas society does still appear, in many respects, to be the most male-focussed of the polygynous primate societies thus far described. The importance to hamadryas of female-female bonding is now being argued, and long-term observations of the ontogeny of undisturbed groups would help us to understand the dynamics of female and male roles in this social system. I would argue that the best explanation for most polygynous primate societies can be found in the suggestion by Eisenberg et al (1972) that uni-male, multi-female societies in primates are the result of a male attaching himself to a group or "core" of bonded, possibly related, females — although hamadryas might prove to be an exception.

Theoretical considerations of the evolution of mating systems in animals have led to arguments that monogamy is usually most beneficial to the female, because she and her offspring receive exclusive male attention, while polygyny is most beneficial to the male because he can inseminate a number of females (for example see Orians 1969; Downhower and Armitage 1971). Pursuing this line of thought, Downhower and Armitage argued that in polygynous societies females will be very competitive and aggressive to one another, while the male will attempt to recruit females and attempt to pacify interactions between them. They suggest that this is true in marmots and many avian species. However it is not at all an accurate prediction of relations in primate polygynous societies, where the females are often closely and affiliatively bonded and may well be female kin groups.

As shall be partially demonstrated in this chapter, primate species which live in polygynous groups also typically show the following characteristics:

1. closely bonded adult females, who will readily coalesce against the male;
2. a somewhat peripheral or socially aloof reproductive male;
3. strong intolerance by the reproductive male of other, potentially reproductive males;
4. leadership shown by at least some females in many aspects of group life, while the adult male shows an outward-from-the-group orientation;
5. some turnover in reproductive males, so that during his life-cycle a male usually spends some years in heterosexual groups, some years in all male groups, and possibly some years as a solitary individual.

All of these features would appear to confirm the contention that polygynous primate societies are not simply the result of individual males bringing together and maintaining harems of females. Rather they result primarily from individual males forming attachments to groups of females. Although males are competitive with other males in such social systems, they may be quite gregarious in certain situations, forming strong male-male bonds in the all-male groups, and taking part in higher levels of social organization, as in hamadryas and gelada societies.

THE LIFE-WAY OF PATAS MONKEYS (Erythrocebus patas)

Description, Distribution, Diet

Patas monkeys are slight of build with slender trunks and long limbs and tails. They are not small monkeys, being 50 - 70 cm. from head to rump, but rather are finely built for speed on the ground, like cheetahs and racing dogs in the carnivore world. They show adaptations of the limbs, hands, and feet for running, and are the swiftest runners of the primate order, having been clocked at 55 km. per hour.

Sexual dimorphism in size is conspicuous, as males weigh on average 10 kilograms, or nearly twice the weight of females who average 5.6 kilograms. Adult males also have much larger canines than females and are sometimes described as being slightly brighter in color, although fur coloration in both sexes is variable, being brick red to reddish brown on the dorsum or back of the body, and fawn to white on the underside or ventrum. The adult males have a longer mantle of fur around the shoulders and neck, and the fur around their rumps is described as conspicuously "pure white." The scrotum, clearly visible from the rear, is bright blue, so adult males are easily distinguishable from the others of their groups. Both females and males have the famous white moustache on a gray-to-black face, which along with their russet coats and stately bearing, may have given rise to their nicknames, the "hussar monkey" and the "military monkey."

The distribution of patas corresponds to that of the savannah belt of Africa north of the equator. They range from Senegal in the west to the Sudan in the east, as far north as Northern Nigeria, and as far south as the forest belts of Uganda (and possibly Northern Tanganyika in the east and the Cameroons in the west). While the closely related vervets are sometimes described as forest or woodland dwellers which can range out onto the savannahs, the patas are fundamentally semi-arid savannah and steppe dwellers who seldom penetrate forested areas. They are terrestrial primates par excellence, and are quite independent of trees, requiring only a few widely-scattered small trees or bushes for sleeping. Quite unlike other nonhuman primates, patas run *down* out of trees, and across the ground, when frightened. Although there is no evidence they have physiological mechanisms for water-conservation, they range further into arid zones with few waterholes than do common baboons or vervets. The patas feed largely on the ground, being omnivorous although mainly vegetarian in their diet. Grasses predominate, followed by seeds, fruits, fungi, leaves, flowers, insects, lizards and bird eggs.

Social Organization

Patas social groups are of two sorts, all-male groups usually made up of adolescent and adult males of mixed ages, and heterosexual groups or reproductive units consisting of 1 adult male, 4 to 12 adult females, and their offspring. Solitary males are also occasionally seen. Reproductive units range in size from 9 to 31 with a mean of 15 group members. Although the formation of new groups, or the fissioning of established groups has not been observed, the presence of solitary males and all-male groups would indicate that males emigrate from their natal groups while females probably remain in their home groups and natal areas throughout their lives. Kinship patterns have not yet been documented through longitudinal research; however, in captivity mothers and offspring defend each other.

Rowell and Hartwell (1978) suggest that patas social organization may be thought of as concentric with a core group of adult females and infants, an outer circle of other females and juveniles, and an outermost circle consisting of the almost "peripheral" adult male who is very aloof from his group except for mating, which is brief and "business-like."

Patas exhibit a mating season and a birth season, although these periods are not as discrete as in macaques, and it is assumed that all mating takes place within the reproductive unit, between the females and the one adult male. There is no direct evidence that outside males approach reproductive units during the mating season; indeed patas groups are normally spaced widely apart, with reproductive units being very dominant, or readily displacing all-male groups, should they venture into the same area around waterholes.

However, the presence of scars and wounds in all-male groups, and the strong intolerance of adult males in reproductive units for the proximity of any other adult male, led two researchers (Struhsaker and Gartlan 1970; Gartlan 1974) to suggest that reproductive unit males are sometimes challenged and perhaps defeated by outside males. Hall (1967) thought that the initiative in keeping a male within the group or not came through the female hierarchy. Direct rather than circumstantial evidence is still needed to settle this question. The data from hamadryas baboons and geladas (see Chapter 12 and below) demonstrate that there are several ways to achieve the formation of new groups, and to increase genetic exchange in polygynous systems whithout the "overthrow of harem males," made popular by the langur data from southern India (see Chapter 17).

Subsistence Factors

Patas are unusual monkeys in many respects. Although they will rest together in one big tree during the daytime (as most monkeys do at night), in the evening they disperse widely and almost singly into scattered trees or bushes. It is speculated that perhaps this is an adaptation to pressure from nocturnal predators, such as leopards. In addition, data from a captive colony demonstrated that patas regularly give birth in the daytime rather than at night as do most nonhuman primates (Chism et al 1978), which might also be an adaptation to the pressure of nocturnal predation.

Patas groups have very large home ranges and they seldom sleep in the same area on consecutive nights. An adult female always initiates group movement and is responsible for the direction and timing of the travel, with the rest of the group following behind her and the adult male often bringing up the rear. The characteristic adult male activity is best described as that of "look-out," as he is continually alert to the possibility of extragroup events. At times he may travel at some distance from the rest of the group and occasionally he may appear to act as a "scout," climbing a tree and surveying an area before his group moves into it.

The adult male never herds his group and in times of fright the females do not seek him for protection. If the group is greatly alarmed they all flee, the longer-limbed and less encumbered male usually putting the most distance between himself and the danger. However patas generally rely on adaptive silence and crouching motionless in the grass to deal with lesser hazards. It is frequently stated in the literature that the patas male performs a diversionary display as an anti-predator tactic, noisily leaping and bouncing off branches and bushes before galloping away with his blue-and-white rump flashing, while the group hides in the grass. Struhsaker and Gartlan (1970) state that during their admittedly short field study in the Cameroons, they saw what might be classified as a diversionary display only once, and that it would take a considerable stretch of the imagination to consider it an effective anti-predator mechanism.

On the other hand Struhsaker and Gartlan did observe adult males chasing jackals on three separate occasions after the jackals had attacked young juvenile patas around a waterhole. In one of these events three males chased one jackal with a juvenile in its mouth, until the jackal dropped the monkey who was then retrieved by a female. This observation is noteworthy for at least two reasons: firstly, only one of the three adult males can have belonged to the same group or have been the father of the juvenile, and secondly, our traditional view of the patas male role as a "sentinel" rather than a "defender" of his group members appears to be contradicted.

It is not clear whether or not patas groups are territorial in the strict sense of defending a defined area; however groups either avoid each other or are very antagonistic when they do meet at waterholes. Females and juveniles actively participate in such encounters between reproductive units, which are interactions involving threats and display rather than physical contact. The adult male's primary concern is sometimes described as that of keeping all other males away from his group. In a similar manner to vervets, it appears that the vigilance and the displaying of adult male patas is directed at least as much, or perhaps more, toward other males as toward possible predators and dangers. Struhsaker and Gartlan (1970) conclude that the adult male plays the major role in driving away outsider males, while the adult females play the major role in maintaining space between reproductive units.

In sum, males are mobile between groups and they may act as a diversion or even a defense against predators, as well as keeping other males away, spending much of their times while travelling and foraging acting as a look-out. Females spend their adult lives in one area which they presumably know well, they determine the direction of group movement, and their usual response to danger is to run, carrying clinging infants, or to remain still and silent in the grass.

Ontogeny and Socialization

Not very much information has been published about the socialization processes within patas societies. It is known that mothers are very protective of their young and that the adult male of the group generally does not interact with infants or juveniles. As is the case with vervets, the patas male is threatened by coalitions of females aiding the mother should the male either threaten a young group member, or simply be in the vicinity of a frightened infant. Adult females seem strongly attracted to newborns in captive groups and allomothering is common, especially by other adult females and sometimes by other mothers or juvenile females and males. Oddly, infants have only been seen to be handled or carried by allomothers once in field studies in Africa, indicating that the amount of "aunting" may be variable between groups, or may be an artifact of observation conditions. Even in captivity, mothers usually resist the taking of infants by others and when they are taken, sometimes have difficulty in retrieving them (Chism 1978). However, close associations between infants and certain allomothers persist as the infant matures. One researcher (Gartlan 1975) suggested that a mother may seek to remain close to a male in order to keep potential alloparents away from her infant, while another researcher (Hall 1967, 1968a) stated that mothers-with-infants and the adult male stay well apart from each other. Clearly the question needs further research.

Juvenile patas are very playful — chasing, wrestling, and bouncing off bushes in a style highly reminiscent of the adult male's "display." The literature does not discuss sex differences in the juvenile years but it does suggest that young males emigrate from their natal groups between 1.5 and 4.5 years of age. In a captive group, Hall (1967) noted that a young male of the group began to be severely attacked shortly after his scrotum turned from its infant coloration to the adult blue, possibly signalling sexual maturity. Thus it has been proposed that young males may be driven from groups by the hostility of the adult male. The strong intolerance of adult males living in reproductive units for all other mature males would tend to support this suggestion. On the other hand, young males in many other species of Old World monkeys are reported to roam at or near the time of puberty, even without receiving apparent hostility from anyone in the natal group.

Social Dynamics

Agonism within patas groups is very rare, and when it does occur, is usually started by females and directed at other females, or secondarily at the male. The male for his part, almost never initiates an attack on a group member, although he may become very agitated when mating with an estrous female and threaten anyone who comes near. A frequent result is that females will coalesce against the male and counterthreaten him, even being joined by the female with whom he has recently been mating. In spite of their larger body and canine size, adult males never enter into physical fights with females, but rather may slap defensively at them to keep them away. Females generally initiate sexual behavior with an elaborate courtship, to which the male does not always respond.

In various studies the adult females have been ranked in a hierarchy according to feeding order, direction of agonistic signals, proximity to the male, or supplantations. The adult male never interferes or participates in disputes between females, and it appears the female hierarchy is established independently of, or without reference to the male. However, one of the behaviors of the top-ranking female in captive groups may be to monopolize proximity to the male. A tentative "leader" role for the top-ranking female has been described by Hall & Mayer (1967) which involves leading the routine group activities, having a privileged relation with the adult male, and being a focus of friendly attentions from the other adult females. Another study suggests that females with new infants are also a great focus of attention. While female-female relations do not alter during the mating season, "the interaction pattern of the females was greatly modified by the arrival of small infants" (Rowell & Hartwell 1978: 163), with a very large increase in both friendly and avoiding behaviors.

Earlier, I noted that the adult male of each reproductive unit is uniformly hostile to other mature males. However, males in all-male groups and solitary males are very tolerant and sociable with one another. As their ages are mixed, the males of all-male groups may indeed be siblings. It appears that during his lifetime a patas male may experience natal group life, solitary phases, periods of sociability in all-male groups, and periods of heterosexual group life when he becomes intolerant of other mature males.

Synopsis

Patas males are probably the most socially detached of primate males found living in heterosexual groups. As adults their only contacts are with the adult females of the group, with whom they groom, mate and occasionally dispute. These males make no efforts to interfere with or to influence the intragroup social life. Females (perhaps one particular female in each group?) direct group movement and regulate intragroup life. They rear the young of the group, engage in between-group encounters (possibly territoriality) and readily coalesce against the physically larger male of the group should he threaten or frighten one of their members.

THE LIFE-WAY OF NILGIRI LANGURS *(Presbytis johnii)*

Description, Distribution, Diet

The Nilgiri langur is an arboreal monkey, found in certain forests in a limited area of South India, and probably closely related to the langurs *(P. senex)* of Sri Lanka. Nilgiri langurs have glossy black hair with an attractive headdress, and whiskers of longer, buff-brown hair. Sexual dimorphism is slight, with males measuring about 78 cm. from head to rump and weighing on average 11.2 kg., while females measure 68 cm. and weigh on average 11.1 kg. Males have larger canines than females, while the latter can be distinguished by a patch of cream-colored hair on the thigh. In both sexes the tail is longer than the body, measuring from 68.5 to 96.5 cm.

These langurs are found only in southern India, from their northern limit in the Coorg district of Mysore, to the Cape Comorin district in Madras state, their southern limit (Oates, 1979, recently reported an even more limited distribution). Even within this limited geographical range, Nilgiri langurs only occupy certain upland forests in the hills, known as *sholas*; a distribution which tends to minimize encounters with the other species of langurs and the macaques found on the same hills. These forests, or sholas, are surrounded by grasslands or cultivated fields and usually have a narrow stream running through them, with evergreens, ferns and mosses being common plants.

Nilgiri langurs sleep in the upper canopy of the trees, forage in the understory and shrubstory, and seldom come down to the forest floor except for short bouts of foraging and playing, or to flee. Their diet consists largely of leaves, but fruits, berries, buds, flowers and stems are also eaten. Langurs are part of the family, Colobidae, all of which are leafeaters or folivores and show specializations of the digestive tract, particularly a sacculated stomach to help process the bulk cellulose of such a diet.

Unlike the rhesus macaques and the common or Hanuman langurs of India, some of which live in and around urban areas, Nilgiri langurs are very shy and wary of humans who do prey upon them, using their flesh and glands for food and medicines, their skin for drum heads and their glossy fur for decoration. Possibly these monkeys have survived by retreating into the recesses of the dense, high altitude sholas.

Social Organization

Nilgiri langurs are commonly found in small groups, averaging about 9 individuals and consisting of 1 adult male, 1 to 5 adult females, and the young of the group. All-male groups, solitary males, all-female groups, and multi-male, multi-female groups are also occasionally seen, a variety reflecting perhaps the structural fluidity of Nilgiri langur societies. The sex ratio in the population of an area (including both reproductive units and all-male groups) averages 1.2 females to 1 male. Changes in group membership seem rather common, with both males and females leaving or transferring groups, although males do so more often and females emigrate in small groups which may be kinship units.

Birth "peaks" have been noted, although these are not as concentrated nor as discrete in time as birth seasons. In spite of the fact that observations have taken place in all seasons and at all hours of the day and night, copulation has not been observed. It is assumed that copulation takes place between the one male of the group and the females, a polygynous system, since the male works dilligently to keep other males away. In southern Hanuman langurs (*P. entellus*) social stimuli such as group changes (entry of new males, etc.) appear to stimulate sexual behavior. Although similar reconstructions were observed in Nilgiri langurs, and the possibility of resultant sexual activity was monitored, none has been seen.

As small as the reproductive units are, they are commonly broken down further into subgroups within which individuals interact more frequently. These subgroups are structured by age and sex (e.g. "adult females") but also possibly by matrilineal factors, as an adult female may be surrounded by certain young ones of various ages, presumably her offspring. The adult male of the group is socially and physically peripheral and the females are described as showing little concern or interest in the comings and goings of males. Poirier described the social organization as female-focal, and noted that adult females are the cornerstone of the society.

Subsistence

The adult male of the reproductive unit does not lead the group's progression, but Poirier (1970a), was of the opinion that group movement may be keyed to his travelling pattern. On the other hand, adult females tend to stay, with the young of the group, in familiar core areas while males range widely and even penetrate the home ranges of neighboring groups. The adult male maintains a constant alert while travelling, climbing high up in the tree tops and scanning the environment. It seems that this vigilance is directed toward spotting other males rather than protecting the group from predators, because the Nilgiri langur response to fright is rapid, arboreal escape with the adult male usually in the lead. Once when Poirier saw a dog catch a young male, the group watched silently and then fled. Although the adult male does not seem to defend his group, the group does appear to pay more attention to his alarm calls in that they react more strongly to them.

Nilgiri langur mothers do not share food with their offspring and are not very cooperative when their infants attempt to obtain scraps from them, turning their backs on the infants or moving away. The majority of the food is obtained in the trees, and the monkeys are very nervous when foraging on the ground, often attempting to obtain the food item quickly and carry it back into the trees to eat.

Nilgiri langur males are highly territorial, but unlike the gray langurs of Sri Lanka (Ripley 1967), the female Nilgiri langur seldom engages in territorial encounters. Only adult males give a special booming "whoop" vocalization while displaying in the trees. This is given every morning, whenever another reproductive unit is encountered, or after any exciting disturbance such as overhead planes or falling trees. While the vocalizations would allow groups to avoid each other easily, males of some groups deliberately seem to seek each other out, even temporarily deserting their own groups, to take part in territorial encounters involving noise, ritual chasing and locomotor displays, but rarely physical contact. Interestingly, certain groups consistently tolerate each other's close proximity, while others are consistently hostile. Perhaps tolerant groups represent divisions of past groups? Hostile encounters between all-male groups and reproductive units have not been observed.

On one occasion, Poirier was able to observe a group forced to move into the territories of neighboring groups because every tree in its home range was being felled. The monkeys were very reluctant to move, especially the adult females, remaining until the last tree was down. When they did move into a neighboring territory and received threats from the resident group, an adult male did attempt to defend his group members against the other group, illustrating that the full potential of female and male roles is not always completely demonstrated in the course of normal day to day life.

Ontogeny and Socialization

Mothers of newborn infants are less active and move slowly, so that mother subgroups tend to be formed, which feed, sleep and travel more closely together. Certain juveniles and subadults of both sexes are drawn to the mother, and are perhaps her prior offspring. Adult males show little interest in newborns. Very occasionally they will retrieve an older infant which is being weaned and is vocalizing in protest, and hold it in their laps, or they may threaten the mother until she retrieves the noisy infant. At three or four weeks of age an infant begins independent locomotion and the mother's protective role flags. By 14 weeks of age, the infant makes its own way over the arboreal pathways and from this time mothers are reported to be curiously inattentive to the needs of, or possible dangers awaiting, their infants.

Adult females often transfer infants but only when the young are from one to seven weeks of age (data from birth to one week were not obtained). A mother may simply deposit an infant on a branch near another adult female, or an allomother may groom an infant and comfort it after its mother departs. Infant transfer occurs only between adult females. Juveniles and subadults are not involved, and friendship bonds, rather than strong interest or dominance on the part of the allomother, seem to be the basis of the transfer. Adult males very occasionally are the recipients of infant transfers. Females willingly nurse infants other than, or as well as, their own and sometimes attempt to hold and carry two or more infants at once. On the other hand babysitters occasionally "walk off" on the infants that have been left beside them.

Sex differences in the young around the time of weaning, or during the first year of life, do not appear very noticeable, and mixed-sex play groups are common until the late juvenile and pubertal stages of development. At this time, male play becomes more unisexual and more vigorous, while female play gradually diminishes. As subadults, females and males play separately. The peripheralization of males from the natal group has not been observed, although the prevalent pattern of all-male groups and one adult male only per reproductive unit, indicates that this must occur some time during adolescence.

Social Dynamics

Agonistic encounters within groups are rare and even amicable interactions such as grooming and play are reported to be less common than in the terrestrial baboons and macaques. Dominance interactions are not an important feature of Nilgiri langur life either, and when they do occur, are usually expressed through supplantations. Poirier scored the individual which gives the most aggressive, as opposed to submissive gestures, in an agonistic encounter as "dominant." By this criterion males rank over females. However almost all agonism occurs within sex groups and separate hierarchies for the sexes are thought to exist. The female hierarchy in each group is judged to be stable; however the top-ranking female is not the object of attention and has no special priorities to infant transference. Alliances during agonistic encounters are rare, evidence which may mitigate against the suggested importance of kinship in Nilgiri social life, as alliances between kin are important in other species. Or it may be simply that alliances are not a feature of Nilgiri social behavior, particularly as dominance encounters themselves are rare.

Adult females are decidedly the more sociable sex in this species, and they groom each other and adult males more often and for longer periods than males groom. Curiously, they rarely groom their infants. Adult males are generally hostile to other males. Even in the occasional multi-male, multi-female group, agonism is more frequent and severe between males. Forming an exception to this pattern the males of all-male groups appear to be mutually tolerant.

In summary I should note that Poirier (1970) sees similarities between aspects of patas and Nilgiri langur societies, despite the fact that their niches are quite different. In both species the groups are female-focal, with a semi-attached, socially-aloof male who acts primarily as a breeder and watcher for external disturbances, especially the presence of other males.

THE LIFE-WAY OF GELADAS *(Theropithecus gelada)*

Description, Distribution, Diet

Geladas are large, ground-dwelling quadrupedal monkeys. Adult males weigh around 20.5 kg. and are approximately one and one-half the weight of females, who average 13.6 kg. Males also have larger canines, and the brown hair on their head and shoulders is longer than that of the females, giving them a striking "cape" which enhances their apparent size and their display-like chases up and down precipices. Both sexes have a patch of naked red skin in the shape of an hourglass on the throat and chest, which presumably gave rise to the nickname "Bleeding Heart Baboons." In mature females this area of bare skin is surrounded by small, globular swellings of tissue, which look somewhat like beads on a necklace. Similar beads appear on the abdomen and perianal region, and appear to indicate active reproductive cycling of the female.

Today geladas are found only in the high mountains of central and northern Ethiopia, although during the Pleistocene, species of the genus *Theropithecus* were found throughout the grassy plains of eastern and southern Africa (C. Jolly 1972). Geladas have a very specialized niche, showing extreme adaptations to a terrestrial mode (they almost never climb trees) in montane, semi-arid grasslands. They occupy cliffs and high plateaus (2000-5000 metres) and eat a largely vegetarian diet of small items growing at ground level; grass blades and seeds, bulbs, tubers and rhizomes. Cliffs are an obvious source of security; geladas never range far into the valleys, or onto the plateaus, and run back to the cliffs when frightened. They also shelter on cliffs and overhangs at night and during rainstorms.

Geladas are sometimes said to live in a habitat that is marginal for primates, with sparse foods of low nutritive value. Some speculate that they have been "out-competed" and pushed into these areas by the wider-spread Papio or baboon species. However, seeds, roots and rhizomes are highly nutritious foods and the Dunbars (1974a,b,c, 1975, 1977, 1979) who have conducted the longest field study of geladas to date, dispute the notion that these monkeys live in an arid, impoverished environment for primates in which extreme competition for food and water has structured and honed the present gelada life-way (see also Kawai 1979). On the contrary, the habitat has a high carrying capacity for geladas, and water, rhizomes, and sleeping sites are abundant even in the dry season. Relations with hamadryas and common baboons, where the species' ranges overlap, are amicable, with the different species occupying quite different niches. The only creatures which appear to be pressuring the geladas and endangering their continued success as a species, are humans.

Social Organization

The social organization of geladas is quite complex, and is described as being structured into three levels. The first level consists largely of small heterosexual or reproductive units made up of one mature male, several adult females (average about 6) and their younger offspring. All mating between adults occurs within these units. Also at this level are all-male groups (sometimes called "bachelor" male groups), juvenile peer groups and "solitary" (or "free-lance") males. Approximately 20% of reproductive units have a "follower" or semi-attached peripheral male.

Thus far the social organization resembles that described for both patas and Nilgiri langurs. However, in geladas, several of these small units share a common home range and a close relationship; individuals obviously are familiar with one another and probably are linked by kinship ties, although this suggestion awaits longitudinal data for confirmation. This larger grouping of reproductive and all-male units has been referred to by the Dunbars as a "band" and constitutes the second level of gelada social organization. Females within a band appear to exhibit synchronized birth peaks; however births and mating are not consistently seasonal in different parts of the gelada range. It can be difficult to distinguish the reproductive units within a band during foraging because the females do not cluster around the male, and units may become somewhat dispersed. However reproductive units do move cohesively into their sleeping areas in the evening and can be distinguished at that time, while individual recognition of geladas can allow the observer to "pick out" the somewhat obscure outlines of the foraging reproductive unit within the larger bands. Because the units within a band are not always cohesive and may travel independently within the same range, the Dunbars suggested that the gelada band is somewhat like the fluid "community" or "regional population" of chimpanzees, rather than analogous to a tightly structured troop of baboons or macaques. Finally, at the third level, geladas may be found in "herds" of up to 600 animals. Herds are described as not really being a social group, but rather temporary congregations or aggregations of animals under good grazing conditions. Where the ranges of two or more bands overlap, the neighboring bands or portions of them form temporary herds for foraging or sleeping on cliffs. Herd progressions are not organized, and herds subdivide readily and often. A team of Japanese primatologists (see Kawai 1979) also described three levels in gelada society, but appear to have labelled them differently.

It can be seen that the constraints on independent movement in gelada bands and herds are not very rigid, resulting in a highly complex kaleidoscope of grouping patterns in which the small reproductive units and all-male units form the basic bits of colored glass which can come together in many different ways. And, like a kaleidoscope, certain structures or principles, possibly kinship, familiarity, and resource distribution, delimit the patterns, such that the configuration of bands and herds is not random but dynamic in pattern.

Subsistence Factors

The travelling pattern of reproductive units is initiated and determined by a particular female within the group, without reference to the adult male. Young males in early prime may attempt to "herd" their group to keep it together (see below), but females are primarily responsible for the cohesion of the group and may give a particular vocalization ("wails," Dunbar & Dunbar 1975) when they wish the RU to join them, or when they are lost from the group.

Geladas are not territorial, although the individual units, especially females in the units, attempt to maintain social and spatial integrity by threatening away most other units. However certain units show tolerant and proximal relations and may be recently fissioned or linked through close female kinship. The adult male of reproductive units is intolerant of other males and is strongly antagonistic to all-male groups.

Predator pressure upon geladas is not heavy today, except by humans and domestic dogs, although it may have been so in the past before larger carnivores such as leopards became scarce. The common reaction to fright is to run for the safety of the inaccessible cliffs and males are not known to protect or defend their groups against predators. Males do however aid and defend their reproductive unit members in antagonistic encounters with other groups. The large size of the gelada male is commonly ascribed to sexual selection rather than to selection for a special role as group defender.

Ontogeny and Socialization

Infants and yearlings are generally found in their natal reproductive unit where they are the object of much attention from group members, especially females. Although mothers are careful with their infants and resist attempts to remove them in the first days of life, juvenile females in particular are attracted to infants and yearlings and sometimes do succeed in carrying newborns. Mothers will swap infants and will nurse infants other than their own. Juvenile males and follower males also are interested in infants and will carry, aid and play with their young charges. Such infant care on the part of alloparents is not considered kidnapping (see Chapter 12), as the infants are free to come and go. Interestingly, the only group member who does not exhibit great interest in infants is the adult male, the putative father. However, in captivity males who are defeated from the position of reproductive male and return to the role of "follower" or "bachelor male" may exhibit a resurgence of interest in infants and yearlings. Bernstein (1975) suggested this may be a form of agonistic buffering or manipulation of immature animals in order to bolster the male's own prestige.

In their second year of life males begin to wander further from their natal units and to form peer groups, especially for playing, but also for feeding and travelling. Around three years of age males are first found in all-male groups and by five years they are invariably in all-male groups. Some males attempt to join established all-male groups and receive a considerable amount of antagonism. Other juvenile peer groups may simply develop into all-male groups as the males mature. All-male groups are very real social units in the sense that they exhibit cohesion, bonding and support between members and a certain degree of stability in membership. It is possible that in many cases the members of all-male groups are related, having come from the same or related reproductive units. Little agonism is observed within these groups, and a hierarchy is not detectable, although a "control" or "leader" individual is said to exist. The adult males of reproductive units are highly nervous of all-male groups, and in some areas, males from such groups are believed to cooperatively harass reproductive males.

Certain young males may become "followers" of reproductive units, although it is not certain if this phase follows a period in an all-male group or is an alternative to it, or both. It does seem clear that young male followers are not associating with their natal units, but rather with units which offer prospective mating partners. "Follower" males may groom and sit with certain adult females of the unit, but seldom, and only surreptiously mate with them. The relationship between the adult male and follower male is variable, ranging from antagonistic to mutual support. Follower males may obtain their own mating partners in several ways: by attracting away the maturing female offspring of the group; by attracting away a subset of the group, probably a matrilineage; or by gradually taking over the entire unit as a former unit male passes his prime.

Yet other young males live for a time as "solitaries," or as one research team called them, "freelancers" (Mori and Kawai 1974) who live in the band, but without unit affiliation. Such males may occasionally attempt to obtain mates by fighting with the adult males of reproductive units. In both this case and the cooperative attacks of all-male groups on reproductive units, it is thought to be the females' decision to go or to stay with one of the males, and not the physical outcome of the fighting which determines the success of males (Dunbar & Dunbar 1975; Mori & Kawai 1974). Females may reject a male who has physically defeated their unit leader, or they may present to, groom and copulate with a male who is attempting a take-over.

Young females often initiate the process of new group formation. After a juvenile period of interest in yearlings and mothers-with-newborns, pubertal females turn their attention to potential mating partners. In her first estrus, a female may solicit the adult male of her natal group, who is possibly her father; however such males seldom respond to the sexual advances of young female group members. Young females often turn to the follower of the unit and may mate with him. Alternatively, she may approach and solicit a male from an all-male group or she may attempt to join another established reproductive unit. In the latter case, the young female attempting to attach herself to a new unit pays a great deal of attention to the top-ranking female of that unit, grooming and following her (Dunbar & Dunbar 1975; Kummer 1971a). It appears that adult males of reproductive units readily accept a new young female from another group, but older females of the group do not.

The development of gelada individuals and gelada groups would appear to exemplify the two interactive principles of sexual selection; male-male competition and female choice. Males compete with each other for access and proximity to females and to attract their attention, while females exercise a very real power in their choice of mates and group companions.

Social Dynamics

Agonistic interactions within reproductive units and all-male groups occur at very low levels and are described as very ritualized, with injuries even in captivity being quite rare. Galadas do have extremely long, sharp canines which they seldom if ever use on each other. Even hostile interactions between units consist of vocalizations and bluff charges and vigorous but ritualized chases in which no one is caught.

Dominance relations *between* units are complex and difficult to establish, but dominance hierarchies are described as stable *within* units. Using the criterion of direction of agonistic signals, Bernstein (1975) states that the adult male of the unit ranks over any one adult female. However the male responds subordinately to a coalition of females, and coalitions are very common in at least some units. The Dunbars suggest that males from reproductive units in their early prime are more domineering and intolerant than older ones. They more often attempt, and are successful, in herding their females and breaking up fights between them. Unlike hamadryas baboons, where herding may be an aggressive act on the part of the male involving attacks and neck bites, so-called gelada "herding" or "rallying of the female" (Mori 1979a,b) consists of "appeasing vocalizations" and defensive, or at the most mildly threatening, facial expressions (Mori & Kawai 1974; Bramblett 1970; personal observation). Even this rather gentle form of herding is not often performed by unit males past their early prime, and when it does occur, usually results in a coalition of females attacking the male.

It is the Dunbars' suggestion that females in older, established reproductive units have formed close and stable bonds, possibly with the emergence of a leader female, and thus achieved a balance of power with the much larger but lone male. There exist intriguing references in the literature to a top-ranking or alpha female, who may act as a leader (Kummer 1971; Spivak 1971; Dunbar & Dunbar 1975). Such a female may become very powerful, determining group movement, influencing the acceptance of new group members, interfering in fights, and even herding stray females herself. She may even suppress reproductive success in lower ranking females. The Dunbars suggest that one such female determined the group movement of an entire herd. This is the most powerful female role described thus far in the nonhuman primate literature.

Gelada females are more socially active than males in all areas except play, and do not concentrate their activities on the group male. They huddle and groom and aggress with each other more than with males. For their part, males are described as somewhat peripheral to these tight social bonds, rather like, but not so extreme as, the patas male. Gelada males exhibit different behavioral profiles, according to the social circumstances in which they find themselves, sometimes being very tolerant of and closely bonded to other males, other times being highly antagonistic to other males; sometimes very interested in infant care and play with juveniles, other times being rather detached and aloof from intragroup life.

In many respects geladas offer one of the most complex and varied pictures of social life in monkeys we have to date, and yet they have been relatively little studied in field or captivity. In the past, gelada society was often likened to that of hamadryas baboons, who also live in polygynous groups, but as the Dunbars note, the similarities are only superficial, primarily because of the differences in the roles females and males play in the social life of the two species.

REFERENCES

Patas Monkeys *(Erythrocebus patas)*
Chism 1978; Chism et al 1978; Gartlan 1974. 1975; Hall 1967, 1968; Hall & Mayer 1967; Hall et al 1965; Loy 1975b; Struhsaker 1969; Struhsaker & Gartlan 1970; Rowell 1974b, 1977; Rowell & Hartwell 1978.

Nilgiri Langurs *(Presbytis johnii)*
Horwich 1972, 1976, 1980; Oates 1979; Poirier 1968a,b,c, 1970a,b; Ripley 1967; Roonwal & Mohnot 1977; Tanaka 1965.

Geladas *(Theropithecus gelada)*
Alvarez & Consul 1978; Bernstein 1970, 1974b, 1975; Bramblett 1970; Crook 1966; Crook & Aldrich-Blake 1968; Dunbar 1978, 1979; Dunbar & Dunbar 1974a,b.c, 1975, 1977; Fedigan 1972a; Iwamoto 1979; Jolly, C. 1972; Kawai 1979; Kummer 1971, 1974; Mori 1979; Mori & Kawai 1974; Ohsawa 1979; Spivak 1971.

Chapter SIXTEEN

Monogamous Societies

Titi
Callicebus moloch
The male (middle) and female sit in a characteristic "hunched" posture with intertwined tails. Titis are arboreal New World Monkeys, showing little or no dimorphism.

Because all Euroamerican societies have prescribed monogamous marriage rules, one must be most careful in describing monogamous mating systems in nonhuman primates. Particular care must be taken to avoid hasty extrapolation from superficial similarities between animal mating systems and human marriage institutions. This is especially so because human marriage rules and human mating practices are not always found to correspond, and because human societies other than those of modern Europe and North America have shown, and still do show, diverse marriage patterns.

A mating system in which one female and one male breed exclusively with each other, and remain together during the rearing of their young is uncommon in mammals; only 3 % of mammalian species are monogamous (Kleiman 1977). In the Primate order, a relatively small number of species live in mated-pair groups, and none of them are very closely related to the human species. To date we only have evidence for monogamy in one species of prosimian (*Indri indri*) and one species of Old World monkey (*Presbytis potentziani*). However this mating pattern is widespread in the families Hylobatidae (gibbons) and Callitrichidae (marmosets and tamarins) and also occurs in several Cebid New World Monkeys (see table in Kleiman 1977).

Those species classified as "monogamous primates" appear to have more features in common than do multi-male, multi-female living and polygynous-living primates. This is a striking finding because hylobatids, (the lesser apes) and the New World monkeys are widely separated in a phylogenetic and taxonomic sense. Perhaps the apparent uniformity of various aspects of morphology and behavior in monogamous nonhuman primates is partly due to the small number of species studied, or to the still superficial view we have of their lives, one which will develop more complexity as field data and longitudinal research become more abundant. But, it is also possible that the similarities result from monogamy being a specialized adaptive pattern for nonhuman primates, which only occurs under certain limiting environmental conditions, or when primates occupy a particular niche.

First I shall examine the similarities among monogamous species and then I shall turn to possible adaptive functions of monogamy, The monogamous primate species tend to exhibit the following features:

1) lack of sexual dimorphism in size or coloration;
2) lack of specialized defense role against predators by adult males;
3) highly developed territoriality in both sexes;
4) extensive care of young by the adult male;
5) close synchrony of all activities by the adult female and adult male.

In addition, Kleiman (1977) suggests that monogamous nonhuman primates may show delayed sociosexual maturation of the young such that older siblings remain in the group and care for younger siblings. Also, higher levels of female aggressiveness than are found in females of other primate species are suggested.

What the limiting conditions, and the special adaptive functions of monogamy in primates might be, is uncertain. Wilson's list (1975) of ecological conditions which bias animals toward monogamy is difficult to assess for primates. The explanations given for the prevalence of monogamous mating in bird species concern the food supply and feeding of young birds in nests and does not transfer easily to primate life-ways. Specialists in animal mating systems as I noted in the last chapter, suggest that monogamy offers the greatest reproductive advantages to the female, and one plausible hypothesis is that monogamy may be favored when more than one individual is needed to rear the young, in a habitat unable to support more than one female and her offspring. This suggestion has been put forth in slightly different forms by Kleiman (1977), Eisenberg (1978) and Leutenegger (1973), all of whom have noted that in monogamous primates the birth weight of the newborn infant is very high relative to the mother's body weight. Thus the mother may literally, as well as figuratively need help with the "reproductive burden."

According to this hypothesis, extensive male care is a key aspect of monogamous primate societies. How the other features that go with primate monogamy — territoriality and lack of sexual dimorphism — fit into a causal chain is not yet clear.

In spite of the similarities discussed here, monogamous primate species do show variability in the character of their social lives and even in the nature of the pair bond. The readings that follow will show that the adult female and adult male of callitrichid societies appear to dominate separate female and male hierarchies and to concern themselves greatly with interactions with their own sex. In contrast, Callicebus adult pairs are highly bonded and interact in a very exclusive fashion with each other. Siamang society seems to contain a rather socially and physically aloof adult female and a highly social, central adult male who actively creates bonds with his young and his mate. In many respects these species offer interesting alternatives to the types of social behavior more familiar to us from the well-studied multi-male, multi-female and polygynous primate societies, and it is unfortunate that the mated-pair species are all difficult to observe and still relatively little studied.

THE LIFE-WAY OF MARMOSETS *(Callitrichidae)*

Description, Distribution, Diet

At the risk of obscuring important differences between species, this section will attempt to describe the life-way of Callitrichidae, or the family of primates including marmosets and tamarins. The description will focus upon common marmosets (*Callithrix jacchus*), which have been quite well studied in captivity, with some supplementary information presented from tamarins (*Saguinus* species) which have been the subjects of field studies.

Callitrichids are extremely small monkeys, the common marmosets weighing only 250 grams (.25 kg.) and measuring approximately 20cm. in body length. For most of the callitrichid species there is no significant sexual dimorphism in body size, pelage or dentition; however the common marmoset female is often said to be somewhat larger in size and greater in weight than the male. Reports are contradictory on this subject. For example Schultz (1969) says common marmoset female weight is 115% of male weight, while Napier and Napier (1967) report it as 93%. In any event the only readily apparent differences between female and male callitrichids are the genitalia, and the circumgenital glands which are bigger in females.

The different species of marmosets and tamarins are variously adorned with striking tufts, crests, manes and colorations of fur on both females and males. Perhaps it is the beauty and small size of marmosets and tamarins that has led to their frequent use as pets, for they strike many people as exquisite and miniaturized monkeys. However their small size, their rapid movements and their proclivity for dense brush, also have made them very difficult to study in the wild.

Marmosets and tamarins are distributed over the tropical forest areas of South and Central America from the coast of Brazil west and north to the slopes of the Andes in Columbia and, for one species, up into Panama. Dawson (1978, 1979) describes the niche of the cotton-top tamarin, from Columbia to Panama, as being "edge" vegetation of riverine forest or secondary growth, and several researchers suggest the common marmoset occupies the same niche in the non-overlapping geographical range of the coastal forests of northeastern Brazil (Coimba-Filho & Mittermeier 1973)

Like all New World primates, callitrichids are exclusively arboreal, rarely, if ever, coming to the ground in the normal course of their lives. They locomote quadrupedally, jumping from branch to branch, and their diet is omnivorous, consisting of fruits, insects, new leaves, stems, epiphytes and, for several species with procumbent lower incisors, tree sap, which they suck after perforating bark with their teeth. Eisenberg (1978) describes marmosets and tamarins as occupying a squirrel-like niche.

Social Organization

Preliminary field observations and extensive experience in putting together successful reproductive groups of marmosets in captivity led primatologists to conclude that the standard composition of callitrichid societies is a mated pair and their dependent offspring. Further research suggested that callitrichid groups may often be quite large, up to 20 individuals, and that mature offspring may even remain in the group, reproductively inactive but caring for younger siblings. This is somewhat similar to the social organization of wolves, for example (Mech 1970), and marmosets are now most often described as living in "extended family groups," with a monogamous pair as the nucleus of the group.

Recent field research on cotton-top tamarins by Dawson (1978, 1979) and Neyman (1978) reports that these animals live in large groups of unstable membership, in which the younger individuals in particular, often change groups. Dawson characterizes or entitles the social organization of cotton-tops as an "age-dependent, one male, one female dominance system" and notes that a stable, reproductive pair is still the central structure of the social group. Some researchers consider this data and some information on regular group splitting in other species of tamarins as challenging the notion of "extended families" in callitrichids (e.g. Kinzey 1979). A major drawback to that challenge is that kinship information is not known in the field studies. Kleiman, in her major review of monogamous mammalian species, still referred to monogamy as a characteristic of all Callitrichidae, although she notes that different species show variations, or modifications, and that a species may "stray from what is considered its modal social system in an optimum environment" (1977:40).

Callitrichids have a very high reproductive rate compared to other primates, females usually giving birth to twins and often producing two sets in a year. The common marmoset appears to reproduce at any time of year, while the cotton-top tamarin is reported to be seasonally polyestrus. Thus, it is not difficult to see how one mating pair could produce a large group in three or four years, especially if older offspring remain in the group.

256

Subsistence Factors

All marmosets and tamarins are highly territorial, employing vocalizations, scent-marking and physical encounters to defend their areas and to maintain exclusive access to them. Callitrichids use olfaction a great deal in many forms of communication (sexual, agonistic encounters, etc.), and from the descriptions in the literature it appears their territorial behavior involves more scent-marking and less vocalizing than the territorial behavior of gibbons and siamangs. Both sexes participate actively in territorial patterns, although some workers report a tendency for females to scent-mark more often and for males to be more involved in actual physical encounters with outsiders or neighboring groups. In captivity, callitrichids show great hostility toward strangers of the same sex.

Neyman reported that there are no obvious leaders of group movement when cotton-tops forage, and Dawson noted that any group member may act as sentinel, watching out and warning of predators. It is not uncommon for a whole group to "mob" a predator, and tamarins react especially strongly to raptorial birds. When young monkeys fall accidentally to the ground, they may be retrieved by mature group members of either sex. In the common marmoset both parents will share food with their offspring. Thus callitrichids do not appear to exhibit specialized sex roles with regard to foraging and group defense patterns.

Ontogeny and Socialization

Rothe (1973, 1974, 1978) observed several births in common marmosets in captivity. Although he noted that the artificial lighting needed to allow observation probably had some effect on the group, he reported that all group members exhibit strong interest in the process of parturition. The female about to give birth moves somewhat away from others in the group, but they groom, greet, huddle beside, sniff and lick her. Earlier studies (Langford 1963) reported that adult male marmosets may "assist at births" by tugging on the emerging offspring, but Rothe reported other group members as watching the birth in great interest but not interfering. Immediately after birth, neonates may be licked, manipulated, groomed and held by the father or older siblings. Other group members participate with the mother in eating the placenta.

Rothe described the common marmoset mother as performing only a minimal and somewhat "reactive" or restrained caretaking role. She does not retrieve newborn infants, even those being injured by other group members, unless they vocalize; and much of the carrying, feeding and cleaning she performs is initiated by the infant. Box (1975a,b, 1978) and Epple (1970b) also have studied many common marmosets in captivity, and they describe a more active caretaking role for the mother. In fact, striking variability between individuals and between groups is a characteristic of descriptions of marmoset behavior.

It has been said often that adult males, or the fathers in marmoset groups do all the caretaking, only passing the infants back to the mother for nursing. Now researchers suggest that this is an oversimplification; mothers do perform other caretaking behaviors, but more importantly, older siblings also often carry and care for young. It is quite remarkable to see juveniles of less than a year carrying days-old siblings, even transporting them when engaged in rough and tumble play with other juveniles. Nonetheless, the fascinating data on juvenile caretaking does not alter the striking and strong participation of adult male callitrichids in infant care. They carry infants from birth, share food with them, play with them and protect and defend them against intruders. Some adult males even regularly huddle up against the mother and the nursing infant.

Marmosets and tamarins are also unusual in the primate world in that most individuals grow up with a peer who is a twin of the same or opposite sex. In their early development, twins may not actually spend very much time together, because they are often carried separately by different family members. Of course this depends on whether they are the first, second, or third set of twins born to the parents. Once twins begin to locomote independently at about five to seven weeks of age they spend much time together and coordinate activities such as feeding, resting and playing. Twins are usually favorite play partners. They usually are described as being very non-competitive (but see Kleiman 1979), and they may direct initial sexual behaviors toward each other. Young callitrichids seem to be very playful animals and at times even adults participate in play. Box, in her intensive study of play in common marmosets (1975b), reported no sex differences in any categories of play, including soliciting play, chasing, and rough and tumble play.

Infant marmosets are very precocious and no significant sex differences appear in their growth patterns, although Hearn (1978) suggests that males reach sexual maturity somewhat earlier. They begin to play and to scent-mark branches at four weeks of age, to groom others at seven weeks, to be weaned at thirteen weeks, and are quite independent by the time a new set of twins is born, generally when they are only five or six months of age. Sexual maturity occurs around 14 to 18 months of age, but as I will explain below, callitrichids do not reproduce while they remain in the parental group.

What happens to twin-bonds, and by what process maturing offspring leave groups and form their own mating bonds is not well documented. However Box (1975b) has suggested that from about 14 months of age, twins become very involved in infant sibling care and "grow apart" as they come to interact individually with their siblings.

There are no generally reported sex differences in juvenile patterns of infant sibling care, and several researchers have noted that both female and male juveniles who are experienced with siblings grow up to have the greatest success in raising their own offspring. However, Ingram (1977, 1978) noted some sex differences in her study of common marmoset socialization, namely that female infants seek the proximity of their parents more, and that mothers wean male infants more abruptly than female infants. She suggested that females at all ages are more interested in infant care, and that adult males tend more toward a protective role and less toward a parental role, especially as older offspring come to participate in infant care. Although these particular sex differences in behavioral development are not reported by other researchers, this might be because they were not looking for such sex differences or they might be the result of intraspecific variability.

Social Dynamics

In most captive studies, callitrichids are described as aggressive animals, engaging in frequent within-group and between group agonism. While neither sex is reported to dominate the other, and indeed the relations between adult females and males are very peaceful, separate dominance hierarchies for each sex are said to be an important aspect of social group life. In fact in every social group, whether artificially constructed of strangers placed together, or naturally developed from the reproduction of one pair, an alpha male and an alpha female actively inhibit the sexual activities of other members of their own sex. Because they are not always successful in this, subordinate individuals do sometimes copulate, but subordinate females do not become pregnant. Thus it appears that every group, no matter how large, has only one reproductively active pair. Hearn (1978) found hormonal differences between subordinate and dominant marmosets, and determined that subordinate females fail to ovulate. It is possible that pheromones play a role in the inhibition of reproductive capacities in subordinate marmosets. Epple (1973) found that dominant individuals scent-mark more often, and that other marmosets can distinguish perches or branches that have been marked by dominant individuals. The alpha male and alpha female also are described as performing a "control role" (see Chapter 8) toward others of their own sex.

Epple has even suggested that in comparison to titis (*Callicebus*) who pair-bond and form monogamous societies through mutual permanent attraction, a so-called pair-bonded marmoset female and male mainly come together through aggressive competition for mates, rather than through strong mutual attraction. Similarly, Rothe (1975a) noted that in large artificial groups of common marmosets the fact that only two individuals breed may not indicate a pair bond, but may rather result from the dual dominance hierarchy in which one female and one male suppress the activities of everyone else. However, mutual and permanent social and sexual attraction between pairs of female and male marmosets is reported in some captive studies. In the wild, where adults are not artificially introduced to each other as strangers without choice, it may well be that more evident social bonding occurs between the reproductive female and male.

As long as mature offspring remain in the same group as their parents they care for younger siblings and do not reproduce. If they are separated from their parents and placed with a mate, they do begin to reproduce. This led Epple (1975a) to speculate that the large groups encountered in the wild might occur when mature offspring leave the group, pair-bond, reproduce and return on occasion to the parent group to travel in extended families. I noted earlier that the extended family hypothesis is currently undergoing some reappraisal, primarily as a result of the instability of groups of *Saguinus* species observed in the wild. However, in terms of kin selection theory and what we know about kinship bonding in other species of mammals with similar social structures (e.g. wolves), it does seem likely that the mature group members who participate in infant care and who do not themselves reproduce, will be found to be in some way related to the parents and the infants of the callitrichid group. If this is found to be the case, "extended family" is probably still an appropriate description of callitrichid society.

THE LIFE-WAY OF DUSKY TITI MONKEYS (*Callicebus moloch*)

Description, Distribution, Diet

Of the three species of titi monkeys in the genus *Callicebus*, only the dusky titi (*C. moloch*) and the yellow-handed titi (*C. torquatus*) have been studied in the field. As more information is available on the dusky titi, the present description will focus upon this species, with some supplementary data about yellow-handed titis introduced where appropriate.

Titis are small, tree-dwelling, New World monkeys, which weigh from .75 to one kg. and show no significant sexual dimorphism in size or coloration. Their fur is long and very thick, gray to reddish brown in color, and the tail is long and "bushy" in appearance, but not prehensile in function although it is often intertwined with the mate's tail. The face is small, round, black and fairly free of hair. Dusky titis have a contrasting, colored "headband" of fur, and yellow-handed titis conform to their name with lighter colored white or yellow hands and feet. Titis exhibit a highly characteristic hunched sitting posture in which all four limbs are held close together to grasp the horizontal branch below the body, while the tail hangs straight down below the branch (see illustration, p. 253). The overall impression is of a soft, rounded form with a teardrop shape and blurred outline. Perhaps it is not surprising then that the titi reaction to predators is that of a camouflaged animal, that is, it becomes silent and motionless.

Callicebus monkeys are found in South America, from the Rio Paraguay basin in Brazil and Paraguay north into Columbia, Venezuela and northwestern Brazil. They are also found in parts of Bolivia, Peru, Ecuador and extend to the base of the Andes in the west. Titis typically inhabit low, thick forests which are seasonally flooded, and Moynihan referred to them as "the closest thing to a real swamp monkey in the New World" (1976: 76). Their preferences for the thickest part of the vegetation in dense native swamps and riverine forests, their "retiring" nature and their tendency to become motionless when approached, should explain why they have not often been the subject of field studies. The titis eat a variety of foods, such as fruits, leaves, insects and nuts, and can be classified as primarily frugivores or fruit-eaters. Dusky titis supplement fruit with leaves, while insects are the second most important food source for yellow-handed titis.

Social Organization

Titis live in monogamous groups consisting of a mated-pair and their dependent offspring, and groups contain from two to five individuals. Mason (1968b, 1971, 1974, 1975) conducted extensive experimental research on captive individuals, as well as some field research, to demonstrate that the social organization of Callicebus is a direct function of the actively affinitive bonding between female and male titi monkeys. In free-ranging titis, the group members remain very close together and closely coordinate all activities such as feeding, travelling, sleeping and territorial encounters. Mates also frequently show positive contact behaviors toward one another such as tail-twining (sitting close together so that one tail can wrap around the other), hand-holding, food-grasping, grooming, lip-smacking. Both members of a pair exhibit distress when separated. The formation of new groups has not been observed in the wild, but Mason's research with captive titis shows that adult females and males, originally strangers to one another, may develop pair bonds after an initial period of "depression." The depression is presumably due to the unfamiliar environment and the rupture of the original pair bond. Pairs formed in captivity however have not yet exhibited the territorality characteristic of free-ranging titi pairs.

Subsistence Factors

Dusky titi monkeys typically have very small home ranges (3200-4700 square metres), and are highly territorial. Both sexes participate in territorial displays and frequent encounters with other groups, largely consisting of vocalizing, posturing, tail-lashing, rushes, counter-rushes and vigorous chases. Titis are noted by several researchers, especially Moynihan (1966) and Robinson (1979), to rely greatly on vocal communication.

Actual physical contact and wounding is infrequent. According to Moynihan, titis pay little attention to potential predators and act at times almost as if they were immune to attack. Their typical response when frightened is to freeze and become totally silent. While specialized roles for females and males have not been noted for predator defense or resource defense (territoriality), it has been suggested, from one field study of yellow-handed titis, that the female leads the group in searching for new food resources, while the male and infant follow behind (Kinzey 1977a). Another report on this same group notes that males, but seldom females, may "share," or transfer food to offspring (Starin 1978).

Ontogeny and Socialization

Very little information has been published on the development and socialization of young titi monkeys. It is known that the male regularly carries the infant on his back and that mobile offspring tend to remain closer to the father than to the mother. Young titis typically beg food from the male rather than the female parent, and the father is tolerant, allowing foods that are difficult to find or to process to be taken from his hands. Immature titi monkeys engage in solitary play or direct play toward the male parent, whose response is again one described as "tolerance." Aggression directed toward subadults by the parents is considered to be important in the peripheralization of maturing young; however nothing is known about the process by which new social units are formed in the wild.

Social Dynamics

The titi monkey was described by Mason (1975) as an extreme example of a "contact" animal. Daily life is marked by behaviors indicating attachment between group members, by close coordination of activities, and by lack of intra-group agonism. Dominance relations between group members are not reported. Lack of sex differences in behavioral patterns is also a notable characteristic. Mason suggested from his experimental research that the female titi is somewhat, but not significantly, more attached to her mate and her territory than is the male. However experiments in which an "intruder" of the opposite sex is placed close to one of the mates, show that males rather than females are more likely to respond with agonism toward the intruder and enhanced attraction to the mate (what the researchers called "jealousy behavior").

In the limited field observations reported, sexually receptive females have occasionally been seen to copulate briefly with a neighboring male. Her own male mate's reaction to this is variable, ranging from mild aggression, to courtship and attempts to restrain her, to no reaction at all. The response of paired titi monkeys to separation is extreme, consisting of increased locomotion, increased vocalizations and increased heart rate. This is similar to the response of infant rhesus monkeys when separated from their mothers (see Chapter 12), and Mason likened the marked pair bonds in *Callicebus* to those of fillial attachment (mother-offspring) in other primate species. Unlike females in many, if not most primate societies, female titis show no attraction toward other females.

Many fundamental questions about social processes in titi monkeys remain to be answered, including adequate descriptions of initial bond formation, and the socialization patterns by which young group members are reared and eventually brought to independence. What we do have good descriptions of, for *Callicebus* species, is a bonding between an adult male and female reproductive pair which is more intense, stable, and complete than that described for any other nonhuman primate species.

THE LIFE-WAY OF SIAMANGS *(Hylobates syndactylus)*

Description, Distribution, Diet.

Siamangs are the subject of some taxonomic dispute. Expert opinion ranges from considering them so different from the various species of gibbons that they should be placed in a separate genus *(Symphalangus)*, to considering them so closely related that they are only a "race" of gibbons. Here I shall follow Groves (1972), who has published a detailed analysis of gibbon material and has placed siamangs within the genus *Hylobates*. Information from the well-studied *H. lars* or white-handed gibbon will occasionally be used to supplement what is known about siamangs, but important differences between the two species will be noted also.

A long, shaggy, black coat, a naked and inflatable air-sack beneath the chin, and a "gangly" body with short trunk and long arms, are the characteristic physical features of both female and male siamangs. Hylobatids in general show very little sexual dimorphism. The male siamang weighs around 11 kg., and while the female is sometimes said to be somewhat lighter (approximately 92% of male weight) and at other times said to be equal in weight, it is agreed that she is fractionally larger (102%) in body length. Canines are long and slender and the same size in females and males. Both sexes have black, relatively immobile faces and are said to exhibit few obvious facial expressions or gestures. Both sexes also exhibit the graceful fluid locomotor pattern which makes the hylobatids the brachiators par excellence of the primate order. The flowing movement of their swinging progression through the trees and the breath-taking impetus that results, puts the observer in mind of dancers or gymnasts. Perhaps it is partially this behavior which earned gibbons an exalted, mystical place in Chinese poetry and Buddhist lore. The characteristic duetting or coordinated calling between the female and male in the group has also led to an association of gibbons with music and musicians in China.

Siamangs are found in Sumatra and Malaya and are completely arboreal, preferring the middle canopy of the trees on hilly slopes in the tropical rain forests of these islands. They are primarily leaf-eaters or folivores, preferring new young leaves, although fruit is their second most important food item and smaller quantities of flowers, buds and insects are also consumed. Siamangs occupy small territories of about 23 hectares on average, and it may be that their territories are smaller than those of white-handed gibbons because the latter are primarily fruit-eaters or frugivores and require larger areas to adequately feed themselves.

Social Organization.

Like all species of gibbons ("hylobatids"), siamangs live in small, mated-pair groups, consisting of one adult female, one adult male, and their dependent offspring. Members of siamang groups are remarkably cohesive in space and synchronized in group activities. The pair bond appears to last for the lifetime of the individuals, or at least to be very stable, as one group has been followed for nearly a decade. Mating occurs between the adult pair, and appears to be seasonal, taking place on an average of once every other day for a "breeding period" of about three of four months during a year. However the interbirth interval is about 2 to 3 years and probably female hylobatids become receptive only every other year at the most. Although all members of the group are closely bonded, the oldest offspring is the most peripheral group member, and as she or he reaches adulthood, may leave the group around the time of birth of the youngest sibling.

Subsistence Factors

Siamangs are highly territorial in comparison to most primate species, but their territorial disputes are less frequent and less intense than those described for the white-handed gibbons. Siamangs maintain their territories in three ways: (1) by regular ranging near boundaries; (2) by calling at, or chasing, neighboring groups about to intrude; and (3) by loud group calling displays (Chivers 1974:52). As well as their brachiating skills, hylobatids are reknowned for their impressive vocal abilities. Siamangs are no exception, and while some researchers are of the opinion that the siamangs' "barks," "booms" and "screams" are less melodious than the duetting of white-handed gibbon pairs (e.g. Fleagle 1977), both species do exhibit complicated, ritualized patterns of coordinated calling.

The territorial vocal display of siamangs consists fundamentally of a rhythmic bark series by the female punctuated by male screams, and group bark-chatters and booms (using the inflatable throat sack) by both sexes. An individual female's call pattern is quite distinctive and can be used by observers to distinguish between groups; however captive studies show that both a female and a male must be present to coordinate and complete the group calling pattern. Siamang vocal displays show similarity to those of white-handed gibbons in that the female plays a prominent role, but differs both in the actual calls produced and in the fact that the whole group, including offspring, participate in siamang territorial calling, while only the adult pair participate in the case of white-handed gibbons.

Group movement over arboreal pathways is usually single file in siamangs, with the female in the lead position most of the time (65% of the time in Chiver's major study, 1975). The adult female is the first to leave the night sleeping tree in the morning and the first to enter the sleeping tree in the evening. Although the female usually moves ahead or "leads" the group, she waits for the others and will backtrack if the male and their young do not follow. Neither sex exhibits special sentinel behavior, but all group members are described as habitually alert to changes in the surrounding environment.

Siamangs typically react to initial encounters with humans by silent retreat, and there is some indication that the adult female may habituate first to the continued pres-ence of observers. The common reaction to predators (possibly large cats, raptors and snakes) is an initial alarm grunt often given by the female (who, remember, is usually in the lead) followed by deafening booms, barks and screams given by everyone in the group. On one occasion when an infant was observed to fall, the adult male retrieved it while the female gave alarm vocalizations.

Ontogeny and Socialization

During the first eight to 10 months of its life, an infant siamang is carried by its mother, but beginning at this age the adult male or father begins gradually to take over the carrying of, and probably much of the caring for, the infant. (Apparently white-handed gibbon males do not show this paternal behavior, Ellefson 1974). Young siamangs con-tinue to sleep close to their mothers at night, but as they gain locomotor independence, travel close to their fathers. As weaning progresses, the mother becomes increasingly intolerant of the close presence of her offspring and even begins to situate herself at night so the young one cannot reach her. As the father does not carry the infant back to the mother at night, but merely settles down in the same sleeping tree, the young siamang tries to reach the mother on its own, and if it cannot, returns to sleep beside the father. The description of the adult female's relations with her older offspring borders on "a-social" as she is said to avoid contact with them much of the time. However, in the first year of her offspring's life, the mother is the primary caregiver and even engages in grappling play with the infant.

Offspring in these monogamous groups do not have peers with whom they can play, and much solitary, and "acrobatic" play is described for siamangs. Infants also play hanging and grappling games with older juvenile or subadult siblings who must inhibit their strength and enthusiasm if they are to avoid hurting the smaller, less coordinated infant. Older siblings will also remain closer to an infant in the absence of both parents.

The adult male of the group interferes in play bouts if the infant appears distressed, and regularly supports or sides with the younger offspring in any dispute which may arise between siblings. As infants reach the juvenile stage more and more of their interactions are with the father, but not all of these interactions are positive. The adult male both grooms and has agonistic interactions with his older offspring more often than does the adult female.

It was suggested by Carpenter (1940) and often repeated in the literature, that in gibbons, older offspring are slowly peripheralized by intolerance from the same-sexed parent; that is, mothers are intolerant of daughters and fathers of sons. The recent research on white-handed gibbons and siamangs suggests that subadults are peripheralized by parental intolerance, regardless of sex of the parent or of the offspring (Ellefson 1974; Chivers 1975). The common context for parental intolerance seems to be feeding situations, and secondarily, sleeping locations.

First as juveniles and increasingly as subadults, offspring are threatened for approaching a feeding parent too closely. Gradually a subadult travels, eats and sleeps at a greater distance from the rest of the group. She or he may also begin to vocalize separately from the group, and researchers believe the sex-specific (recognizably female or male) vocalizations act as an excellent advertisement for a prospective mate.

Indeed, one new siamang group was seen to be formed in just this manner: a subadult moved to an area adjacent to his parental group, gradually attracted a young female to his area, probably through his solitary calls, and then set up a new territory and group with her. As both female and male subadults are reported to be peripheralized from their groups, it is equally possible for a female to attract a male to her area with her recognizably "young female" voice. One case is recorded of a recently peripheralized male seen travelling on the edge of several different groups, some of which contained young females.

Social Dynamics

"Co-dominant" is a word often used in respect to the female and male pair of a hylobatid group. Chivers states that it is not helpful to think of one siamang as dominating another, even in specific situations. Agonistic episodes are very rare within siamang groups, occurring on average one and one half times a day. When they do occur, they mainly involve threats directed toward older offspring. During the breeding period of the adult pair, the male in particular becomes increasingly aggressive toward the oldest offspring, whether it be female or male, and it may be a common pattern for the subadult offspring to break with its group during the gestation or birth of the newest offspring. Agonism between the mated pair also rises during the breeding period.

The positive aspects of bonding between the adult female and male pair of siamangs is not emphasized in the literature as it is for *Callicebus* or titi monkeys (tail-twining, body contact, etc.). But this may be an artifact of different perspectives by different researchers, or the absence of offspring in many *Callicebus* studies, rather than reflecting weaker or less obvious pair bonding in siamangs. Indeed it is often remarked for the latter species how well and how peacefully the group members do everything in synchrony, with so little obvious communication. The adult female siamang does approach and groom the male, but the male engages in more grooming interactions and spends more time in grooming. Perhaps it is because the male interacts with his older offspring very closely as well as with the female, that he appears the focus of group activities, while the female appears more socially aloof. The male also appears to shift his attention from his offspring during part of the year, and pays more concentrated attention to the female during the breeding period. He solicits sexual activity and she either responds positively or moves away. Although female and male behavior patterns in this species are not as sex-differentiated as in macaques or baboons for example, such differences as do exist offer an interesting contrast to our more familiar picture of primate males as socially-aloof, externally-oriented, and leaders of group movement, and primate females as caregivers and social foci.

REFERENCES

Marmosets *(Callitrichidae)*
Abbot and Hearn 1978; Box 1975a, b, 1977, 1978; Dawson 1978, 1979; Eisenberg 1978; Epple 1970, 1973, 1975a,b, 1978; Hearn 1978; Ingram 1977, 1978; Kleiman 1977, 1978, 1979; Langford 1963; Nelson 1975; Neyman 1978; Rothe 1973, 1974, 1975a, b, 1978; Stevenson and Poole 1976; Voland 1977.

Dusky Titi Monkeys *(Callicebus moloch)*
Cubicciotti and Mason 1975, 1978; Kinzey 1972, 1977a, b, 1978; Kinzey and Gentry 1979; Kinzey et al 1977; Mason 1968b, 1971, 1974, 1975; Mason and Epple 1969; Moynihan 1966, 1976; Robinson 1979; Starin 1978.

Siamangs *(Hylobates syndactylus)*
Aldrich-Blake and Chivers 1973; Carpenter 1940; Chivers 1972, 1973, 1974, 1976, 1977; Chivers and Chivers 1975; Chivers et al 1975; Ellefson 1974; Fleagle 1977; G.T. Fox 1972; Groves 1972.

PART V

SEX DIFFERENCES
IN EVOLUTIONARY THEORY

Chapter SEVENTEEN

Sexual Selection:
Female Choice or Hobson's Choice?

Gorilla
Gorilla gorilla
*These are the "knuckle-walking," great apes of Central Africa. They are
extremely dimorphic with males showing great robustness and enlarged
sagittal crest.*

hob-son's choice — [after Thomas Hobson 1631 Eng. liveryman fr. his
practice of requiring every customer to take the horse which stood nearest
the door] 1: an apparent freedom to take or reject something offered when
in actual fact no such freedom exists: an apparent freedom of choice where
there is no real alternative. a: the forced acceptance of something whether
one likes it or not, b(1): the necessity of accepting something objectionable
through the fact that one would otherwise get nothing at all (2) the necessity
of accepting one of two or more equally objectionable things 2: something
that one must accept through want of any real alternative.
Webster's Third New International Dictionary

In many species of animals sex differences are not limited to the generative organs directly necessary for reproduction (primary sexual characteristics), but also include male-female distinctions in size, color, plumage or fur, ornamentation, and behavior. These latter differences are called secondary sexual characteristics and usually appear at reproductive maturity. Examples in primates are the mantle or cape of the male hamadryas baboon, the sagittal crest on the head of the male gorilla, and the swollen perianal skin of the estrous macaque female. Occasionally in the animal kingdom secondary sexual characteristics have become so spectacular and extreme, as in the case of the peacock's tail or the antlers of certain deer, that natural selection does not seem to some to be an adequate explanation for these features; or, in more modern idiom, would not predict these features. It appears, for example, that an individual peacock could not only survive, but might be better off without such a tail.

In order to explain this intriguing aspect of sex differences, Darwin developed (although he did not originate) the theory of sexual selection. This theory holds that certain (but not all) features of dimorphism serve not to make an individual more likely to survive, but rather to render the individual more attractive to members of the opposite sex, and/or more frightening to members of the same sex. Darwin concluded from the data he had amassed, that when the sexes of a species differ in external appearance it is the male who has been modified, because the female generally retains a close appearance to the young of the species. There are reversals to this rule, where females are larger or more differentiated, but examples are rare. Darwin did not accept the suggestion, now documented for some bird species, that the physical appearance of females (apart from males) can be modified for camouflage and protection against predators (Selander 1972). Therefore, he saw sexual selection as acting mainly upon males, selecting for ornaments and behavior useful in competition with other males, and features effective in the attraction of females for mating. Thus, he wrote sexual selection "depends on the advantage which certain individuals have over others of the same sex and species solely in respect of reproduction" (Darwin 1936: 568); and sexual selection consists of "a struggle between individuals of one sex, generally the males, for the possession of the other sex. The result is not death to the unsuccessful competitor, but few or no offspring" (1936: 69).

Darwin held that among almost all the animals there is a struggle among males for the possession of females, and females have the opportunity of selecting one of the "strugglers." This competition would be most severe if there were a numerical preponderance of males, and this conformed with his observation that species with polygynous mating systems also exhibit the greatest degree of sexual dimorphism. Nonetheless, he was aware of the fact that some monogamous species display marked secondary sex differences, while some polygynous species do not, and that some aspects of sexual dimorphism are related to niche differentiation between the sexes rather than to sexual selection. The theory of sexual selection has two major principles, later entitled by Huxley (1938) "epigamic selection" (interactions between the sexes, usually female choice), and "intrasexual selection" (competition within one sex, usually between males). Translating this into the world of primates, it is sometimes said that male savannah baboons threaten off other males, but attract females with their hypertrophied canines and their belligerent behavior, features which are thus considered to be sexually selected. We will consider suggestions such as these in a later section of this chapter.

THE PROBLEMS

Probably none of Darwin's ideas have generated as much criticism among scientists as the theory of sexual selection. Indeed, one of the more curious aspects of the application of sexual selection principles in the behavioral sciences is that is spite of widespread discussions of the shortcomings of this theory, the major tenets still operate as hidden assumptions, or even axioms, in much of the writing on social behavior in animals. To try to understand this paradox I shall discuss what these researchers consider to be problematic, or simply wrong, with this theory; the kinds of data that have been used in its support; and some of the reasons for the continued acceptance and application of the theory in the face of serious criticisms.

A. Fitness

We have seen that Darwin conceived of sexual selection as being quite different from natural selection. He said:

> Sexual selection depends on the success of certain individuals over others of the same sex, in relation to the propagation of the species; while natural selection depends on the success of both sexes, at all ages, in relation to the general conditions of life (1936: 916).

This means that Darwin distinguished between characters which help an individual survive and characters which give an individual the cutting edge in competition for mates. The former features render an individual more fit, the latter are ornaments which render an individual more attractive. As Mayr (1972) pointed out, Darwin's concept of fitness was significantly different from the concept of fitness currently in use by most evolutionary scientists. Darwin saw fitness as a characteristic or general property of an organism which improved its chances of survival. The extensive development of the science of genetics in this century has defined fitness in terms of the individual's contribution to the gene pool of the next generation. In other words, an individual's fitness is now measured not merely by viability (some would even say not at all), but by reproductive success; and these qualities are seen as part of the same process, natural selection. This new definition obviates the need to postulate a special process, that of sexual selection, to explain secondary sex differences.

Both concepts of fitness, with their apparent criteria of survival of the organism or survival of offspring, are seen by some writers as tautological or circular, and as a result, to constitute a major flaw in the theory of natural selection (eg. Peters 1976). If natural selection is defined as the survival of the fittest, and the fittest are defined as those which survive, then natural selection would be simply, and tautologically, the survival of those which survive. However, "survival of the fittest" is not today, and probably never was, an adequate description of how evolutionary theorists conceive of natural selection. It has been argued by evolutionists that differential survival may be a

271

result of differential fitness, and thus a measure of fitness, but there also exist criteria of fitness independent of survival. Further, the survivors are not always seen as those which are fittest; for example, random or genetic drift may bring about differential survival. This sometimes highly abstract, yet very important, debate is too complex to pursue here; however, good discussions of it can be found in Lewontin (1974); Peters (1976); Gould (1977); G.L. Stebbins (1977); Caplan (1977); Castrodeza (1977); Bethell 1976.

The redundancy of a special process to explain sexual dimorphism was argued in Bastock's (1967) analysis of animal courtship. She maintained that sexual selection (competitive situations and female choice) need not be invoked in order to explain the evolution of male ornamentation and displays. She saw courtship displays as being adaptive and arising through natural selection because they function to bring the female and male together at the right time and place, as well as to synchronize physiological processes necessary for copulation, and to prevent cross-breeding. She concluded: "Hence sexual selection, in Darwin's terms, may not exist" (p. 92). Apparently deciding that adaptation is not a sufficient explanation for the extremes of male ornamentation and display sometimes observed, she then returned to sexual selection for an explanation, seeming, like other biologists, to accept the theory for want of a better one, even when she had shown it to be unsatisfactory.

B. Context for Development of the Theory: Race

Although Darwin cited sexual selection to explain sex differences in a wide variety of animals, he only developed the concept in detail in his book, *The Descent of Man* (1871) in order to explain the differentiation of the human races. Racial characteristics appeared to Darwin to be non-adaptive or non-functional in terms of survival, in the same manner as he saw male embellishment in other animals as unrelated to fitness. That he was aware of the problem may be seen in the following quotation:

> Although with our present knowledge we cannot account for the differences of colour in the races of man, through any advantage thus gained from the direct action of climate; yet we must not quite ignore the latter agency, for there is good reason to believe that some inherited effect is thus produced. (1936: 554).

Some of Darwin's critics even claimed that he invented sexual selection because he was unable to explain many of humankind's attributes using only natural selection (see Mayr, 1972). He may indeed have felt uncomfortable with the implications of applying natural selection to humans, but he did not invent the concept of sexual selection, already discussed briefly in chapter four of *On the Origin of Species* (1859); and he attempted to document extensive examples of sexual selection in many animal species, as well as in humans.

Today, of course, the entire study of human races remains extremely controversial. Some physical anthropologists think that human races, at least as traditionally conceptualized, do not, in fact, exist. Those others who still find value in the study of race, ascribe the differences between the races largely to the operation of natural selection, quite the opposite of Darwin's views. For example, two major theories in physical anthropology relate variability in skin color to environmental differences in solar radiation, therefore viewing skin color differences as adaptations resulting from natural selection.

So, in both cases, secondary sexual differences and human racial differentiation, the apparently non-functional phenomena which Darwin was attempting to explain with his theory, now are explained in other ways as the result of conceptual shifts in the thinking of biologists.

However, rather than abandoning the theory altogether, many scientists today think sexual selection is a type of natural selection. Since, unlike Darwin, most scientists now believe natural selection in its modern definition is sufficient to explain male ornamentation and racial differentiation, we are confronted with the paradox mentioned earlier: if the theory of sexual selection is no longer satisfactory or needed, why do scientists contine to afford it more or less qualified support? Three reasons suggest themselves:

1. We continue to be intrigued by the different, sometimes widely different, characteristics seen in females and males of the same species. It is postulated that there must be different selection pressures operating on females and males, and much discussion of sexual selection today deals with the causes and consequences of the variations between females and males.
2. More recent theories of differential female and male reproductive contributions (e.g. Fisher 1958; Williams 1966; see Chapter 18) and differential parental investment (Trivers 1972) can be seen to be reformulations or revitalizations of the theory of sexual selection. This area will be discussed in the following chapter.
3. Sexual selection, and in particular its principle of male-male competition, appears to some biological and social scientists to be a very good explanation of phenomena observed in the animal world. To others it is simply the best explanation to date, even though certain aspects remain controversial or unclear.

C. The Principles: Epigamic and Intrasexual Selection

What was the relationship between epigamic and intrasexual selection postulated by Darwin and how did he actually envision the operation of male-male competition and female choice? These are not easy questions to answer, at least partly because Darwin equivocated in his several statements on female choice. An example is his use of "sometimes" in the following quotation: "We shall further see, and it could not have been anticipated, that the power to charm the female has sometimes been more important than the power to conquer others in battle" (1936: 583). Certain scientists, Selander (1972) for example, interpret the word "sometimes" as indicating that Darwin did not ascribe the greater importance to female choice. Others argue to the contrary, that Darwin thought of female choice as being so important, it even suggested the term "sexual" selection to him (Mayr 1972). Whichever way we decide to interpret the word "sometimes" in the passage above (and it probably depends on our own biases) this is obviously the sort of word an author uses when attempting to be extremely cautious or deliberately vague. On this topic Darwin was both. He seems to have been uncomfortable with the implications of female choice, and his ambiguous treatment of the principle set a trend which continues to this day.

Why uncomfortable? First of all, consider Darwin had concluded that female animals actually exercise an aesthetic sense akin to our own human sense of beauty, for this is the basis of female choice, and indeed of all sexual selection. Thus we read:

I cannot here enter on the necessary details; but if man can in a short time give beauty and an elegant carriage to his bantams, according to his standard of beauty, I can see no good reason to doubt that female birds, by selecting, during thousands of generations, the most melodious or beautiful males, according to their standard of beauty, might produce a marked effect. (1936: 69).

Just as man can give beauty according to his standard of taste, to his male poultry, or more strictly can modify the beauty originally acquired by the parent species . . . so it appears that female birds in a state of nature, have by a long selection of the more attractive males, added to their beauty or other attractive qualities. No doubt this implies powers of discrimination and taste on the part of the female which will at first appear extremely improbable; but by the facts to be adduced hereafter, I hope to be able to show that the females actually have these powers. (1936: 570).

Darwin probably had reason to feel defensive about this idea, because the notion of female animals discriminating and appreciating beauty for itself, or of having a sense of aesthetic values is antipathetic to the way most biological scientists view animals, and few have supported this suggestion. It seems to imply more than just a neurologically simple stimulus response to color and form, and Darwin himself said that a sophisticated taste for beauty such as this, implied certain "mental powers" to have been developed in the same manner and to the same degree in various and widely distinct groups of animals. Although many writers accept that human aesthetic values are derived from simple, naturally occurring phenomena, inquiries into the subjective consciousness of animals have not been encouraged in the disciplines of either ethology or behaviorism, the two most influential schools of animal behavior. (See Shafton 1976; Griffin, 1976.)

But perhaps there is something even more profoundly distasteful in the idea of female choice based on beauty. As Bastock (1976: 78) phrased it: "He [Darwin] held that the whole advantage of displays favored by sexual selection lies in the satisfaction of female whims." Female whims? Whims are silly or irrational passing fancies, and are traditionally associated in Euroamerican culture with spoiled and foolish females. Could male animals be at the mercy of female fancy, like so many medieval knights battling and composing songs for fickle, exigent ladies? This appears to many people to be an exceedingly trivial foundation on which to base so important a phenomenon as secondary sex differences. But if female caprice does not seem adequate, on what other basis might males be chosen? Darwin suggested a possible answer to this while speculating how sexual selection might also operate in monogamous species. If the most attractive males are also the most "vigorous" (by which he meant "fit"), females might be most enticed by "the more ornamented, the best songsters, or those that play the best antics" (1936: 573), but they actually would be choosing the most fit. This suggestion tends to meet with more approval, especially after later genetic theorists demonstrated that a genetic linkage between general fitness and extreme ornamentation would be a feasible mechanism in selection (Fisher, 1930; O'Donald 1962).

One major drawback to this idea, noted by some authors (Mayr 1972; Bastock 1967) is the assumption that there is such a unitary characteristic as overt vigor or fitness in males, and that it correlates with attractiveness, strength and being "well-armed." Mayr refers to this idea as naive, and it certainly does imply a very one-dimensional notion of fitness. There is in fact no necessary correlation beween these features, and the concept of "general vigor" is so vague as to be virtually untestable. Bastock demonstrates the multifactorial nature of "vigor" very nicely in describing a set of experiments in which albino peacocks were preferred by females to normal peacocks because the albinos courted and displayed more vigorously. Because albinism is always considered to be maladaptive, Bastock refers to this as an example of females choosing vigorous courters of low fitness. Further, modern biologists suggest that this model of females choosing the fittest males and thus ensuring that their offspring inherit the fittest genes, implies an unrealistically high, and simplified, heritability of fitness (Williams 1975; Maynard-Smith 1978).

Where does male-male competition enter into this picture of female choice? It is at this conceptual point that I think Darwin equivocated to the degree that actually undermined his own principle of female choice, writing that: "It is certain that amongst almost all animals there is a struggle between the males for possession of the female. Hence, the females have the opportunity of selecting one out of several males . . ." (1936: 571). From this and other statements (see below), it appears that Darwin assumed that male competition is prior to female choice and that males which win out in the struggle with other males will be most attractive to females. He had earlier argued (without presenting evidence) that "a hornless stag or a spurless cock would have a poor chance of leaving offspring" (1936: 69). This is an interesting hypothesis, but not a generally proven one; then or now. As a matter of fact, there are some intriguing data relating to this from a study of red deer, which showed that mutant forms without antlers collected and maintained large harems and did well in fights, but the mutant strain failed to spread in the population (Darling 1937). This exemplifies again that courtship vigor, and fighting vigor and size of "weapons" and other aspects of fitness need to be considered separately.

In many bird species males compete for territories rather than fight for possession of females as Darwin supposed, and females choose not the strongest or most beautiful male, but an adequate territory for raising young (Searcy 1979). In males there is not automatically a correlation between strength, beauty and the acquisition of the best territory, and there is no evidence whatsoever that female choice is determined by the size, or quality, of those attributes which have first helped a male to be victorious over other males (Mayr 1972). Contrary to what has often been supposed, there is no general rule that female animals will be "observed like the ladies in the times of chivalry, to attend the car of the victor" (Erasmus Darwin, *Zoonomia* 1874, in King-Hele 1968).

Nonetheless, the idea that winners in male-male competition will automatically be chosen by females is probably the most ubiquitous assumption in the sexual selection literature. Consider the implications: if the females choose only those males which first have been the "conquerors" in male-male competition, that is, if they must choose the already victorious males, then they have no real choice at all. This is a form of "hobson's choice"; a choice of what is offered or nothing at all. At one point Darwin seemed to recognize this deductive sleight-of-hand by which female choice disappears and he cautioned:

But in very many cases the males which conquer their rivals, do not obtain possession of the females, independently of the choice of the latter. The courtship of animals is by no means so simple and short an affair as might be thought (1936: 573).

Unfortunately, shortly thereafter he essentially ignored his own caution by concluding that females will select those males which are vigorous and well-armed (p. 578), and finally he diminished the importance of female choice even further when he wrote:

. . . the female, though comparatively passive, generally exerts some choice and accepts one male in preference to another. Or she may accept, as appearances would sometimes lead us to believe, not the male which is most attractive to her, but the one which is least distasteful (1936: 579).

This is indeed an example of hobson's choice, and it is a long way from a view of females as being the creative force in sexual selection, using their powers of discrimination and choice to add to the beauty and attractive qualities of males. So, perhaps we can see now why scholars have disagreed over the importance which Darwin ascribed to female choice, for he was definitely inconsistent on this topic.

In spite of the fact that Mayr (1972) and Zihlman (1974) and others recognized the "catch 22" nature of female choice, the idea that the same features which first allow a male to win in competition are necessarily the features which also attract females, continues to be almost axiomatic in ethological research. We shall see how often this has been assumed in primate studies. And I think it is a measure of the relative importance attributed to the two principles that we have a fairly good idea of how male-male competition operates in many animals, but we still really have no general understanding of the bases and the mechanisms of female choice. For some authors, so passive is the female role in selection that female discriminative powers have been reduced to the mere ability to recognize and choose males of the appropriate species (e.g. Lack 1968).

If brilliant ornamentation and extravagant displays are necessary before females even recognize males of their own species, then the importance of the female, and female discrimination in sexual selection is indeed minimal. It is perhaps small wonder that some nineteenth century evolutionists have described the human brain as being essentially a male adaptation, leading them not only to attribute lower intellectual powers to females but even, on occasion, to wonder why females are as intelligent as they are! Darwin himself was party to such speculation:

Thus man has ultimately become superior to woman. It is, indeed, fortunate that the law of the equal transmission of characters to both sexes prevails with mammals; otherwise it is probable that man would have become as superior in mental endowment to woman, as the peacock is in ornamental plumage to the peahen (1936: 874).

A careful reading of Darwin on sexual selection indicates that what truly concerned him was an explanation of "ornamentation," which appeared to be non-adaptive in terms of his understanding of fitness. Once this special explanation was rendered unnecessary by subsequent developments in evolutionary theory, what we were left with was a theory which emphasized males and their strategies for reproduction. One of the goals of future research should be to redress this imbalance. If sexual selection is to be a viable explanatory theory in the study of animal behavior and evolution, even as "merely a special form of natural selection" (Lack 1968), then our understanding of female choice must be greatly augmented. Bastock (1967) pointed out that male discrimination or choice has been documented in many species (contrary to the persistent expectation that males will mate with anything), and perhaps what is needed today is a major shift in emphasis to a focus upon how and why *both* females and males choose their sexual partners and compete for access to them, and how this influences the evolution of secondary sex differences. The assumption that a male has a unitary characteristic such that he is either totally fit or unfit, and that he has access to females or they "choose" him on this simple basis, does not do justice to the complexity of courtship, or of animal behavior in general. Neither does it do justice to the best expressions of the complex interplay of forces and processes which make evolution a powerful and productive theory of great intellectual and aesthetic value.

APPLICATION OF SEXUAL SELECTION THEORY TO NONHUMAN PRIMATES

A. Faces and Hinder-ends

Darwin himself considered the application of his sexual selection principles to primates, and in so doing, gives us some insight into the difficulties of explaining exceptions to the theory. For he recognized that in many primate species (as in some bird and many mammal species) beautiful ornamentation occurs in both sexes. For example, almost all the forest-dwelling species of the genus Cercopithecus have elaborate patterns and colors of fur and skin on the face and body, involving brightly-tinted moustaches and beards and bibs and crests of hair. Some species (see p. 278) even have a great colored patch in the middle of the face, looking for all the world like the exaggerated painted nose of a circus clown. Both sexes partake of these mardi-gras appearances (although some think the male colors are slightly brighter), and Darwin concluded that this is an example of "equal transmission of ornamental characteristics." This, he believed, occurred when features first acquired by males are then transmitted to females. He offered no evidence (other than analogy to sex differentiated species), and there exists no convincing evidence of which I am aware, that males were the first to acquire these features; rather Darwin seemed to have difficulty conceiving of females as being directly modified through sexual selection (however, see Huxley 1938; Winterbottom 1929).

This is another example of what we might call the "coattails" trend in evolutionary thinking about females, illustrated earlier by Darwin's comments about mental endowment in human females (and see also "Man the Hunter," Chapter 19). Much as unimportant politicians are elected on the coattails of the powerful politicians who endorse them, so Darwin suggested that females are carried along on the coattails of males. Or to choose a Christian metaphor with which Darwin would have been well-acquainted, much as Eve was made from Adam's rib, so Darwin believed that beautiful and conspicuous ornamentation in females could only be the result of a secondary transference from males.

Mutual display involving ornamentation in both sexes, is now thought to have a very important courtship function in many animal species (e.g. Bastock 1967). For a

Cercopithecus
An array of Cercopithecus faces showing the variety of patterns which,
combined with colors, are characteristic of both females and males of each
species. *Clockwise from top right:*
C. *neglectus* (de Brazza's), C. *pogomias* (golden-bellied), C. *aethiops* (vervet),
C. *diana* (diana), C. *cephus* (moustached), C. *alboguris* or *mitis* (Sykes'),
C. *ascanius* (of the spot-nosed group), C. *hamlyni* (owl-faced).

modern and detailed analysis of the particularly decorative colors and patterns of female and male Cercopithecus monkeys, see Kingdon (1980). He suggested that the patterns may serve three possible functions: (1) to augment facial expression; (2) to signal social messages; and (3) to act as a mechanism for social isolation. While outlining these hypotheses, Kingdon also called for further research on this topic, which clearly is needed.

In a supplemental note to his *Descent of Man* (reprinted from *Nature*, Nov. 2, 1876) Darwin offered a lovely set of musings on the bright red "hinder-ends" of monkeys. He was intrigued as to why monkeys display their hinder-ends to each other, and while he realized that he might be ridiculed for suggesting monkeys find these rather grotesque hinder-ends attractive, he did indeed think they were a form of sexual ornament developed through sexual selection. A zookeeper had informed him that monkeys display their hinder-ends to each other, and Darwin concluded that perhaps this was no odder than the habits of many "savages" who greet by rubbing their bellies or their noses together.

The displaying of hinder-ends was described in Chapter 7, in the less evocative modern primatological language as an "anogenital present." And while the functions of such presentations are still arguable, I would support the proposition that they are gestures of social affiliation used in contexts as varied as greeting, agonism, and interactions with infants. Probably Darwin was correct in speculating that monkeys do, in some sense, find each other's bright red hinder-ends attractive, for they do use the presentations of them as a cohesive gesture in many different types of situations.

B. The Priority of Access Model

In terms of the general impact of sexual selection theory on primate studies it seems fair to say that the major result has been a strong emphasis on male-male competition and male reproductive strategies (see Goss-Custard et al 1972; Crook 1972). This is true to such an extent that Caspari, a non-primatologist reviewing the literature on sexual selection in primates, was led to say:

> One factor mentioned by Darwin is assumed to be missing in nonhuman primates: female choice. Reproductive success is based on social interactions between males. But the possibility that the females, at least in some species, can express preferences cannot be dismissed. It is rather characteristic of the present state of primate behavior studies that new and unexpected observations are frequently made (1972: 345).

It is not in the least correct to suggest that female choice plays no role in primate reproduction and that reproductive success is based (only) on interactions between males. However, it is certainly correct to say that a reading of the literature could give one this impression.

This bias is nowhere better exemplified than in the "priority of access to estrous females" principle (discussed in Chapter 7), and ubiquitous in the primate literature. The original suggestion is often traced to C.R. Carpenter (1942b), the development of a predictive model to S. Altmann (1962), and the most sweeping (and apparently premature) confirmation of the idea to Devore (1965). However, numerous primatologists have taken part in its elaboration. This theory basically holds that in multi-male, multi-female troops of primates, the adult males compete for a position in a stable dominance hierarchy (female hierarchies often being seen as unstable, if existing at all), and the males then have access to estrous females in direct relation to the rank they hold. In such a system, dominant males will father the most offspring, and thus propagate the genetically-based characteristics determining male dominance.

Since the model and the problems surrounding the concepts of dominance and innate behavior have been discussed in Chapters 3 and 7, several deficiencies should be apparent. There is no conclusive evidence that in primates the more dominant are the most genetically fit, or that dominance shows high heritability in the narrow sense, or that males form long-term, stable hierarchies, or that dominant males in multi-male troops father the most offspring. And look what has become of female choice in this model − it is no longer even hobson's choice; the perspective has become entirely male-centered, because "priority of access" implies the *right* to approach or use, an incentive, in this case, estrous females. Small wonder Caspari concluded that female choice plays no part in primate reproduction!

The model then is inherently biased, because it assumes that males alone determine the patterns of mating. Even worse, the model, rather than simply generating hypotheses to be tested in the field and laboratory and then accepted or rejected on the basis of the results obtained, appears in an unscientific manner (though in a manner not uncommon in science) to have become reified. That is, the model has become accepted as the reality, and so one can read discussions of sexual behavior and other aspects of primate behavior in which the statement that dominant males have first access to estrous females and thus, propagate the trait, is considered axiomatic rather than hypothetical (e.g. R. Fox 1972; Crook & Gartlan 1966; Wilson 1975).

Because it is not possible to administer good biological tests of paternity (biochemical blood protein analyses) in most of the multi-male, multi-female groups studied, researchers have been forced to rely on the indirect evidence of mating activities to infer male parentage. This is a questionable, if necessary practice, since female primates may mate in the nonovulatory phases of their cycle. Such studies all ask a version of the questions: is there a relationship betwen high dominance in males and access to estrous females; or, is there a relationship between male rank and sexual activity?

To date, the answers truly seem to be "six of one and a half-dozen of the other." A number of studies did conclude that male rank determines mating success (e.g. Carpenter 1942b; Hall & DeVore 1965; DeVore 1965; Kaufman 1965; Struhsaker 1967a; Tokuda 1961-2), while a number of other studies concluded either that there is not a correlation (e.g. Baldwin 1969; Conoway & Koford 1964; Drickamer 1974; Eaton 1974; Reynolds 1966; Saayman 1970), or that many factors including dominance rank are at work in mating success (Tutin 1979; Taub 1980; Packer 1979b; Hanby et al 1971; Hausfater 1975). In one case two research efforts working consecutively (Conoway & Koford 1964; Kaufman 1965), with the same group of monkeys came to different conclusions, although some of the males had changed groups between the study periods (reported in Rowell 1972a). Lindburg (in press) found for rhesus in India, that the alpha male was the most sexually active by various measures, but that below the alpha position rank did not correlate with mating activities. Tutin (1979) described a similar situation for chimpanzees.

Although it is likely that the variability in the answers reflects to some extent actual variability in primate behavior, it is probably equally true that these contradictory results are a consequence of different measures of dominance and different measures of reproductive success by different primatologists studying different groups at different times. I suggested earlier that dominance and reproductive behavior are actually complex sets of phenomena which can be measured in many different ways. For example, "priority of access" to a female may be measured as the ability to approach and remain close to a female, which is not necessarily the same as actual acceptance or active participation in mating on the part of the female. In addition, mating success in males may be, and indeed has been, measured with such varied standards as: simple ability to court and mount a female without interference or harrassment; overall rate of copulations; number of copulations which actually terminate in ejaculation; or rate of copulations which occur during the probable conception period of the female.

A recent study by Drickamer (1974) added yet another confounding variable to this already confusing mosaic. He pointed out that previous research had been based on the assumption that all males in a multi-female, multi-male troop were equally visible to the human observer. In fact, the males are not equally visible and researchers bias their results unless they control for this effect. Drickamer then presented data which at first appeared to show that high ranking rhesus males mated significantly more often, but when the data were corrected for the lower visibility of low ranking males, the significant differences vanished. One is led to wonder if many earlier results would lose their significance if other researchers corrected for visibility.

In a similar vein, Altmann & Altmann (1977) in a recently published paper of methodology, also noted that we cannot answer questions about rates of behavior unless we control for sample bias. Clearly the relationship of rank to mating success cannot be conclusively determined until studies correct for the lower observability of mating activities in peripheral and low ranking animals. Even more conclusive would be biological tests of paternity that would allow us directly to determine which males are in fact fathering the most offspring. Some biochemical tests were completed on rhesus monkeys at the Yerkes Primate Field Station (Duval, Bernstein and Gordon 1976), which indicated there is no direct relationship between rank and reproductive success. These tests found that some low ranking and even some adolescent males fathered as many offspring as did high ranking males.

Hausfater (1975) carried out the most detailed testing of the priority of access model, and from his results one can see that the more carefully one attempts to answer the question, the more complex it becomes. In a study of baboons he found that the first-ranking male did not do the most mating and that access to estrous females did not follow the dominance hierarchy. In fact, both females and males discriminated in their partner choice on some other basis than dominance, so that high ranking females did not necessarily choose high ranking males, and the top ranking males did not always mate with an estrous female even though she was the only one available.

On the other hand, Hausfater did find that high ranking males were more likely to mate with females during the latter's probable conception days. (The significance of this may be somewhat lessened by the fact that sperm may be viable in the female for several days.) Also, as Hausfater noted, because males change ranks often, short-term differences in reproductive success may "even out" or vary extensively in the long-term perspective of a male baboon's lifetime. He suggested that males who live the longest and initially come from the highest ranking lineages are probably the most reproductively successful.

The most important point to make in relation to the priority of access model is that even if we *were* able to determine that in general a high ranking male mates more often with estrous females than a low ranking male (and remember that we have not), it is not thereby proven that a high ranking male has the right to any female as a *result* of his dominance or his aggressiveness. A correlation does not prove the direction of causation. In point of fact, it is possible to develop a model which postulates the reverse causation from the traditional assumptions. When male-male competition takes place in a social matrix, as it does in multi-female, multi-male troops, it is possible to argue that females, through aid to males or prior alliances with males, help to determine the outcome of male-male competition. For example, a peripheral male monkey may be repulsed from access to a social group by females in at least some species (see Yamada 1971; Packer & Pusey, 1979), or he may gain access to the interior of a new group and possibly enter into a male dominance hierarchy after certain females or female lineages form mutual support alliances with him (Bernstein and Gordon 1974; Yamada 1971; Fedigan 1976). He may then move up the dominance hierarchy, maintaining his coalitions and perhaps increasing his social skills in forming alliances, and eventually he may become a high ranking or an alpha male. If we then find that he mates more often than low ranking males, his reproductive success and his competitive success (his rank) may be seen to be at least partially the result of his attractiveness to females, rather than vice versa, as would be premised by the priority of access model.

While I have indicated that it is possible to argue that female choice is in some respects prior to male-male competition (and Janetos 1980, has done the same), in fact I would suggest that a better model, one which accords with the best expressions of sexual selection principles, would be one in which female choice is seen to act in concert with male-male competition. Bernstein (1976) made a similar argument in a paper examining (among other things) the priority of access model. He concluded that alpha males have highly developed social skills and social skills are related to success in attracting mates as well as success in competition. This would indicate that it is not dominance itself which causes mating success, but that both attracting mates and success in competition are interacting with a third factor, social alliance skills. In such a model, the problem of which comes first, male-male competition or female choice, disappears, to be replaced by a contemporaneous interaction of male-male competition, female-female competition, and male and female choice. Choice here becomes real choice, not a hobson's choice.

C. Female Choice

Caspari was forced to the conclusion that female choice plays no role in primate mating. Is there, in fact, any research on female choice in primates? Female selectivity as part of the mating patterns of primates has been alluded to since Carpenter (1942a) and Yerkes (1929), although female choice as such has not been discussed until recently. There exist a few studies which relate directly to this topic and it does seem to be of growing interest.

In the laboratory setting, female primates may be experimentally offered a choice of males, with male competition and choice held constant. For example, in one set of experiments, Eaton (1973) conducted choice tests on pigtail macaques in which adult females could release cage doors allowing a neighboring male to join them for social interactions, often including copulation. Eaton found that the females showed definite preferences for certain males, and that they chose males who were not aggressive (i.e. intent to injure) toward them. The females, with one exception, eventually stopped releasing one particular male who attacked them.

Herbert (1968) and Michael et al (1978) also found that rhesus females preferentially seek out (i.e. release from neighboring cages) males who are not aggressive to them, and groom and copulate with them. That females choose non-aggression in males is especially noteworthy in light of a recent model of human evolution involving female choice for cooperative, sociable males, described in Chapter 19. Also, Kleiman (1978) found that given a choice, captive female lion tamarins selected unrelated males or more socially and sexually active males. Although lion tamarins, like all callitrichids, mate exclusively with one partner, extra males who might stay in the vicinity while not sexually active, are well tolerated and aid in caring for offspring of the mated pair.

Michael & Saayman (1967) also performed laboratory choice tests on rhesus macaques which led them to conclude that both females and males exhibit preferences for certain types of "personalities" in sexual partners, which may be more important than the reproductive status (e.g. ovulating, producing sperm, hormonal levels) of the partner. This is one study which documented partner choice or discrimination on the part of male as well as female primates. Bachman and Kummer (1980) also conducted choice tests in which they found that both male and female hamadryas show preferences for certain partners. Further, they found that males may be able to asses the females' preferences, and this (but not the males' preferences) may determine whether a "rival" male respects, or attempts to break up, a previously heterosexual pair.

Turning to non-laboratory research on female choice, Seyfarth (1977) concluded that female baboons in his study group preferred the alpha male, and Hausfater (1975) also found female preference for higher ranked males (which the authors recognize is different from saying that the alpha male has priority of access to them). However, Jay (1963) found that the langurs in her study group did not prefer the alpha male. Saayman (1970) noted that chacma baboon females choose males who are not aggressive to them and who have intervened on their behalf in fights. Scott (1978) suggested that female choice in common baboons operates through female initiation of sexual interaction and cooperation with males, or through avoidance and lack of cooperation with unwanted males. She suggests that female baboons choose males on the basis of age, genetic relationship, previous interactions, and male response to female solicitation.

Tutin (1975, 1979) described how in Gombe chimpanzees, female cooperation is essential for the maintenance of special consort relationships and thus females exercise choice. In Tutin's opinion the selection criteria appeared to be social and caretaking abilities of the male chimpanzees and not their dominance status. Concerning macaque species, Taub (1980) found that female Barbary macaques initiated and terminated the majority of mating interactions, which he termed "consociations" (very brief consorts). Females appeared to choose to mate with as many different males as possible, rather than preferring one male to another. In contrast, Lindburg (in press) stated that male-male competition is primarily responsible for the selection of mates in rhesus macaques, although female choice may also play a substantial role. Lindburg found that female rhesus, in at least one study group, preferred the highest ranking male, and he suggested this is because the alpha male has a protective and leadership role in the group, and is attractive to all group members.

Our own study of consort partner selection in the Arashiyama West Troop of Japanese monkeys (Fedigan and Gouzoules 1978) indicated that it is very difficult to determine who initiates a consort relationship; however, both females and males chose and/or accepted partners on the basis of age (similar age group) and kinship (outside of the kin group), and females may even choose to consort with other females. Because there was no lack of available or even actively courting mature males, we interpreted homosexual behavior as part of a larger pattern of female sexual initiative and as part of female choice, rather than as stemming from a lack of male partners. In addition to these general criteria which have some influence on choice of partners, we think that a large part of the variability may be attributed to more individualistic or idiosyncratic characteristics. However, because they *are* idiosyncratic features, it is very difficult to carry out good, quantitative studies on them.

It was once suggested to me, only partly in jest, that the most popular male sexual partner in a multi-male troop had the prettiest tail, since males of this species display their tails very prominently during mating season. Darwin would certainly have thought it a resasonable basis for choice, and I would not dismiss the suggestion lightly. I do think however, that behavior of the male toward the female is probably the most important factor in determining her choice, and the previous experience of a satisfactory consortship with a male may influence a female to choose him again.

But what behavioral features may initially attract a female to a particular male in a multi-male, multi-female primate society? It has been suggested, for baboons in particular, that females may choose males as consorts if they have previous friendship bonds with them, and males may also prefer such females (Rowell 1972b; Ransom & Ransom 1971; Saayman 1971; Packer 1979; Hausfater 1975; Scott 1978). Unfortunately for the sake of simple, definitive answers, this appears to be widely documented for baboons, but does not seem to hold true for Japanese monkeys. Our studies (Fedigan & Gouzoules 1978; Baxter & Fedigan 1979), and those conducted by Enomoto (1974, 1978) found that female (and male) Japanese monkeys do not choose their consort partners from among the individuals with whom they have year-round friendship bonds (see Chapter 10).

Eaton (1977) offered another possible answer in a study of "display behavior" in Japanese monkeys. Displays are defined as stereotyped locomotor activities or routines accompanied by noise and/or motion generation, which serve to call attention to the displayer and to transmit signals, of which tree and branch-shaking were the most common in his study group. Displays were performed significantly more often by males and their frequency increased during mating seasons. He found that the frequency of male display did not correlate with individual male rank or attack frequency, but *did* correlate significantly with ejaculation rank (his measure of reproductive success in males). Thus, it seems possible that female Japanese macaques are attracted to males that perform these displays in the classic sense that Darwin saw female animals as drawn to those males that "perform the best antics" (1936: 573), although I doubt, as does Eaton, that this is more than a partial explanation for female choice in this group of monkeys.

In addition to displays, Japanese macaque males have elaborate courtship routines which vary from individual to individual and from troop to troop (Stephenson 1973) and females might well make their choices, at least in part, on the basis of courtship skills.

Female Japanese monkeys also have variable courtship gestures and skills, which in turn may give males a basis for choosing partners. Unfortunately, female courtship patterns have not thus far been studied in the same detail as those of the males.

In all of the studies described above, and in all of the nonhuman primate species I have observed, it is obvious that females are seldom sexually responsive to, or cooperative toward, males who harm or frighten them. Indeed, Hinde (1974) said that aggression interferes with mating rather than promoting it.

In summary, we truly know very little about what does attract primate sexual partners to one another. But the aggressiveness which many researchers believe helps male primates to win out in male-male competitions does not seem to be an attribute females choose.

The studies described above are only the beginning of what should prove to be an intriguing investigation of sexual attraction in primates. To succeed in drawing a multi-dimensional and balanced picture of intra-specific competition and choice, a better model than that of priority of access is needed, since the latter has the effect of minimizing both male and female choice and discouraging their study. Those few studies carried out so far indicate that both females and males are discriminating in their choice of mating partners and that the basis for choice is probably multifactorial, with social skills playing an important part. Among those qualities with some support from research studies, positive findings are reported for: previous successful consort; age; rank; lack of relatedness; personality; and idiosyncratic features. Some negative findings are also reported for rank, and previous friendly bonds.

Thus, some of the findings are contradictory unless we are willing to accept the possibility that there may be great variability among the non-human primates, as there is among human primates, and that a quality which may attract mating partners in one species or social group of primates, may not attract partners in another. Given the findings that personality and idiosyncratic characteristics may be important to successful consort relations, we need to be cautious in generalizing about all primate species. With these caveats in mind, it is to be hoped that we shall see in the near future some exciting research on the dynamics of sexual attraction and social bonding.

Chapter EIGHTEEN

Sociobiological Theories and Sex Roles

Celebes macaque
Macaca niger
The macaques are found only on the island of Sulawesi (ex Celebes). The female (left) is smaller than the male and here shows swollen estrus skin.

Evolutionary biologists have long puzzled over the reasons why animals appear to help each other, and why they often live in social groups, when evolutionary theory, especially the principle of natural selection, seems to emphasize individual competition. Proffered answers to this question from the time of Darwin until the 1960's have never completely satisfied researchers, and in the past decade a new set of explanations, related through a refocused view of selection and fitness, have been suggested, under the general label of sociobiology. The theory, or more properly, theories, of sociobiology are linked by an effort to explain social behavior entirely in terms of self-maximizing biological processes, most importantly the process of self-replication, and the achievement of reproductive success. Since these are viewed as the only and ultimate goals of life, they take on a power, almost an "animus" of their own, becoming the driving force behind all behavior, and providing the meaning of individual and social existence. E.O. Wilson, a leading proponent of sociobiology, stated this view succinctly: "The organism is only DNA's way of producing more DNA" (Wilson 1975a: 3).

One of the tenets of sociobiology is that individuals are at all times helping to ensure the ultimate maximal reproduction of their own genetic material, even when they may appear to be sacrificing themselves to help others. Such an assertion is not without philosophical, ethical and political implications, and the development of sociobiological theory has brewed a major storm of controversy in the scientific and lay community. This is especially true when the theory is applied to the explanation of human behavior, as most of its proponents believe it should be, and as many willingly do. An adequate discussion of the epistemological arguments surrounding sociobiology, is beyond the scope of the present, or any single, chapter, but the interested reader can find stimulating discussions of the numerous aspects and ramifications of sociobiological theory in the following sources: Caplan 1978; Clutton-Brock & Harvey 1978; Chagnon & Irons 1979; Gregory et al 1978; Ruse 1979; Gould 1977; Dawkins 1976; Daly & Wilson 1978; and Sahlins 1976.

In this chapter I would like to restrict the discussion to a brief description of and commentary on three principles of sociobiological theory directly relevant to the study of sex differences in behavior: kin selection, reciprocal altruism, and parental investment. Each description is followed by an example of its application in primate studies. Finally, we will turn to the implications of the three suggested principles for sex roles, dealing particularly with the charge, by some critics, of "sexism" in sociobiology. In order to introduce the reader adequately to kin selection, reciprocal altruism and parental investment, it is helpful first to consider an issue in evolutionary biology known as "the level(s) of selection."

THE LEVEL(S) OF SELECTION

The principle of natural selection is derived from the observations that organisms are highly variable in all physical and behavioral features and that these organisms have differential rates of survival and reproduction ("fitness") in different environments. The premise is that those variants which are best suited to the environment are the most reproductively successful: they are "naturally" selected. Further, if the variation has a heritable aspect, it will be preserved in the offspring, as an "adaptive" trait. This powerful theory, like many great ideas in science, is both succinct and open-ended, so that evolutionary biologists may be united in their adherence to the theory, while diverging in their interpretation of it.

One point of divergence concerns the understanding of "competition" in differential survival and reproduction. Does the observation that more organisms are produced than do survive (the "struggle for life") mean that individuals are continually in active combat with each other for limited resources, or simply that the variants best suited for a particular environment will leave the most surviving offspring? In other words, is natural selection pushing all organisms to their limits so that the need for constant optimization leads to an extreme efficiency and adaptiveness in all traits? Or is there, at least for much of the time, a lot of "slack" in natural systems, meaning that natural selection is largely the process of *de-selecting* grossly maladaptive traits, while allowing the survival of a wide variety of traits, some of which could be neutral or even marginally maladaptive in themselves. Lewontin (1970, 1979) and Dawkins (1976) are examples of evolutionary biologists with widely different answers to this question.

Another point of divergence, and the most relevant for this chapter, lies in answers to the question of what it *is* that natural selection acts upon, or, what is the level of selection. It is commonly thought, and indeed widely held to be true by most evolutionists, that selection acts upon individuals. It is, after all, individual organisms which die or survive and reproduce even if they are conceived of as simply packages of DNA. However it is also possible to think of groups, populations, species and even eco-systems as exhibiting differential variability, survival and reproductive success. We are all aware that many, many species have become extinct in the history of life on earth, and we often think of different species as being in competition with each other for survival. Yet, although the differential survival, reproductive success or extinction of any unit greater than an individual may be referred to as group selection, it is essential to remember that what lives and dies in a literal biological sense, and not a metaphorical sense, are individual organisms, and not groups of individuals. Gould's (1977) use of David Hull's aphorism, "genes mutate, individuals are selected and species evolve," illustrates well these three levels. It is not possible to discuss the vast body of literature dealing with group selection in this chapter, but the reader is referred to the following sources offering a variety of perspectives: Darlington 1972; Lewontin 1970; Stanley 1975; Williams 1966; Wilson 1975; Wynne-Edwards 1962; Maynard-Smith 1978a; and Packer 1978.

It has been assumed by many scientists that individuals may do things for the good of the group or the good of the species, even if these things are detrimental to themselves as individuals. Groups which contain such helpful individuals may be more successful than other groups. This idea was carried to one possible conclusion by Wynne-Edwards (1962), who argued that individuals may regulate and suppress their own breeding capacity or individual reproductive success for the good of the group, and thus avoid overpopulation and over-use of resources. It was in criticism, response, and indeed contradistinction to this particular conceptualization of group selection that some of the principal aspects of sociobiological theory were formulated.

One major criticism of Wynne-Edward's theory is that there cannot be selection for individual self-sacrifice of reproductive capacity for the good of the group, because, by definition, such individuals would be reproductively unsuccessful, while other individuals who did not regulate or suppress their breeding capacities would leave more offspring.

Thus the heritable basis of the self-sacrificing behavior would disappear. Nonetheless, we do know from the empirical data that in the animal world individuals often do behave in ways that give every appearance of diminishing their chances of survival and reproductive success. This brings us full circle to the question of why do animals help each other? Various theorists converged on the idea that what gives the appearance of self-sacrifice *must* in some way ultimately promote an organism's own reproductive success.

This is easiest to envision if individuals are conceptualized as promoting the replication of their genetic material rather than themselves as whole entities. Thus the focus of selection which some scientists had directed toward clarifying differential *group* and *species* survival, is redirected by sociobiological theorists to bring into relief selection of the basic units of heredity: the genes. It is possible to argue that ultimately genes are the units naturally selected (e.g. as in kin selection, see below). Also, it is possible to argue from mathematical modeling of processes in population genetics, that while selection may occur at many levels — for example selection of genes, individuals, and groups — such selection would occur at different rates, making each process differentially significant. Thus group selection might occur, but not rapidly enough to equal or override the effects of gene selection (for an opposing point of view see Wade & Breden 1980). We shall see in the discussions to follow that some principles of sociobiological theory may be interpreted to involve the selection of genes, individuals, and/or groups.

KIN SELECTION

Individuals not only share genetic similarities with their parents and their offspring, but to varying degrees with all their biological relatives. Thus it is possible for an organism to have at least some of its genetic material well represented in succeeding generations, even if it never directly reproduces, as long as its relatives are reproductively successful. This is the key to Hamilton's (1963, 1964) theory of kin selection, which suggests that individuals may help their kin, giving the appearance of self-sacrifice but ultimately promoting the maximum replication of their genes through their relatives. (Hamilton's ideas were presaged 30 years earlier by two population geneticists, Fisher 1930 and Haldane 1932). Kin selection theory introduces a new conceptualization of fitness — known as "inclusive fitness" — as well as a new way of thinking about reproductive success. *Individual* fitness concerns an organism's ability to produce offspring and thus contribute genetic material directly to the next generation. *Inclusive* fitness involves individual fitness *plus* the organism's effect on the fitness of relatives who are not direct descendants, the latter being the organism's *indirect* genetic contribution to the next generation.

Kin selection involves three important parameters: the degree of relatedness between individuals, which implies the probability of genes shared in common by descent; the cost of a behavior to an individual, in terms of its own fitness; and the benefit to the recipient, in terms of her or his individual fitness. Hamilton's theory predicts that individuals will help each other whenever it costs them little and benefits a relative more. Or stated more formally, individuals will help each other whenever the ratio of the recipient's benefits to the donor's costs is greater than the reciprocal of the degree of relatedness. ($k > 1/r$, where k = benefit/cost; r = coefficient of relationship.)

There are a number of possible ambiguities, or seeming paradoxes in kin selection theory as just described. One concerns sociobiologists' notion of selfish and altruistic behavior, already discussed in Chapter 2. Although I have just described kin selection theory without the use of these words, they are in fact central to all sociobiological theory. Helping another individual under the conditions described above in terms of cost, benefit, and degree of relatedness appears to be, and indeed is called, "altruism," but nevertheless it is considered to be ultimately or "truly" self-serving or selfish. Hence helping others becomes the same as helping oneself in an ultimate sense, and although the two words, selfish and altruistic, are still used, the distinction is collapsed, and is considered paradoxical by some, "special" sociobiological definition by others, and simply mischevious use of words by still others (see Chapter 2).

Another paradox that blurs some discussions of kin selection results from the difficulty described earlier of deciding the appropriate level of selection, since kin selection too can be interpreted as operating at different levels. For example, kin selection has been interpreted by various of its proponents as a form of group selection (Wilson 1975; Hamilton personal communication cited in Kurland 1977; Brown 1975); as selection at the individual level (for example Kurland 1977; West-Eberhard 1975); *and* as selection at the level of genes (Dawkins 1976, 1979; DeVore 1977; Kurland 1977). If the individuals who live in social groups are related to each other and help each other, then it can be argued that kin selection is the same as, or has the effect of, group selection. It is simply kin-group selection. At the same time, however, it is individuals who act to maximize their inclusive fitness and thus kin selection can be said to act upon individuals. Finally, the degree of relatedness is a measure of *genetic* similarity, and so it can also be argued that reproduction of genes, not individuals or groups, is the unit of selection.

APPLICATION OF KIN SELECTION THEORY TO NONHUMAN PRIMATES

In recent years quite a few papers in primatology referred to the mechanisms of kin selection, but so far only two published studies set out to extensively test the predictions of the theory in particular primate groups (but see also Kaplan 1978, 1979; and Chepko-Sade & Olivier 1980, discussed in Chapters 7 and 9). Massey (1977, 1979) analyzed "support" or "aiding" behavior in a captive group of 49 pigtail macaques, and concluded that individuals not only discriminated between relatives and non-relatives, but also chose to aid relatives of a closer degree of relatedness (e.g., siblings, offspring) more often than they did more distantly related relatives (e.g., cousins). She also found that age and sex were significant variables in giving aid — older individuals and females being more likely than younger individuals and males to aid others.

A second and more extensive study of kin selection was that conducted by Kurland (1977) on the Kaminyu group of Japanese macaques. He defined behaviors such as grooming and defense of an attacked animal, as altruistic acts, and interactions such as agonism, agonistic buffering and alloparenting as selfish acts and then tested for a correlation or linear relationship between the frequency or duration of these behaviors and the degree of relatedness between individuals. His findings were many and were detailed in his monograph; however, they can be summarized as containing some findings that offered support to kin selection theory, and others that did not. For example, grooming between individuals *did* increase with increasing relatedness, one of Kurland's predictions from kin selection theory, but agonism *did not decrease* with increasing relatedness, as also predicted from kin selection theory.

Altmann and Walters (1978, 1979, and unpublished manuscript) reviewed Kurland's study in detail. Some of their comments apply to Massey's study as well, and indeed, speak to the problems faced by all attempts to apply kin selection theory to empirical research. Many of these problems were recognized by Kurland and mentioned in his monograph or later publications (e.g. 1979); however they remained unresolved in his study. One major difficulty underlying many others, is that what some scientists see as an elegant and parsimonious theory — kin selection — can become cumbersome and problematic when concepts are operationalized and applied to the actual social behavior of animals. For example, giving aid, which involves an individual in a fight and risks wounding, may seem clearly altruistic, but how do we decide that grooming is altruistic and alloparenting is selfish? This difficulty arises from the near-impossibility of truly determining the "cost" and "benefit" of the vast majority of behaviors or acts in terms of future fitness or future reproductive success (Waser & Homewood 1979, also discussed this).

Most sociobiologists either tacitly leave out the "cost and benefit" measure from their discussions (as noted by West-Eberhard 1975), or else fall back on the calculation of *energy* expenditure as a measure of costs and benefits. Thus, for example, if an individual grooms another, she or he is said to be using up a limited supply of time and energy which could be put to other uses, hence the "cost" of grooming must be balanced by a "return" on the investment price. However this assumes that all animals are on something analogous to a welfare or pension energy budget, and that they must continually evaluate the spending of every "penny" of energy, an unproven assumption of the optimization or hyperselective interpretation of adaptation. It assumes that energy expenditure is a meaningful and adequate measure of cost in terms of fitness, when even those who use it recognize it is not. Rather, energy expenditure is used as a proxy variable, that is, as the next best thing. This brings to mind the adage that it may be possible to calculate the price of everything without knowing the value of anything. Determining the real value of a behavior or interaction to an animal's fitness rather than its theoretical price, calculated from an abstract energy budget, may be the crux of the problem in operationalizing much of sociobiological theory. When attempts are made to move beyond simple calculations of energy expenditure, such as in Kurland's suggestion that grooming is altruistic because a grooming monkey might ingest quantities of hair and die of hair balls in the intestine, the discussion moves from elegant theorizing to not very credible speculation.

On the other side of Hamilton's kin selection equation $(k > \frac{1}{r})$ from the benefit/cost ratio, is the degree or coefficient of relatedness. That individuals should help their kin in linear correlation with the degree of relatedness between them seems straightforward and easily testable. Indeed the testing of this hypothesis provides the most convincing of the primate evidence for kin selection thus far: Massey found that pigtail monkeys aid each other and Kurland found that Japanese monkeys groom and defend each other, according to how closely related they are. Even here there exist complicating and confounding variables, again clearly defined by Altmann and Walters (1978, 1979). Kin selection theory does not necessarily imply that an individual should dole out altruistic behavior to all relatives according to their degree of relatedness. According to Altmann and Walters the optimal strategy, if benefits accrue in constant proportion to the frequency or duration of the behavior, would be to direct *all* of the altruistic behavior to the closest relative, and *none* to others. On the other hand, Alexander (1974) argued that kin selection could also result in individuals treating everyone in the group according to the *average* relatedness of all members or potential recipients of altruistic behavior. Thus it has been argued that theoretically there may be at least three ways in which kin selection could operate: "all or nothing" (Altmann and Walters); "the golden mean" (Alexander and others); or "to each according to its genetic relationship" (Kurland and others). The last has been the most common interpretation of kin selection theory, although Dawkins (1979) has recently acknowledged that it is incorrect.

To my mind a more pressing problem for the present studies is adequate determination of all coefficients of relationship in any multi-male, multi-female primate society, such as the Japanese macaques studied by Kurland, and the pigtail monkeys studied by Massey. As I noted earlier, paternity in this social situation is unknown to the observer and gives every appearance of being unknown to the animals. This means that in our geneaological charts we only record the *maternal half* of an individual's biological relations. Even that maternal half was not completely known in Kurland's study, as some mother-offspring relationships were inferred from proximity, which led to an unfortunate circularity, since proximity was also one of his measured kin effects. If kin selection did occur in the manner Kurland suggested, then it should operate according to the degree of relatedness in a genetic sense, and relatives through the father's line should be treated equally to relatives through the mother's line. Although we cannot test to see if the unknown patrilineal relatives are treated like the known matrilineal relatives precisely because we do not know biological paternity, it seems unlikely to be happening in macaque societies since observers report that individuals of one matriline regularly defend each other against members of another matriline. We recognize siblings born to the same mother who may often be half-siblings because they have different fathers, but we cannot recognize half-siblings who share the same father and different mothers.

This point may be easier to visualize if the reader returns to the matrilineal kinship chart, Figure 3 in Chapter 9. For example, if we find that Pelka 60 regularly aids her sibling or half-sibling, Pelka 65, in any agonistic encounter with Matsu 58, we cannot be certain that Pelka 60 is not just as closely related to Matsu 58 as she is to Pelka 65, but through an unknown common father. To carry this argument one step further, we might note that breeding males in Japanese macaque societies join and leave social groups frequently, perhaps once every four years. If, as sociobiologists generally believe, a very few males inseminate many females and produce most of the offspring during their short tenure in the group, then it would be highly likely that monkeys born in the same years to many *different* mothers, and therefore different matrilines, would have a very few *common* fathers. Patrilineal kinship charts would thus largely crosscut our currently known matrilineal charts.

Indeed, Jeanne Altmann (1979) recently argued that in some cases offspring born in the same year (an "age cohort") might be half-siblings through the father, and thus form a "paternal sibship." The latter, because they are of the same age, and because their life-spans overlap more than do those of maternal sibships, might provide more opportunity for the evolution of sibling altruism or competition, than do maternal sibships. The importance of paternal sibships or age cohort relatedness will be appreciable in cases where only one, or only a few males, father the young by several females and where young are born in very small litters, particularly of size one, and where the reproductively active males are frequently replaced.

Of course, we do not know biological paternity in multi-male, multi-female societies such as Japanese and pigtail macaques, and thus have no confirmed data as to the general pattern of patrilineality. It is possible that a female has each of her young fathered by a different male and it is also possible that the same male returns to the same female for several or many years to father her offspring. The point of this discussion is to illustrate that all we know thus far is that pigtail and Japanese macaques show differential behavior toward their relatives *through the mother*, and I would not be surprised if we continue to find this is true for many primate species. However, this is short of being compelling support for kin selection.

Altmann and Walters also criticized Kurland and other researchers in this area, for not developing and testing alternative hypotheses to kin selection, and for failing to reject a kin selection hypothesis when it is inconsistent with the data. Their suggestion for one alternative hypothesis to explain kin-deployment of behavior is that selection for enduring and persistent mother-infant bonds leads to physical and social proximity between matrilineal kin. If individuals tend to interact affiliatively toward those with whom they have experienced consistent proximity since birth, then recipients of such positive behaviors will be matrilineal kin. Kurland recognized that "because siblings have a common focus of interest, their mother, they may preferentially interact with each other" (1977: 20). But he regarded this as only a proximate mechanism, of little explanatory value, which only serves the ultimate purpose of kin selection. It is not uncommon for sociobiologists to reject or dismiss possible alternate explanations for behavior by relegating them to the realm of proximate mechanisms. However, it is obvious from Altmann and Walter's discussion of selection for extended mother-offspring bonds that, unlike Kurland, they regard this hypothesized selection as an important mechanism of ultimate explanatory power. If future research were to reveal that Japanese monkeys do treat their matrilineal relatives differently from their equally-related patrilineal kin, then extended mother-offspring bonds, rather than any and all genes shared in common by descent, would seem to be the underlying mechanism at work in these social interactions.

In sum, kin selection theory is difficult to test in primate social behavior, even in those few cases where we have adequate longitudinal data to know matrilineal relations. Meaningful measures of costs and benefits of behavior in terms of fitness need to be developed; confounding variables, such as the roles which age, sex and simple proximity play in "altruistic" and "selfish" behaviors, must be considered; and finally, paternity must be known, or at least reasonably estimated, before we can truly measure genetic relatedness in primate social groups.

RECIPROCAL ALTRUISM

Animals not related to each other, or only distantly so, occasionally help each other, and this behavior cannot be attributed to kin selection. Trivers (1971) developed a model he called "reciprocal altruism," to explain how altruism could occur between individuals independent of any genetic relationship — when the behavior is more beneficial to the recipient than costly to the giver, and if some kind of return or a similar act in reverse could be expected in the future. His primary examples are of humans saving each other from drowning, warning calls in birds, and species of "cleaner" fish, which remove parasites from other fish. The theory of reciprocal altruism is largely a mathematical formulation of the Golden Rule: do unto others as you would have them do unto you. Individuals help others in order to be helped in the future. "Cheaters" are eliminated because they are remembered and essentially removed from the altruistic exchange system (cf. the contrary argument that unselfish animals would be selected against, in reference to Wynne-Edward's group selection theory). The major differences between the mechanisms of reciprocal altruism and kin-selection are the mutual or reciprocal aspects, and the independence of genetic similarity proposed for the former.

It has been proposed that reciprocal altruism also can occur between kin, if they regularly return favors to each other, and indeed Kurland attributed one pattern of grooming relations in his study group of Japanese macaques, that did not fit kin selection predictions, to the mechanism of reciprocal altruism (1977: 123).This substitution of a second sociobiological hypothesis for a first one which is not substantiated has been criticized as rendering the body of sociobiological theory unfalsifiable (Allen et al 1976). Ruse (1979) countered that it necessarily takes several mechanisms to explain complex matters in behavior. Notwithstanding his defense of substituting rather than discarding hypotheses, it seems to me that while a specific kin selection hypothesis may by itself be testable and thus falsifiable, combining it, or "backing it up" with reciprocal altruism must render it less so. Reciprocal altruism is almost impossible to operationalize and test because we cannot define within what time period a "return" should occur, or what forms the reciprocation might take. If we predict that individuals help others in accord with some measure of benefit to cost and degree of relationship, this prediction can be tested. If we add, OR individuals may help anyone regardless of genetic relationship in order to be helped in some unspecified manner in the future, then all cases of helping have been described and the testability and explanatory power of the prediction is considerably diminished.

This is not to deny that cases of unrelated individuals helping each other in a reciprocal pattern do exist, as we well know from human ethnographic examples, and Trivers devotes a large part of his discussion to such human examples. However, the original insight into animal relations and the predictive power of this suggested mechanism seems considerably less than kin selection. Although its proponents see it as operating along *with* kin selection to maximize inclusive fitness, it is also possible to argue that it would confound and occasionally negate the effects of kin selection. That is, either one helps one's kin consistently against non-kin, OR one helps anyone that could be expected to return the favor.

APPLICATION OF RECIPROCAL ALTRUISM TO NONHUMAN PRIMATES

According to Packer (1977) in order to test this model we must be certain that the altruism in question is not the product of kin selection. Thus the behavior must occur regularly and reciprocally between unrelated individuals. Packer's study of coalitions among adult male baboons (*Papio anubis*) is to date the only such published test of the model for nonhuman primates. He began by noting that adult male baboons have generally tranferred into his study troop singly from elsewhere, and will be presumed to be unrelated, although the possibility that some males come from the same natal troop cannot be ruled out. Baboon males have often been described in the literature as forming temporary coalitions when one individual solicits another against an opponent.

From the patterns of solicitations and aid in his group, Packer concluded that individuals may have preferences for particular partners based partly on reciprocation. He found it difficult to test whether nonaltruists ("cheaters") or those who refused a solicitation to form a coalition were excluded from future aid. He also found it difficult to determine whether the benefits to the recipient exceed the costs to the donor, as required by the model. Further, he pointed out that a solicited male may sometimes join a coalition when there is a prospect of immediate benefit to himself, rather than simply to give aid.

In general, Packer's careful treatment of the topic points up the difficulties of testing the model. And while his study offers some cautious support for the hypothesis that male coalitions may be partly based on reciprocal aid, it cannot really be said to offer any significant new insights into the mechanisms of male-male interactions in olive baboons, a limitation probably inherent in the model.

TRIVERS' MODEL OF PARENTAL INVESTMENT

The third and last of the principles of sociobiological theory listed earlier, and the one must clearly relevant to sex differences, is Trivers' (1972) model of parental investment. He defined and described parental investment as:

> *any investment by the parent in an individual offspring that increases the offspring's chance of surviving (and hence reproductive success) at the cost of the parent's ability to invest in other offspring.* So defined, parental investment includes the metabolic investment in the primary sex cells but refers to any investment (such as feeding or guarding the young) that benefits the young. It does not include effort expended in finding a member of the opposite sex or in subduing members of one's own sex in order to mate with a member of the opposite sex, since such effort (except in special cases) does not effect the survival chances of the resulting offspring and is therefore not *parental* investment (1972: 139).

This model has been widely used to generate explanations for the different behavior patterns we can observe in female and male animals in relation to reproduction. Since survival of offspring represents the measure of reproductive success for both parents, females and males have often been conceptualized as behaving cooperatively to ensure their *inter*dependent fitness. However, the implications of Trivers' model, and one of the underlying tenets of sociobiological theory, is that females and males experience continual conflicts of interest, and only cooperate in so far as their previous investment in one jointly-conceived offspring, or one set, or litter, of offspring, forces them to do so. Producing and rearing one offspring can be viewed as having energetic costs, and more importantly "fitness" costs, in terms of other possible offspring. Each young is seen as an independent investment and, in a finite world, increasing investment in one tends to decrease the possibilities of investment in others.

In order to further her or his own reproductive success, the premise is that each parent will act so as to bear the minimum cost of the offspring. The "optimal strategy" for each of the pair after copulation and conception, therefore, would be to leave the other partner with the offspring to rear, and to move on to another act of offspring production — in other words, "desertion" of the partner. Since at the moment of conception, the female gamete, the egg, is considerably bigger than the male gamete, the sperm, the female's initial investment in the offspring is always seen to be greater. Futher, the model presumes that a female's reproductive success is fundamentally limited by her ability to produce eggs, but a male's reproductive success is not limited by the number of sperm he can produce, but by the number of eggs his sperm can fertilize. Thus females, with a lesser number of high-cost eggs are seen as the limiting resource, for which males, with a larger number of low-cost sperm, compete.

Perhaps the reader can begin to see why parental investment is considered a modern reformulation of sexual selection theory. In this theory females, with their initial "handicap" (a larger investment) at the moment of conception, and their more limited number of possible offspring, would choose their mates carefully, either choosing for males that give evidence of vigor and fitness in a general sense (good genes), or males that maintain an area with resources adequate for the young, and/or males that give evidence they will stay around to contribute to rearing the young with other forms of parental care (good parent). On the other hand, since males invest less at the initial conception of an offspring, and since they must compete with other males for access to females or female gametes, individual male reproductive success can vary enormously, and a male's best strategy would be to leave the female as soon after conception as possible, and move on to fertilize another female. Males would act so as to produce as many offspring as possible, while females would act so as to successfully rear the offspring they conceive, and these two "strategies" would usually result in antagonism between the sexes.

Although, according to Trivers, females suffer an initial handicap at the moment of conception, they do have and exercise some power of control, or choice, over which male fertilizes their eggs. For example, females could choose males who have no other mates and who aid in rearing the young, as in many monogamous bird species. Or they might choose a male who already has several mates but who maintains a large territory with ample resources, and a polygynous mating pattern would result. However, Trivers believes that the strong check, or counteracting force, on the power of female choice is her initially much greater investment in the young, such that breeding failure would effect her more strongly than the male:

> . . . because of their initial large investment, females appear to be caught in a situation in which they are unable to force greater parental investment out of the males and would be strongly selected against if they unilaterally reduced their own parental investment (1972: 156).

Numerous and varied aspects of female and male behavior, of parental behavior, and of female-male interactions have been described and explained using this model, and the reader is referred to Daly and Wilson (1978) as one extended example.

297

APPLICATION OF PARENTAL INVESTMENT THEORY TO NONHUMAN PRIMATES

While extensions of Trivers' model have been used to explain aspects of maternal care in primates and sibling competition (e.g. Hinde 1974; Trivers 1974) I would like to concentrate on its application to female and male reproductive patterns. The best known and most extended of these applications is Hrdy's explanation of infant deaths, or infanticide, in some groups of Hanuman langurs, described in Chapter 6. In some parts of India, langurs live in polygynous breeding groups which are periodically attacked by neighboring all-male langur groups. In the context of these attacks, suckling infants die and disappear, and the invading males are strongly suspected of killing them, since mortal wounding of infants by these males has occasionally been seen by primatologists, and is often reported by local villagers.

Some langur specialists interpret this pattern as a pathology, resulting from high population density and human disturbance of habitat. They point out that polygyny, all male attacks, and infanticide do not occur in low density, less disturbed habitats, and that infanticide is a well-known result of overcrowding in animal experimental research (Curtin 1977; Curtin and Dolhinow 1978; Boggess 1979). Other langur specialists, particularly Hrdy (1974, 1977a, b, 1979a), who has articulated the position most extensively, argue that infanticide is a facultative response in a high density context and an adaptive male strategy to produce more offspring in a highly competitive situation.

Using as her basic assumption the parental investment premise that female and male reproductive interests are independent and often antagonistic, Hrdy's explanation of infanticide states that a male with a limited breeding tenure kills the young of his rival males, thereby hastening the females into estrus, and quickly produces his own offspring before he himself is subject to an all-male invasion and ejection from the breeding group. The finding that generally only nursing infants die or disappear, is explained by the "pathology" school as resulting from the fact that they cling to their mother's body and are highly vulnerable during attacks on her. It is explained by the "male strategy" school as resulting from the fact that nursing suppresses estrus, and death of a suckling infant may hasten the mother into receptivity and conception of another.

Because males come and go from breeding groups so frequently at Hrdy's study site of Abu in India, she has proposed that males show some capacity to avoid killing their own possible progeny, by taking account of recent copulations with the mother. This allows females one possible counterinfanticide mechanism — they may copulate with several males at the time of conception and soon afterward, especially during all-male invasions, and thus make paternity of future infants highly uncertain. Hrdy argues that this is not a conscious strategy or mechanism, rather a selected capacity, and indeed sociobiologists are frequently forced to point this out since they often speak, or use metaphors, as though animals are conscious strategists.

This suggested capacity of males to associate copulation with birth or future infants, even if unconscious, and the attempt by females to confound the male's ability, does exemplify the degree to which sociobiologists believe natural selection alone can construct complex patterns of social interaction in animals. (Critics of sociobiology refer to this assumption, that every aspect of social behavior is the result of natural selection, as "hyperselectionism" and even "panglossianism," a criticism to which we shall return later.)

In the case of Hrdy's model, infanticide is an adaptive male reproductive strategy if at least the following premises prove to hold true:

1. if males deliberately or in a goal-directed fashion kill infants who are not related to them;
2. if females, because of infant deaths, return more quickly to estrus and mate with the infanticidal male;
3. if the infanticidal male is usually the father of the next infant;
4. if the infanticidal male has thereby meaningfully shortened the time interval before he could have produced an infant with a female without killing her former one;
5. if the average male tenure as the breeding male of the group is such that in an infanticidal population, he will not simply in turn have all his own offspring killed by the next invading male.

These premises are the points of disputation on which the langur controversy turns. Hrdy believes that evidence exists, or will be forthcoming, to substantiate each of these premises, while critics argue the contrary. The model of parental investment states that males will always act so as to produce as many offspring as possible, and Hrdy built quite an elaborate argument to demonstrate that in the case of langur infanticide, males are employing a reproductive strategy which does promote their own success in producing offspring. That the five premises listed above are difficult to substantiate does not in-validate her argument, but it does show that the graduation from plausible theoriz-ing (males will always act so as to produce as many offspring as possible), to empirical demonstration is a difficult one.

FEMALE STRATEGIES

Those who interpret langur infanticide as a social pathology were not long in noting that if infant-killing is a profitable male strategy, it is certainly detrimental to female langurs, who may repeatedly lose infants they have gestated, suckled, and "invested in." In addition, such a strategy is possibly detrimental to the future of langurs at Abu, when the death rate reaches 83% (Dolhinow 1977). Primates are generally a slow-reproducing species; each infant represents a considerable investment, and is singly, lengthily and carefully nurtured. This pattern is known as "k-selected," as compared to "r-selected" species. The latter produce many offspring of small cost each, many of which do not survive. Although early descriptions of langur infanticide (Sugiyama 1965) reported that females easily or passively allowed their infants to be killed by males, Hrdy described females as highly resistant to the male attacks on their infants, and discussed female "counter-strategies" to infanticide.

Female langurs at Abu are usually strongly antagonistic to males outside the breeding group and will attempt to chase them away. Females may also come to each other's aid when one mother-infant pair are the object of an invader male's attacks. Hrdy was not sure whether to interpret this as kin selection or reciprocal altruism, since she did not follow

299

the study animals for long or continuous enough periods to determine kinship, although she suspected that females in one group are related. However, in her view, females are ultimately unable to stop a male who persistently attacks their young, and may elect to abandon a severely wounded infant and start the process of producing another.

Why have female langurs not evolved effective counterstrategies to male infanticide? One obvious possibility for an effective counterstrategy would be to refuse to mate with an infanticidal male, and thus eliminate the inherited aspect of the trait, if such an inherited component exists. According to Hrdy, if infanticidal behavior is advantageous to males, then a female needs to ensure that her own sons will inherit the trait and thus be reproductively successful. Thus females are said to be caught in an "evolutionary trap" (1977: 49) out of which they have not been able to find their way. This is just one example of a rather persistent theme in the literature on parental investment in mammals, and some would say, in sociobiological theories in general — most behavior is interpreted first and foremost in terms of selection upon males, and female "strategies for reproductive success," when they are considered at all, are viewed as less successful due to competition with male strategies.

IS SOCIOBIOLOGY SEXIST?

The Argument
One of the many criticisms of sociobiology is that it is sexist, or androcentric, either inherently, or in application, or both. Sexist, it is said, from the metaphors its proponents choose to describe it (see Chapter 2), to its assumption that the interests of the two sexes are necessarily antagonistic, and that females lost the initial skirmish when they invested more in each individual gamete than males did in each of theirs. Haraway (1978) and Pilbeam (1979) note that sociobiology is not necessarily sexist, in spite of the way some if its proponents interpret it, holding as an example a theory of human evolution by two women, Tanner and Zihlman, in which women are seen as the primary forces in the evolution of unique human capacities precisely *because* they bear the brunt of the parental costs. This theory will be described at length in the following chapter. Similarly, Hrdy (1979a, b) has turned her attention lately to a more female-oriented analysis of the evolution of primate sexuality.

Nonetheless, it is true that sociobiologists devote much attention to developing biological explanations of why females appear doomed to lose in competitions with males. From Hrdy's "evolutionary trap," to Trivers' "females appear to be caught in a situation in which they are unable to force greater parental investment out of the males," to Barash's "Ironically, mother nature appears to be a sexist" (1977: 283), the sociobiological interpretation does not hold much promise for females in competition with males. The charge of sexism has perhaps been best summarized by Michael Ruse (who then found it to be "not proven"):

The whole analysis of sexual relations starts with the initial premise that females begin with a disadvantage, namely that they are literally left holding the baby! And the analysis continues with this theme of women as underdogs, as they work to get around their initial disadvantages, either by trapping

300

the male ("domestic bliss strategy") or by selling themselves to the highest bidder ("he-man strategy"). Through and through females (women in particular) are presented as second best, always trying to catch up with or overcome their disadvantages as compared with males. Moreover in the human case this leads to all kinds of sexist stereotyping: women are the domestic animals, the child-rearers; men are the aggressive humans, dominant over females, by nature polygamous (polygeny) and so on. In short, the sociobiology of sexual relations is male chauvinist through and through (1979: 96).

The Counter

Ruse largely counters the charge of sexism by using the same techniques sociobiologists commonly use to counter criticisms concerning the ethics, and implications for social and political values of their various theories — by denying that they are guilty of the "naturalist fallacy." As I describe it in Chapter 2, this is the name given to the logical flaw involved in reasoning from what *is*, to what *ought* to be. Thus sociobiologists argue they are only describing what *is* operating in the natural world of animal behavior, and are not proposing principles for how one *ought* to behave, especially, of course, in the human case.

I regard the invocation of the naturalistic fallacy here as something of a smokescreen which obscures rather than clarifies some basic issues in evolutionary thinking. First of all, the sociobiological literature is replete with examples of predictions and warnings about what humans *can* do, and can hope to do, given the evolutionary forces which are believed to constrain us, and at a subconscious level, to direct us. Secondly, many sociobiologists do think their theories can both explain and justify human ethical systems. But the main reason I regard the invoking of the naturalistic fallacy as a smokescreen, is that it deflects attention from the theory itself. The issue here is whether sociobiological theory *does* accurately and usefully reflect and predict animal behavior, or does it represent attempts to develop biological justifications of the belief, rather than the fact, that females are in some way inferior. Thus sexism may lie not only in statements by experts as to what *ought* to be, but, more importantly, in the manner in which they decide and determine what *is*.

In Trivers' model of parental investment, there are three pervasive generalizations about what *is*, or three underlying "facts," I would like to consider further. These are: 1. females have already invested more than males in offspring at the moment of conception, and are therefore "stuck" with rearing the young; 2. sexual reproduction is inherently a process of antagonistic competition between females and males; 3. females lost the first round by committing the Original Error of investing in a few large gametes, while males made the winning opening move of investing in many, smaller gametes.

ANOTHER LOOK AT REPRODUCTIVE COSTS
AND FEMALE VERSUS MALE INVESTMENTS

Even if we choose to argue within the sociobiological framework of intra- and intersexual competition, and use the microeconomics and gaming language of costs and benefits, it is possible to offer other explanatory pictures of female and male reproductive

301

behavior. Parental investment theory begins with the premise that females invest more at the moment of conception because the egg is larger than the sperm. Years ago, Simone de Beauvoir (1954) argued that this is a trivial difference, and I am somewhat inclined to agree with her. However, optimization theory, widely applied in biology today, suggests that marginal differences result in major differential effects, and many biologists view the size differences of female and male gametes as highly significant.

But to begin at the beginning of the parental investment model, it is possible to argue that males do not just invest one sperm in a zygote, they invest millions in a copulation, and countless millions in the several copulations which a male usually undertakes with one female before a conception results. More importantly, males invest significantly in terms of energy and risk to future fitness in order to reach the point of copulating with females. Both competition between males for proximity to females, and courtship of females, are enormously costly to males. For example, in many species, males must compete to maintain territories and/or gain acces to female-bonded groups, they must advertise their presence vigorously to females, and/or build nests, and/or provision females, etc. all *before* they copulate. Trivers recognized these necessary expenditures on the part of males (known as "mating costs," Daly 1978), but he ruled out these costs as significant to parental investment because, he argued, they do not affect the survival chances of the resulting offspring. (That is, a mother lactating for the young affects their survival, but the father courting the mother prior to copulation does not.) However, by choosing to start the clock of parental investment at the moment of conception, Trivers effectively eliminated the period of maximum male effort in reproduction. And he decided in his model to consider "reproductive success as if the only relevant variable were parental investment" (1972: 139), when clearly it is not.

Further, while Trivers argued that these sorts of male activities still "cost" less than producing eggs, it is improbable that we can calculate and compare the costs of such activities in any meaningful way. Ralls (1976) has also commented upon the impossibility of realistically evaluating and comparing the "costs" of various female and male activities involved in reproduction. It is clear however, that by the time the moment of birth is reached in primates, both females and males have considerable, if non-comparable, investments in the offspring, contrary to the oft-repeated assertion that males need invest only sperm.

It has also been argued that the cost of infant loss is greater to the female because of her limited opportunities to reproduce in her lifetime, compared to the males' theoretically unlimited possibilities. We do not have any data on how many infants the most successful primate male produces in his lifetime compared to the most successful primate female. Researchers working with polygynous animal species have sometimes found, or argued, that a very few males out of the possible breeding males may produce a majority of the offspring, but this is almost all data from short term studies. Application of most evolutionary models suffers from the lack of a life cycle perspective, and yet the models, by their very nature, require life history data to determine adequately the reproductive success of individuals.

For the primates, it is highly likely that a male may produce heavily in a few peak breeding years, and not at all in others. Thus his total lifetime reproduction may not be generally greater than that of a female. At any rate, if competition is as strong between males, and variability in success as extreme as believed, then males must work very hard indeed to become one of the minority of successful males. Further, I would argue that whatever it has cost a male to reach the moment of conception, it is still important to remember that the *survival*, and not the mere production of young, is the criterion of success for males, as well as females.

At least two related objections to parts of Trivers' model have been made: (1) what an animal does is a function of the options available to it; and (2) if animal behavior is based on sound economics, then present action should be based on future gain, not past investment. Grafen (1980) has given a good explanation of the first point when he argues that a male's optimal strategy may not be to desert a female (thereby leaving her to rear the young), *unless* there exists a better alternative use for his time and energy. The cost of performing an act must be evaluated in the light of what opportunities to perform alternative acts exist (and would therefore be missed). Grafen has called this the "opportunity cost" of a behavior, and while it certainly adds substance to the sociobiological concept of "cost," it also renders the concept, if anything, more difficult to measure.

Perhaps more damaging to Trivers' model is the second point, that natural selection should favor parental behavior based on potential future gains rather than past investment. Although a number of researchers have recognized this flaw in the model (e.g. Boucher 1977; Maynard-Smith 1977; Robertson & Biermann 1979), it was, as usual, Dawkins (1976, 1979) who produced the most flamboyant metaphor, labelling it the "Concorde fallacy." Following these theorists, parents should care for their young according to what benefits will accrue to them in the future, and not according to how much they have already invested in the past (but see Dawkins & Brockmann 1979).

In sum, the "economics" of male and female "investments" in reproduction would appear to be far more complex and better balanced than Trivers initially assumed. And many of the inferences widely drawn from his model (e.g., that males are better off deserting their female mates and that male parental care should be unwilling and based soley on certaintly of paternity) would also seem to be unwarranted.

HOW ARE WE FALLEN?

It may have occurred to the reader that nowhere in this discussion have we begun at the beginning of female and male biological differences — if female and male gametes differ in size and abundance and must unite in order to reproduce themselves, how and why did this come about? Explaining the origins of sexual reproduction is a major problem in current evolutionary thought, and all attempted explanations to date are highly speculative, as their authors readily admit.

If the purpose and driving force of all life is to reproduce itself as often and as perfectly or as completely as possible, then the optimal pattern would seem to be for all organisms to produce clones, or to reproduce asexually. Sexual reproduction involves an organism only passing on half of its genetic material to an offspring, or what some refer to as a 50% loss, when compared to asexual reproduction in which organisms reproduce *all* of their genetic material. The classic adaptive or functional explanation for sexual reproduction has been that it results in increased variability and thus increased potential to adapt to changing environments. Sexually, as compared to asexually, reproducing populations can achieve more rapid adaptations to environmental changes because of the variability which exists in the gene pool. However, this explanation involves selection and adaptation at the level of *group* or populations or species, and is not acceptable to those who believe that selection operates only at the level of the individual or at the level of genes. Thus there has been much discussion in recent years as to how sexual reproduction can be adaptive for competing individuals.

One such explanatory attempt by G.C. Williams (1975) argues that *a*sexual reproduction is the most successful pattern in a stable environment, while sexual reproduction is most successful in highly fecund species in an unstable environment. Some species of plants and animals (e.g. aphids, rotifers) have the capacity for both forms of reproduction, and exhibit asexual reproduction while in a stable environment, but switch to sexual reproduction in times of "ecological crunches" and environmental change. Daly and Wilson (1978) interpret this as using sexual reproduction in the face of an uncertain future, or in anticipation of dispersal. How such species determine when to switch to a sexual mode is unclear, and the implications of prescience on the part of the organism in the description above ("anticipation," "in the face of an uncertain future" (1978: 42), is unfortunate. Many species have lost the capacity to switch between the two modes of reproduction and have become exclusively sexual.

Williams (1975) concludes from his examination of the topic that except for a very few species with extremely high fecundity, sexual reproduction is *not* an adaptive pattern. Perhaps many, many species have simply become "stuck" with this less successful pattern, and if the capacity for asexual reproduction should reappear through mutation, it would replace sexual reproduction. Both Williams and Maynard-Smith (1978) who also treats the question of advantages of sexual versus asexual reproduction, state clearly that they are "thinking out loud" about difficult and controversial questions, and the work of both authors takes the form of presenting ideas for discussion, rather than emphatically drawing conclusions.

Most discussions of the evolution of sex deal briefly with the question of differential size in female and male gametes, or "anisogamy," since sexual reproduction almost always involves not just the union of two gametes, but the joining of two asymmetrical gametes. Maynard-Smith (1978b), Williams (1975) and Daly and Wilson (1978) all describe favorably a speculative model for the origin of anisogamy developed by Parker et al (1972). This model suggests that originally all gametes were of the same size, but there was selection for *greater* gamete size to maximize cytoplasmic resources for the zygote. Once such large gametes were available there could be selection to exploit the situation by producing many small gametes which would compete for access to the large gametes, in a sense parasitizing them. The large gametes, represent proto-ova (female gametes) and the smaller represent proto-sperm.

At first, proto-ova might join preferentially with other large gametes, and discriminate against the smaller proto-sperm, but there would be more intense selection on the many proto-sperm to overcome the discriminatory devices of the fewer proto-ova, since the former's genetic survival absolutely requires a large partner. Williams concludes: "This primeval conflict between the sexes was resolved in favor of the males, because of more intense selection for male functions" (1975: 113). Daly and Wilson underscore the significant implications of this theoretical model by stating: "In this and the following chapters we shall trace some of the paths of causality by which femininity and masculinity derive from a mere size distinction" (1978: 48).

There are, of course, objections one can offer to this speculation; for example, there is no explanation for the proposed selection for gametes very much larger than they need be, except that it needs to have happened in order to then allow selection for much smaller gametes. Such an explanation is little more than a fable, or Just-so story.

It seems to me that the above model for the origin and maintenance of differentially-sized gametes justifies the charge that this area of modern evolutionary thought *begins* with the assumption that females lose in competition with males, and then reasons backwards to a plausible biological origin for female inferiority (the primeval conflict and loss by larger gametes). The fact that this model is everywhere presented as speculative, while oft repeated, does not exempt it from criticism, and in fact it is probably in their more free-wheeling speculations, rather than the carefully-argued research of scholars, that we may best observe the underlying assumptions of their thinking.

CONCLUSIONS

I have tried to describe, in the preceding sections, how parental investment theory, and related investment theories of reproduction, are premised on three assumptions, or tenets, concerning what *is*. These three assumptions are: 1. females have already invested more than males in offspring at the moment of conception and are therefore more likely to give parental care than males; 2. females and males even at the level of the union of gametes are in direct competition; and 3. females are regularly observed to be the losers in competition with males, and this requires biological explanation.

To these three tenets can be added a fourth, central to all sociobiological theory, and alluded to earlier in the chapter in its extreme form of "hyperselectionism." This fourth assumption can be stated thus: any aspect of nature we wish to conceptualize as an entity may be viewed as an adaptive trait (and therefore as the direct result of selection), whose function we can determine through induction, inspection and/or the creation of plausible explanations.

The extreme view that every aspect of living organisms is an optimal (adaptive) solution to problems set by the environment, and therefore the direct result of natural selection, has been criticized extensively by Lewontin (1977, 1979). Lewontin variously referred to this approach, which he says is typical of all sociobiology, as a "caricature of Darwinism," "panglossian," and more conventionally, an "extreme adaptionist program." One example of the assumption that all that exists must be adaptive, is perhaps seen in the following statement by Williams concerning the question of the function of sex:

> If it can be shown, for a variety of organisms that can reproduce asexually, but use sexual reproduction in special situation s, the answer [to the question of the function of sex] arises from inspection. Sex is an adaption to special situation s. This is a valid conclusion regardless of possible ignorance as to why it should be adaptive in relation to s (1975: 3).

In spite of this, Williams concludes his book with the suggestion that sexual reproduction is actually *not* adaptive in most cases where it does continue to exist; in other words, in situation s.

Lewontin (1978, 1979) stressed that there are many alternative evolutionary forces besides natural selection and adaptation which may help bring about the establishment of physical and behavioral traits. These include genetic drift (sampling error or random fixation of genes), allometry (differential growth of different body parts), pleiotropy (multiple effects on the organism traced to one gene or genetic change), multiple selective peaks (alternative paths of evolution when more than one gene influences a character), and developmental noise (the organism is subject to various random influences as it develops). Thus sociobiology has been described and criticized by Lewontin as the view that each aspect of an organism is an adaptive, optimal solution to some problem, a solution achieved by natural selection of the heritable trait. Ruse (1979) countered for sociobiology that evolutionists must necessarily and logically begin by looking for the functional or adaptive value of traits. Like many others, this controversy will not soon be resolved, and one's sympathies in this issue are probably affected by an underlying aspect of world view — do we see nature as operating most efficiently, perhaps perfectly, according to a known set of strict rules, or rather as setting out a flexible, imperfectly understood, structure for behavior?

As to the first sociobiological assumption, females have invested more than males once offspring are conceived, I suggested that females and males have invested differently, and perhaps non-comparably. The second assumption, that female and male gametes are in competition at the moment of their union to form a jointly conceived offspring, does not seem thus far to lead to very helpful conclusions about the adaptive advantages of sexual reproduction to *individuals* or to their *genes*, compared to asexual reproduction. Neither Williams nor Maynard-Smith seem themselves convinced they have indeed determined any such generally adaptive advantages for individuals or genes. Perhaps some acceptable advantages will be forthcoming as more evolutionary theorists put their minds to the problem, or perhaps the widespread presence of sexual reproduction in the biological world will remain anomalous, challenging the sociobiological paradigm, and bringing forth alternative theories of natural processes.

Finally, the assumption that female are the losers in competition between the sexes, although widespread in sociobiological literature, does not seem to me to be inherently necessary to evolutionary theories, precisely because it *is* a cultural assumption rather than a biological given. And it is to be hoped that more theorists in sociobiology will follow the lead of Tanner and Zihlman, and Hrdy, in examining the processes and repercussions of selection upon both sexes, rather than viewing female traits as merely a by-product of selection upon males.

306

Chapter NINETEEN

Theories
of the Evolution
# of Human Social Life

Chimpanzee
Pan troglodytes
*The female of this moderately dimorphic African great ape uses a tool (twig)
to draw termites from a termite nest, watched by her infant and an adolescent
male.*

The planets will move as they always have, whether we adopt a
geocentric or a heliocentric view of the heavens. It is only the
equations we generate to account for those motions that will be
more or less complex; the motions of the planets are sublimely
indifferent to our earthbound astronomy. But the behavior of
men is not independent of the theories of human behavior that
men adopt. (Eisenberg, 1972: 123)

In the film, "The Man Hunters," a section from an old silent movie never fails to bring delighted laughter from the audience. A stooped, heavy-browed, fur-clad man sits at the entrance to his cave idly manipulating a club-like branch, while other creatures similar in appearance move around the area. He puts the end of the branch into the hole of a doughnut-shaped stone and absent mindedly bumps himself on the head with it. Shortly thereafter his group is attacked by an obviously ferocious enemy group of ape-men and appear to be losing until the hero, with a "eureka" expression on his face, clobbers one of the invaders over the head with the newly-invented tool (read: weapon) and thus routs the enemy. In some ways this film clip can be seen as epitomizing the dual nature of attempts to explain the origins of those things which may make us unique as human beings. Part-science and part-imagination, empirical and mythical, plausible and ludicrous, univeral and ethnocentric, the theories seldom fail to engage our rapt attention. For there really *were* early hominids, and whatever they really were like, our continuing urge to recreate them may tell us as much about the way we are, as the way we were.

In Chapter 1, I pointed out that the social behavior of living nonhuman primates is one line of evidence now used to reconstruct the life of early humans. While most primatologists would probably prefer that models for early peoples not be the *major* reason for studying prosimians, monkeys and apes, as these animals are eminently worthy of study in and of themselves, still we cannot entirely avoid the implications of our research for the study of humans. Speculation or reconstruction of early social life was common long before the advent of primate behavior studies. For example, Darwin and Freud both thought that primates, including early humans, lived in polygynous social groups, then called "primal hordes." But the first data were no sooner filtering back from field studies of primates which proliferated in the late 1950's, than this information was happily incorporated into models of "early man." Indeed, at least one of these field studies was conducted explicitly to gather information for early man analogies. In this chapter I would like to review several of these models from the point of view of a primate behaviorist, and from the perspective of postulated sex role differences. I think these models generally involve myth-making as well as scientific reasoning but do not consider this an adequate reason to ignore them or to reject the process. For we are intensely curious about our past, and we will never see a film clip showing us exactly what did take place. As long as we do not reify our models, we might as well enjoy trying to satisfy our intellectual curiosity by playing this scientific version of "Clue," a creative game of deduction, detection, and reconstruction.

"BABOONIZATION" — THE COMMON BABOON MODEL

Reasons for the Model
While we saw in Chapter 11 that the rhesus macaque is the monkey of preference for laboratory studies of biology and behavior and from which extrapolations are often made to humans, the favorite primate prototype in the study of human evolution has no doubt been the common baboon. Although neither of these two kinds of primates are any more closely related to us than any other Old World Monkey species, they have been more accessible for study, and they are terrestrial or ground-dwelling animals, as we are. The

macaque-baboon (*Macaca-Papio*) genera also happen to be the most aggressive of the non-human primates in terms of daily levels of self-assertion and overt conflicts, and the idea that this is due to their terrestrial niche is an oft-repeated, but untested assertion, whose relevance for humans will be described shortly.

A number of reasons are offered for the use of the common baboon as the prototype for early man, but the one proffered most often is that these baboons, like the early hominids, have evolved a terrestrial lifestyle adapted to the open savannahs of Africa. Ecological similarity is thus the basis for the analogy. Other reasons suggested are that baboons were the first nonhuman primates to be extensively studied in the field, and that the aggressive male-dominated, rigidly-structured picture of their social life, drawn for us by the early field workers, appealed to Euramerican folkbeliefs concerning innate human and primate nature (Reynolds 1976; Martin & Voorhies 1975).

The former suggestion is not strictly true; three excellent field studies on the howler monkey, the spider monkey, and the gibbon, had been completed by Carpenter (1934, 1936, 1940) nearly 20 years before the first of the post-war baboon field work. But it *is* true that the data from the savannah baboon studies, and interpretations thereof, were much more widely disseminated among scientists and lay people, achieving wide publicity in various media, and even forming a substantial component of an extensively-used elementary school social science curriculum. Why was baboon behavior so interesting to everyone when howler and gibbon behavior had not been? The validity of the suggestion that baboon social life, was seen to conform to our cultural worldview is difficult to test of course, but there is the problem of the conflicting pictures of baboon behavior painted for us by different researchers discussed in Chapter 4. If baboon behavior is not as male-oriented and militarily-structured as was first described, whence did these ideas originate? When one primate species out of at least 200 species, is chosen as a model for early humans, the explanation for the choice probably lies as much in the nature of the argument being developed, as in the behavior of the animal upon which the argument is to be based.

The Model

There are several versions of the model for early humans which make use of the baboon analogy (Pfeiffer 1972; Morris 1967; Tiger 1969; Tiger & Fox 1971; DeVore & Washburn 1963; Ardrey 1961; Washburn & Lancaster 1968) but they share in common the basic formula which we might write as follows: take one common baboon-type society, add hunting and its consequences, and the sum equals early human social life.

The underlying theme to the several versions is that early hominids, like common baboons, moved down out of the trees and out of the forest onto the grassland savannah, where they were to become both predators, and themselves more subject to predation by the large African carnivores. In order to protect themselves from all the new dangers of living in the open, the humans, like the baboons, must have given up the relaxed social life characteristic of forest-dwelling primates, and adopted a rigidly controlled, hierarchical social structure, in which males cooperated to protect the group and ruled over the females and young as part of their necessary dominant and aggressive role. Although they cooperated in defense against outside attacks, males of the group competed internally for female mates, the most dominant being the most successful, and thus the important male trait of pugnacity was promulgated. The genetic bases of a tendency to form male hierarchies were incorporated into the hominid gene pool, and manifest themselves even today in all aspects of human behavior, including for example, the relations between the sexes, politics, and even mental illness (Tiger 1969; Pfeiffer 1972).

However, hominids on the savannah made one adaptation to open-country life which was different from that of the baboons, and this is considered in the present model to be the most significant and influential of all human behaviors — they began to hunt and to make meat a regular part of their diet.

> Hunting is the master behavior pattern of the human species. It is the organizing activity which integrated the morphological, physiological, genetic, and intellectual aspects of the individual human organisms and of the population who compose our single species. Hunting is a way of life, not simply a "subsistence technique," which importantly involves commitments, correlates, and consequences spanning the entire biobehavioral continuum of the individual and of the entire species of which he is a member (Laughlin 1968: 304).

There are a number of traits which anthropologists believed set human behavior apart from that of the other primates, and these have all been attributed at one time or another to the influence of hunting. Because most readers should be familiar with some version of the "Man the Hunter" theory of human evolution, and because various authors have suggested several different sequences in which hunting gave rise to one trait, which then gave rise to another, and so forth, it seems easiest and clearest to list the human behavioral characterisitcs which have been traced to a hunting way of life:

1. the reduction of canines and increased skill in the manufacture and use of tools and weapons;
2. bipedalism (two-footed locomotion);
3. growth of the human brain and human intelligence;
4. sharing of food;
5. sexual division of labor such that males provide sustenance as well as protection, and females provide sexual and reproductive functions;
6. the human nuclear family;
7. continual sexual receptivity of the female in order to attract and hold a provider-male permanently;
8. the incest taboo, to avoid disrupting nuclear families;
9. exogamy or the exchange of females;
10. cooperation replacing competition among males;
11. male bonding and prominence in social, especially political life;
12. language, in order to stalk and bring down prey cooperatively;
13. territoriality, larger home ranges and increased mobility;
14. aesthetics, developed from the appreciation of beautiful tools;
15. pleasure in killing.

The model suggests, then, that a prehuman ape came out of the African forest and "baboonized" (a term coined by Pfeiffer 1972: 315) its social life in order to survive the dangers of the savannah, and developed extensive cognitive, technical, and social skills as part of its adaptive complex as the best hunting species the world has ever known.

Comments

Many people find this a satisfying story of our origins, especially because it offers plausible explanations for such a wide range of apparent human traits, from our technological skills to the economic dependence of females. However, the shortcomings of this model have been noted from numerous points of view, which I shall characterize, for the sake of discussion as the ethnographical, paleontological, archaeological, primatological and ecological criticisms, although they overlap in some respects. Also, the gelada-based and chimpanzee-based models of human evolution, described in the following sections, have been developed and proffered as direct alternatives to this one. They offer many criticisms of the common baboon analogy, and alternative explanations of the origins of human traits such as those listed above, so these will not be repeated in detail here.

Ethnography

It is interesting that in the same symposium and resulting book (*Man the Hunter*, 1968) wherein are found the two most famous expressions of the "Man the Hunter" school of thought (Washburn & Lancaster; Laughlin) we also find the ethnographic data and resulting interpretation that were to turn the minds of many researchers to an alternative explanation of human origins; that is, to the significance of human gathering, carrying and sharing of vegetable foods. When the experts on all the hunting-gathering peoples existing today came together for presentations and discussions, it was their unexpected conclusion that on average, 60-80% of the diet of these peoples comes from vegetable foods gathered primarily by women, and only 20-40% of the diet on average consists of meat.

This has given rise to several expressions of a "Woman the Gatherer" theory of human evolution which will be described below. What is of greatest note here, is that the idea of dependent females trading sexual favors and infant care for protection and support has been pervasively projected, and yet *economic* dependence is not generally characteristic of human females any more than it is of other primate females (see for example Martin & Voorhies 1975 on the roles of women in horticultural, agricultural, pastoral and industrial societies). Apart from considerations of subordination or submission to males, a different and thorny question discussed several times in this book (see in particular Chapter 7), the point is that primate females are often able to take care of themselves and their offspring with little or no help from male mates.

Paleontology and Archaeology

The evidence from paleontological studies indicates that the first hominids were indeed descended from an ape-like creature which, of the living primates, would most closely resemble the chimpanzee. The reduction of the canines, and possibly the upright stance, preceded the "move out of the forest habitat," and probably preceded the use of tools by several million years, insofar as we have any evidence from tool remains. This makes it unlikely that canine reduction, or bipedalism, resulted from, or even occurred at the same time as, the hunting adaptation. As regards tool use, the archaeological evidence indicates that hominids were using some resources of the mixed open and woodland area of Africa for 4 to 5 million years before they had the technological skills for big-game hunting, which is what most people think of when "Man the Hunter" is described to them. As will be reiterated below, the earliest tools were probably not preserved, and the "uses" or functions of the first stone tool remains we do have (river pebble tools, choppers) are not at all clear, although it is more likely they were used for processing foods than for bringing down animals. In sum, this evidence indicates that a number of the human traits which were thought to *follow* upon the hunting adaptation, in all likelihood *preceded* it.

Ecology

Several different researchers have pointed to conspicuous distortions of ecological data, and of ecological principles, involved in the baboonization model, three of which are mentioned here. The first is the assumption that the habitat in which we find a species today is the habitat in which the animal evolved, and for which it is adapted. For example, Rowell (1966b) suggested that baboons today live in forested areas throughout much of their range, and those which do live on the savannah may have moved there recently, as human burning practices opened up more of Africa to grassland savannah. A number of ecologists proposed that the majority of African grasslands today are the relatively recent result of burning and clearing by agricultural societies (for example Rattray 1960; Hopkins 1965). Thus, baboon social organization may not be specially adapted to savannah life.

The second misconception is the idea that forest and savannah are two discontinuous ecological zones in Africa. In spite of the consistent implication, seen for example in Morris' (1967) statement that early man was "jettisoned" into his new life, it is improbable that humans, or baboons, ever popped out of the dark forest, blinking and frightened, into the sunlight of the open, short-grass savannah. The areas of Africa over which baboons range today, and over which early hominids probably moved, consisted then, as now, of a mixture of intermediate types of woodlands, riverine areas, open areas interspersed with bush, large lakes and lake edges, and so forth. If the ecologists cited are correct, there is more open savannah in Africa today than there was during the evolution of hominids or baboons, but even now savannah and bushland are not always sharply demarcated.

Third and finally there is no evidence that ground-dwelling, open-country primates suffer more from predator pressure than arboreal or terrestrial forest primates. This assumption is part of the terrestrial-bias in the study of nonhuman primates described in Chapter 1. The large African cats which often hunt on the savannah may impress us more as predators, but there is no evidence that they take more primates as prey than do the large monkey-eating raptors of the tropical forests.

Primatology

This model vastly oversimplifies the data from the study of primate behavior in several respects, but perhaps especially in relation to baboons, It has already been noted that Rowell does not think the common baboon evolved its social patterns in adaptation to the open African savannah, and that later field workers did not find the baboon to be nearly as hierarchical, male-focused and aggressive as early research suggested. Thus, even the validity of the baseline data for the model are in some doubt. Some proponents of baboonization have countered that because savannah-dwelling baboons are said to be more hierarchical, army-like, etc. than forest-dwelling ones (and this has not yet been adequately substantiated), and because there is a report of chimpanzees becoming more cautious, subdued, and cohesive as they move out onto a savannah, this proves the significant influence of the savannah habitat, and justifies the baboon-human analogy. On the contrary, if valid, these examples underline once again the flexibility of primate behavior, and the primate ability to adapt to different environments without special inherited proprensities to hierarchies or aggression.

Although the chimpanzee is the living nonhuman primate most closely related to humans, it is often rejected as a model because it seldom ventures into the open savannah. However there *is* another nonhuman primate which, although semi-arboreal, semi-terrestrial, also ranges over the same parts of Africa and in similar habitats as the baboon, including the grassland savannah. This is the vervet monkey. Good field studies of the vervet have been conducted (Struhsaker 1967a,b,c,d; Lancaster 1972; Gartlan & Brain 1968) and the data show that in comparison to the savannah baboon data, vervets are less sexually dimorphic in morphology and less sexually differentiated in behavior, have lower frequencies of agonism, and females are not generally submissive to males. In spite of the fact that they have been well-studied and can dwell on the ground in open areas, and that one of their popular names is the "savannah monkey" (Gartlan & Brain 1968), we do not as yet have a "vervetization" theory of human evolution.

The continual willingness of the human female to mate has figured largely in explanations of human social cohesion, from the time when it was thought to be a trait characteristic of all primate females and underlying all primate societies (Zuckerman 1932), to now, when it is often used to contrast and explain the unique human version of monogamy and exogamy (e.g., Morris 1967; R. Fox 1972). One reads often that human females were the first to gain enough cognitive control over sexual motivation to be freed from a hormonally-driven estrus, and that this figured largely in their early "bargain" with the men they needed for protection and economic support (a sort of Original Prostitution myth). However, I demonstrated in Chapter 11 that social factors are major determinants in the sexual behavior of, at least, the Old World anthropoids, and this beginning emancipation from strict hormonal controls cannot be considered a uniquely human state. Further, Beach (1976a) calls these the theories of the "Ever-Ready Vagina," and points out that the two assumptions, 1. human females are constantly receptive; and 2. human males are constantly needful of, or ready for sex, are both untenable. In the third section of this chapter I will discuss theories which argue against the idea that early hominid females had to trade sexual and reproductive functions for support and protection.

In recent years some surprising data have come in from field studies concerning the catching and eating of animals by both common baboons and chimpanzees. (A number of other primate species eat insects and/or very small animals such as fledgling birds, which may be important sources of protein.) Although neither chimpanzees nor baboons are found to make meat a substantial part of the group's diet, there are enough observations of predation to make some generalizations. These provide both good and bad news for the "Man the Hunter" theory. In both the group of baboons (Strum 1975; Harding 1975) and the group of chimpanzees (Teleki 1973a) in which predation has been best studied, it seems to have developed as a learned or protocultural tradition, and is primarily confined to males, although females can and do hunt.

Thus the data available so far do appear to indicate that primate males are more prone to hunt than females. The reasons for this are obscure, but are not those usually offered by the "Man the Hunter" theory. Baboon and chimpanzee hunting does not require greater strength, endurance or even cooperation, the reasons traditionally offered for leaving human females out of the hunt (and human females do sometimes participate in hunting, see Martin & Voorhies 1975). Further, baboon and chimpanzee hunting does not require anatomical specializations (bipedalism, etc), or tools. Prey animals are caught with the hands and torn apart with the teeth and hands.

Chimpanzees do use tools, but they use them primarily for foraging aids; rocks are used as hammerstones to crack hard fruit, and probing twigs are used to bring up termites from underground nests. They will also hurl stones and branches during conflicts. Although chimpanzees use tools, and "weapons," they still have very large canines. It is questionable whether large canines in primate males have been selected for predator defense, and can therefore be replaced by tools and weapons, as suggested in the model (see also Chapter 5 on sexual dimorphism in dentition). The "Man the Hunter" theory tends to emphasize the uniqueness of humans in catching and eating meat, but we do not really know what portion of the early hominid diet was meat; and, what is more, today's hunters and gatherers should properly be called gatherers and hunters, and "omnivory," or the eating of a mixture of foods, is common to many primates.

WHAT OUR ANCESTORS ATE* — THE GELADA MODEL

Reasons for the Model

Criticizing the common baboon model on several accounts, including those of tautological reasoning and distortion of the behavioral and ecological data, an anthropologist named Clifford Jolly developed an alternative model, often labeled the "seed-eating" or "small-object feeding" model (1970). It was his argument that an analogy could be made between the gelada and early hominids on the basis of evolutionary parallelism or functional equivalence in the way these two evolved and diverged from their ancestors. His evidence is mainly morphological, based especially upon evolutionary trends in teeth or dentition, and the core of his argument is that some of the same things happened to geladas and hominids during their evolutionary development because they were both adapting to a grassland environment and to a dietary staple of grain or grass seeds. Ecological and morphological parallels are the reasons given to justify the model.

The Model

Theropithecus gelada, as was noted in Chapter 15, live in open-country areas of Africa and are not at all dependent upon trees or tree products (fruits, nuts, etc.) as are the vast majority of nonhuman primates. They sleep on cliffs and spend their days foraging in small polygynous groups on the ground. Foraging mainly consists of shuffling

---

* "We cannot escape the conclusion that we are what our ancestors ate" (Zihlman 1976a: 6).

along on the haunches, maintaining truncal erectness and picking up small bits of food, seeds, rhizomes, and grassblades, with the best precision grip of the nonhuman primates. The precision grip is the ability to hold a small item between the pad of the index finger and the opposing thumb, and is thought to be a morphological trait which was significant to human tool use and tool making.

How can this gelada pattern of shuffling along on the haunches remind anyone of early hominids? Following a rather ingenious line of reasoning, Jolly argued that hominids, like geladas, moved out onto the grasslands, and developed a small-object feeding complex. To support this, he offered a list of similarities in morphological trends between evolving geladas and evolving hominids, but the most significant are the parallels in reduction of canines and incisors (the anterior teeth, or teeth at the front of the mouth); habitual truncal erectness; freeing of the arms and hands for manipulation rather than locomotion; and development of the significant precision grip.

The traditional explanation in anthropology for the reduction of human canines in size compared to our prehuman ancestors, is that canines were no longer needed when humans developed tools and weapons. This explanation has been criticized on many grounds (see previous section and Chapter 5), and Jolly's model offers the alternative suggestion that hominid and gelada teeth, jaws, and accompanying muscles, are adapted to the grinding of small tough objects between the teeth. Large canines and incisors, useful for peeling, nibbling and tearing fruit or flesh, would only inhibit the rotary motion necessary to seed eating and grinding. For early hominids, it often is not mentioned that incisors as well as canines were reduced, and it is probable that the anterior teeth form a genetic complex or "package."

One important difference between gelada and human evolutionary trends is that gelada ancestors were quadrupedal monkeys, and thus geladas imposed their erect posture on a monkey-like body, and when upright, shuffle along on their haunches; while hominids imposed a similar erect adaptive pattern upon a semi-brachiating, ape-like form (something like a chimpanzee), which already included truncal erectness, forelimb independence and some locomotion on the two hind limbs. Hominids came to move across the grass bipedally and to sit while eating. Jolly noted that geladas and humans are the only two primates to have fatty pads on the buttocks, which he attributes to sitting while eating.

He further postulated that since the three ground-dwelling, open-country primates (patas, geladas, and hamadryas) all have polgynous social systems (see Chapter 15) this might have been the social structure of early hominids also. Following a popular line of argumentation (e.g. Desmond Morris), he postulated that in primates who spend much of their time sitting down, sexual signals will be transferred to the front of the body. He drew an analogy between the cape of the gelada male and the hair on the face and neck of the human male; and a second analogy between the breasts of the human female and small tubercles or vesicles which run down the chest and stomach of the gelada female, and which change in size and coloration with her sexual and reproductive state.

Jolly's model of hominid differentiation has two phases. In Phase 1, an ape-like creature (called a "basal hominid") adapts to seed or small-object eating in the grassland areas of Africa which occur in patches within woodland or seasonal forest zones. Although grass seeds are the staple diet and this foraging adaptation leads to morphological and perhaps social changes, the hominids remain omnivorous like their primate relatives.

In Phase 2, sexual division of labor is developed such that females gather vegetable foods and males hunt for meat. This would involve minor dietary and morphological changes, but major social and technological ones as well. Females would develop better techniques and tools for gathering, and males for hunting. Already living in polygynous societies, individuals would share with mates and offspring. Other human traits such as those listed earlier would follow. Phase 2 resembles the "Man the Hunter" hypothesis with two important differences: it attributes traits to a sexual division of labor, not hunting alone; and the adaptation is built upon a seed-eating gelada, not a hierarchical common baboon.

Comments

A number of anthropologists have been impressed by the line of reasoning used in this model, by the anatomical and behavioral comparisons, and by the synthesis of several types of evidence into an attractive theory with the self-proclaimed intent of moving beyond earlier, circular explanations which explained a set of data by use of that same data. Perhaps most of all, people welcomed a theory which is apparently able to explain familiar information (the hominid traits listed earlier) in a refreshingly original manner. Jolly cautions against scientifc Just-so stories in reconstructing human evolution, and argues that evolutionary models should provide testable predictions, or the closest approach to an experimental situation by examining parallel adaptations. However, the nature of adaptive arguments is, unfortunately, often tautological, and the nature of ecological niches is so complex that our formulations of analogies concerning them always *sound* like Just-so stories. Even Jolly acknowledged that the major test of any evolutionary model, including his own, is its plausibility, which does not take us far beyond Just-so stories.

Just how plausible is this theory, and how well does it account for early hominid traits? The gelada model really pivots on the dental analogy, and anyone who has studied living geladas has some difficulty accepting the notion that their canines are meaningfully "reduced" because of their still impressive size. The paleontological evidence does indicate that anterior teeth reduction has been a trend in gelada evolution, but their canines remain absolutely the longest among primates (Bernstein 1975). Jolly explained that gelada canines and incisors are not as reduced as in hominids because geladas still eat a lot of grassblades, while hominids concentrated and adapted more to grain-eating, which requires the utmost in rotary grinding ability. However, another primate biologist, Frederick Szalay, has taken exception to Jolly's theory and provided a very different interpretation of hominid teeth. He argued that hominid dental changes were an adaptation to "incisivation," the grasping and tearing of meat with the anterior teeth (1975). Ape-shaped canines would interfere with this process, and therefore human canines were "transformed" for their new meat-tearing role, rather than simply "reduced."

Many elaborate, clamorous and long-lasting arguments in the study of human evolution have revolved around differing behavioral or functional interpretations of human and primate teeth, usually the best preserved fossil remains. The study of dentition is an exacting and esoteric science, and one where laypeople often feel they know so little they must simply accept the word of the experts. However, if dental arguments are difficult for outsiders to evaluate, one thing *is* clear—when dental experts have widely differing opinions on the adaptive significance of the same teeth, then obviously inferring function or primary use of teeth from their shape or form is not a straightforward process. Indeed, inferring function from form is a risky, if creative, scientific process, whether we are talking about stone tools, or teeth, or bones. Arguments which are based on structure or morphology have tended to give short-shrift to behavioral variability, as in the study of locomotion where the structure of the body may suggest only one pattern of locomotion (e.g., the quadrupedal monkey), but observation of the living animal reveals multiple locomotion patterns (quadrupedal monkeys may walk on two legs, swing through the trees, cling to tree trunks, hang by their feet, etc.).

Perhaps this argument-from-morphology is why Jolly's theory seems particularly weak as far as primate and hominid behavior is concerned, especially social behavior. Indeed, noting that the fatty padding of gelada and human buttocks is the only evidence that both geladas and early hominids fed while sitting down, Alison Jolly (1972) aptly noted that this is a lot of theory to rest on a couple of species' fat bottoms. As I mentioned in Chapter 17, secondary sexual characteristics (epigamic features) are believed to be important in sexual selection and therefore in choice of sexual partners and sexual behavior. While the human beard and breasts are believed by some to be epigamic features, to suggest that they are visible from the front, as are the gelada cape and tubercles, because these creatures spend most of their time sitting down, is the worst sort of Just-so reasoning. In fact, the fatty padding of the buttocks itself, which is more prominent in the human female and only seen from behind, has been suggested by others as yet another epigamic feature.

Finally, Jolly's model does fall into a similar "ecological-determinist trap" as the common baboon model, even whilst arriving at a different conclusion. He argued that early hominids probably lived in polygynous ("harem") social groups because they were open-country dwellers. Yet, I noted in Chapter 4, that while social organizations in non-human primates are adapted to environmental conditions, they are not determined in some mechanistic manner by a unidimensional ecological category, such as "open-country." Polygynous primate societies are found in forest-dwelling, arboreal species as well as terrestrial open-country ones, and patas, gelada and hamadryas societies are only similar at a very general, perhaps superficial, level of analysis.

Phase 2 of Jolly's theory moves on from the gelada model and shares some ideas in common with both the previously discussed common baboon model and the chimpanzee-based model which will follow. Although the Phase 2 ideas are interesting, they are briefly presented with very little elaboration, and most will be covered in the next section. In sum, while Jolly's theory is invigorating in its presentation of an alternative explanation of human evolutionary trends and traits, and while early hominids may indeed have exploited the small vegetable foods of the African savannahs, the suggested morphological, behavioral and ecological analogies on which the gelada model is based are not indisputable and do not appear altogether convincing under closer scrutiny. For example, Dunbar (1976) suggested that the "small-object feeding" aspect of the model is a useful analogy between geladas and early homininds, but that more care needs to be taken to differentiate the points of departure between the niches of geladas and early hominids.

OUR KITH AND KIN — THE CHIMPANZEE MODEL

Reasons for the Model

The fossil evidence indicates that the first hominids descended from ape-like creatures, known collectively as "Dryopithecines." Variant forms of Dryopithecines were probably also the ancestral stock which gave rise to modern chimpanzees, gorillas and orangutans. Thus the living great apes are our closest relatives in the animal world, and many lines of evidence — anatomical, genetical, biochemical, paleontological, and behavioral — show that chimpanzees are the most closely related of all. This does not mean of course that chimpanzees were our ancestors, rather that we share with them some common ancestors. As a matter of fact, both the biochemical and behavioral data gathered in recent years have startled everyone by the extent of the similarity between us. In terms of molecular evidence, we are as close as sibling species, and in terms of behavior, having been introduced to chimpanzee social data in Chapter 14, you may have decided that there are some striking behavioral likenesses also. Thus, the reasons for the chimpanzee model are our close phylogenetic (evolutionary) relationship with these animals, and our apparent similarities in many behavioral propensities.

The Model

Specifically, the chimpanzee characteristics most often listed as relevant to the study of human behavioral evolution are the following: elasticity and complexity of chimpanzee society; matrifocal kinship units, and movement of pubertal females between groups; tool use; occasional erect posture and bipedal locomotion; food sharing; and predation by males who tend to range widely and locate food sources, although they remain in one community. These traits, along with the fossil and archaelogical data, have been used to produce various chimpanzee models (Tanner & Zihlman 1976; Zihlman 1978a, b; Reynolds 1966, 1968, 1976; Lancaster 1978b; Martin &Voorhies 1975; Slocum 1975). I shall attempt to synthesize these models, and still do justice to the intent of each author.

The models suggest that the first hominids lived in complex, flexible social communities, in which fluctuating subgroups traveled about in search of food (note that this is characteristic of both chimpanzees and modern gathering and hunting societies). Females and their dependent offspring consistently traveled together, and adult offspring often returned to martifocal units for varying periods of time. At other times subadult and adult males, and perhaps a few younger females unburdened by offspring, formed rapidly-moving, mobile groups which located good food sources. Mothers shared some food with their dependent young, although the juveniles had to forage for themselves. Males occasionally killed an animal and the meat was shared with friends and relatives.

These opportunistic, explorative creatures were able to exploit the resources of the savannah areas as well as the woodlands, tropical forests and riverine areas. For reasons upon which researchers do not agree (it does not seem likely that the forest receded significantly at this time, or that hominids were "pushed" onto the savannah), the transitional hominids began to use the grassland resources more and more. Zihlman suggested they collected and carried food from the open areas back to the familiar, safer clusters of trees to eat. Anyone who has seen any of the films showing chimpanzees running off with provisioned bananas will have noticed that these animals have difficulty

318

carrying large amounts of food, and tend to shuffle off bipedally, holding food in the arms and mouth and groin. Whether or not the carrying of food brought about the transition to bipedalism, most chimpanzee modelers agree that the gathering, carrying and sharing of food were critical hominid developments.

Sticks for digging, and containers for carrying were probably the first regularly used tools, although being organic, they were not preserved for us, as were the later stone tools. Chimpanzee females are the most adept tool-users today (McGrew 1977), and the hominid mothers, who shared their food more regularly and had therefore more to gather may well have invented containers. As the hominid pattern of locomotion became more habitually bipedal, the foot shape changed in order to bear the weight of the erect body rather than to grasp and hold. Zihlman and Lancaster both suggested that mothers, faced with the problem of infants who were less and less able to cling of their own accord, may then have adapted gathering-bags into infant-slings, another highly significant human invention, some version of which is used almost universally by humans today.

In this view then, the early tools (both organic and stone) were used not for hunting, but for gathering, carrying, and processing vegetable foods. Females passed their technological traditions on to their offspring as do chimpanzees, and perhaps on to other communities when they transferred for mating. Males occasionally preyed upon animals, although there is disagreement over whether they tended to scavenge them from carnivore kills or hunt them down themselves. As sexual division of labor became more prominent in early hominid life, the males shared the meat they were able to obtain, with their mothers and siblings, the matrifocal unit which had traditionally shared food with them. Males were useful to their groups, but females were not dependent upon them.

Females initiated sexual behavior with males outside of the kinship unit, and often outside of the community. Like other primate females living in multi-male, multi-female societies, they formed temporary, intense sexual liasons, or consorts, and they chose males who were less aggressive and more sociable. Tanner and Zihlman suggest that female choice for male partners who were not frightening to them or their offspring led to reduced canine size in the males. At some juncture, adult males became attached more permanently to matrifocal units containing their consorts rather than their relatives, although modelers disagree as to how, why, or when this might have occurred, and most feel it was late in human evolution.

Reynolds argued for one further, crucial human development which occurred after the social pattern described above had been present for some time — the emergence of conceptual awareness of social forms and relationships. His argument is that hominids became truly unique when they became self-aware and conceptually aware, so that they could classify, regulate and manipulate their social patterns, something that chimpanzees cannot do.

Comments

The principle value of the model may lie in its demonstration that the data of primate behavior and of human evolution may be plausibly explained in such a diametrically-opposed manner to the traditional anthropological theories. In its exposing of the androcentric bias which has permeated all past attempts to recreate early human social life it turns the traditional explanations around and examines what Boulding (1976) has called the "underside" (the female perspective) of human life. Some may think this model swings too far the other way and gives short shrift to the significance of males, but the time is well nigh to stress the importance of the female role in primate life, and here specifically in the evolution of human social life. It is essential to show that females as well as males could have invented tools, chosen mates, constructed social systems, provided for themselves, and generally contributed to the evolution of significant human abilities.

Since, it is usually stressed that it is the females who conceive, nurture, and largely raise offspring, surely they should be placed at the center of any "investment" theory of evolution, as Zihlman and Tanner place them. Even in many of the androcentric models, males are portrayed essentially as insecure, anti-social or male-bonded individuals, seeking only to inseminate all and any females they encounter, and then participating little or reluctantly in the care of offspring, whose survival and future reproduction are, after all, the criteria of evolutionary success. While I doubt that this portrayal does justice to the complexity of either female or male roles in primate societies, the survival of offspring or related genes, and not male reproductive strategies, male dominance or costs of insemination, is the direct criterion of evolutionary success. Thus the behavior of those individuals usually considered to be primarily responsible for the survival of offspring, the mature females, would seem to be the point where any evolutionary theory of behavior should begin.

One might argue that this should be self-evident. Yet anyone familiar with the study of human evolution must have absorbed the implications of ideas such as Darwin's, that the human brain evolved through equal transmission of traits (males evolve and females inherit the traits through them, see Chapter 17); or the "Man the Hunter" postulate that all unique capacities evolved due to the human hunting adaptation, and *only males hunt*. Or consider the distortion of primate data, and the implications for female significance involved in R. Fox's (1972) argument that the essential social change from simian to human was the switch from females as "objects of sexual use," to females as "objects of social exchange."

Having suggested that the chimpanzee model is, in a sense, a needed antidote to earlier models, and does demonstrate how the same data can be variously interpreted, I should add that for this very reason, the model is open to many of the same objections as the previous models. As just one example, there is now disagreement among the experts as to whether chimpanzees can be accurately described as peaceful and cooperative (see Chapter 6). Neither is it necessary, of course, to suggest that either sex alone, or on its own, invented and developed the traits such as tool-use and language, which we use to define ourselves as humans. Finally, there is no one species of nonhuman primate which is the necessary prototype for the human species. For while chimpanzees may be the living species most closely related to us by all current scientific evidence, and while savannah baboons may live in a similar niche to our ancestors, or geladas may eat a similar diet, still these species and our species have travelled separate evolutionary paths for many millions of years. As I explained in the first chapter of this book, chimpanzees can not be regarded as "not-quite-human," as primitive creatures retarded on the path to humanness, because they are moving along their own, unique, evolutionary path. Of course, the purpose of model-building in science, and what these model-builders have in mind, is not to suggest that common baboons or geladas or chimpanzees are primitive versions of ourselves (models are not *reproductions* of the phenomena they represent).

The purpose is rather to develop an interpretive reconstruction or representation of early human life, through the use of extended analogy from nonhuman primates (models, like metaphors, suggest resemblances).

CONCLUSION

The very fact that we may disagree with the suggested resemblances, and that diverse models can be developed from similar information about early hominids, demonstrates the interpretive aspect, the value, and perhaps the danger, of such model-building. Models may satisfy us when they seem to synthesize and explain a body of data, and especially when they provide testable predictions, but they also satisfy us when they simply reflect our underlying assumptions about the data. The three models described in this chapter reflect three different views of what it is to be human, and female or male. The sometimes-disputed empirical data on the life-ways of common baboons, geladas, and chimpanzees are woven into what we might call, but not in any derogatory sense, "evolutionary creation myths." For, like all creation myths, they are valuable statements of self-understanding. Were the first human societies hierarchically-organized and male-dominated like those of common baboons are sometimes said to be, or polygynous seed-eaters like those of geladas, or flexibly structured in cooperative matrifocal kinship units like some descriptions of chimpanzee society? Or were they any or all combinations of the three, or something entirely without resemblance to these portraits of our past? Since I doubt the answer will be forthcoming from paleontological, archaeological, or other forms of empirical data, perhaps the answer lies, in Haraway's (1978) words, on our skills in the construction of mirrors. For surely these models, to a large extent, are reflections of how we see ourselves. Therein, to me, lies their major importance, for I believe that we are, or we become, what we think we are.

Each person has a unique view of the world, patterned by the kaleidoscope of their past experiences and a culturally-shared view of the world developed during social learning. Scientists are not exceptions. All scientific theories spring from human minds, and all human minds are greatly influenced by cultural learning. This realization should not make us value scientific endeavor less, for science and society are best served when scientific theories are treated as dynamic exercises in human reasoning, rather than a quest for immutable truths. In their interpretations of data, scientists present us not with final truths, but with valuable views of their subject matter. In this tradition, this book is an effort to present my view of sex differences in primate behavior.

321

BIBLIOGRAPHY

BIBLIOGRAPHY

Abbott, D.H. and Hearn, J.P. 1978, Physical, hormonal and behavioural aspects of sexual development in the marmoset monkey, *Callithrix jacchus, J. Reprod. Fert.* 53:155-166.

Adler, N.T. 1969, Effects of the male's copulatory behavior on successful pregnancy of the female rat, *J. Comp. Physiol. Psych.* 64:613-622.

Adler, N.T., Resko, J.A. and Goy, R.W. 1970, The effect of copulatory behavior on hormonal change in the female rat prior to implantation, *Physiol. and Behav.* 5:1003-1007.

Akers, J.S. and Conoway, C.H. 1978, Female homosexual behavior in *Macaca mulatta, Arch. Sex. Behav.* 8:63-80.

Aldrich-Blake, F.P.G. and Chivers, D.J. 1973, On the genesis of a group of siamang, *Am. J. Phys. Anthrop.* 38:631-636.

Alexander, B.K. 1970, Parental behavior of adult male Japanese monkeys, *Behaviour* 36:270-285.

Alexander, B.K. and Bowers, J.M. 1967, The social structure of the Oregon troop of Japanese macaques, *Primates* 8:333-340.

Alexander, B.K. and Hughes, J. 1971, Canine teeth and rank in Japanese monkeys *(Macaca fuscata), Primates* 12:91-93.

Alexander, R.D. 1974, The evolution of social behavior, *Ann. Rev. Ecol. Sys.* 5:325-383.

Allee, W.C. 1978, Where angels fear to tread: a contribution from general sociology to human ethics. IN: *The Sociobiology Debate* (A.L. Caplan, ed) New York: Harper & Row, pp. 41-56.

Allen, E. *et al* (Sociobiology Study Group of Science for the People) 1976, Sociobiology — another biological determinism, *BioScience* 26:182-186.

Allpert, S. and Rosenblum, L.A. 1974, The influence of gender and rearing conditions on attachment and response to strangers, *Symp. 5th Cong. Int'l. Primat. Soc.*, pp. 217-231.

Altmann, J. 1978, Infant independence in yellow baboons. IN: *The Development of Behavior: Comparative and Evolutionary Aspects* (G.M. Burghardt and M. Bekoff, eds.) New York: Garland STPM Press, pp. 253-277.

Altmann, J. 1979, Age cohorts as paternal sibships, *Behavior Ecol. Sociobiol.* 6:161-164.

Altmann, S.A. 1962, A field study of the sociobiology of rhesus monkeys, *Macaca mulatta, Ann. N.Y. Acad. Sci.* 102:338-435.

Altmann, S.A. 1968, Sociobiology of rhesus monkeys. IV: Testing Mason's hypothesis of sex differences in affective behavior, *Behaviour* 32:49-69.

Altmann, S.A. 1974, Baboons, space, time and energy, *Amer. Zool.* 14:221-248.

Altmann, S.A. 1979, Baboon progressions, order or chaos? A study of one-dimensional group geometry, *Anim. Behav.* 27:46-80.

Altmann, S.A. 1979, Altruistic behaviour: the fallacy of kin deployment, *Anim. Behav.* 27:958-962

Altmann, S.A. and Altmann, J. 1970, *Baboon ecology: African Field Research*. Chicago: University of Chicago Press.

Altmann, S.A. and Altmann, J. 1977, On the analysis of rates of behavior, *Anim. Behav.* 25:364-372.

Altmann, S.A. and Altmann, J. 1979, Demographic constraints on behavior and social organization. IN: *Ecological Influences on Social Organization* (E.O. Smith and I.S. Bernstein, eds.) New York: Garland Press, pp. 1-26.

Altmann, S.A. and Walters, J. 1978, Book review of "Kin Selection in the Japanese Monkey," Man 13:324-325.

Altmann, S.A. and Walters, J. 1978, Critique of Kurland's "Kin Selection in the Japanese Monkey," Unpubl. manuscript.

Alvarez, F. and Consul, C. 1978, The structure of social behavior in *Theropithecus gelada, Primates* 19:45-59.

Anderson, B., Erwin, N., Flynn, D., Lewis, L. and Erwin, J. 1977, Effects of short-term crowding on aggression in captive groups of pigtail monkeys *Macaca nemestrina, Aggr. Beh.* 3:33-76.

Anderson, C.O. and Mason, W.A. 1974, Early experience and complexity of social organization in groups of young rhesus monkeys (*Macaca mulatta*), *J. Comp. Physiol. Psych.* 87:681-690.

Anderson, C.O. and Mason, W.A. 1977, Hormones and social behavior of squirrel monkeys (*Saimiri sciureus*), *Horm. and Behav.* 8: 100-106

Anderson, C.O. and Mason, W.A. 1978, Competitive social strategies in groups of deprived and experienced rhesus monkeys, *Dev. Psychobiol.* 11:289-299

Andrew, R.J. 1964, The displays of the primates. IN: *Evolutionary and Genetic Biology of Primates* (J. Buettner-Janusch, ed.)

Archer, J. and Lloyd, B. 1974, Sex roles: biological and social interactions, *New Scientist* 64:582-584.

Ardrey, R. 1961, *African Genesis*. New York: Dell Publ. Co.

Ashton-Warner, S. 1963, *Teacher*. New York: Bantam Books.

Bachmann, C. and Kummer, H. 1980, Male assessment of female choice in hamadryas baboons, *Behav. Ecol. Sociobiol.* 6:315-321

Baldwin, J.D. 1969, The ontogeny of social behavior of squirrel monkeys (*Saimiri sciureus*) in a seminatural environment, *Folia primat.* 11:35-79

Baldwin, J.D. and Baldwin, J.I. 1973, The role of play in social organization: comparative observations on squirrel monkeys (*Saimiri*), *Primates* 14:369-382

Baldwin, J.D. and Baldwin, J.I. 1977, The role of learning phenomena in the ontogeny of exploration and play. IN *Primate Bio-Social Development: Biological, Social, and Ecological determinants* (S. Chevalier-Skolnikoff and F.E. Poirier, eds.) New York: Garland Publ., Inc., pp. 343-406

Banton, M. 1965, *Roles, An Introduction to the Study of Social Relations*. London: Tavistock.

Barash, D.P. 1977, *Sociobiology and behavior*, Amsterdam: Elsevier.

Bardwick, J. 1971, *Psychology of women: a study of bio-cultural conflicts*. New York: Harper and Row.

Barfield, A. 1976, Biological influences on sex differences in behavior. IN: *Sex Differences, Social and Biological Perspectives* (M.S. Teitelbaum, ed.) Garden City, N.Y.: Anchor Books, pp. 62-121.

Barnett, S.A. 1968, On the hazards of analogies. IN: *Man and Aggression* (M.F. Ashley-Montagu, ed.) London: Oxford University Press, pp. 18-26.

Basckin, D.R. and Krige, P.D. 1973, Some preliminary observations on the behaviour of an urban troop of vervet monkeys (*Cercopithecus aethiops*) during the birth season, *J. Behav. Sci.* 1:287-296.

Bastock, M. 1967, *Courtship*. Chicago: Aldine.

Bauer, H.R. 1975, Behavioral changes about the time of reunion in parties of chimpanzees in the Gombe National Park, *Contemporary Primatology*. Basel: S. Karger, pp. 295-303.

Bauer, H.R. 1976, Sex differences in aggregation and sexual selection in Gombe chimpanzees, *Am. Zool.* 16:209;

Bauer, H.R. 1979, Agonistic and grooming behavior in the reunion context of Gombe Stream chimpanzees. IN: *The Great Apes. Perspectives on Human Evolution,* Vol. 5 (D.A. Hamburg and E.R. McCown, eds.) Menlo Park: Benjamin/Cummings Publ. Co., pp. 395-403.

Baxter, M.J. 1979, Behavioral patterns of a population of free-ranging spider monkeys (*Ateles geoffroyi*) in Tikal National Park, Guatemala. Masters Thesis, Edmonton: The University of Alberta.

Baxter, M.J. and Fedigan, L.M. 1979, Grooming and consort partner selection in a troop of Japanese monkeys (*Macaca fuscata*), *Arch. Sex. Behav.* 8:445-458.

Beach, F.A. 1965, Retrospect and prospect. IN: *Sex and Behavior* (F.A. Beach, ed.) New York: Wiley, pp. 535-569.

Beach, F.A. 1968, Factors involved in the control of mounting behavior by female mammals. IN: *Reproduction and Sexual Behavior* (M. Diamond, ed.) Bloomington: Indiana University Press, pp. 83-131.

Beach, F.A. 1970, Some effects of gonadal hormones on sex behvaior. IN: *The Hypothalamus* (L. Martini, M. Motta and F. Fraschini, eds.) New York: Academic Press, pp. 617-639.

Beach, F.A. 1971, Hormonal factors controlling the differentiation, development, and display of copulatory behavior in the ramstergig and related species. IN: *The Biopsychology of Development* (E. Tobach, C.R. Aronson and E. Shaw, eds.) New York: Academic Press, pp. 249-296.

Beach, F.A. (ed.) 1974, *Sex and Behavior*. New York: R.E. Kreuger Publ. Co.

Beach, F.A. 1976a, Cross-species comparisons and the human heritage, *Arch. Sex. Behav.* 5:469-485.

Beach, F.A. 1976b, Sexual attractivity, proceptivity and receptivity in female mammals, *Horm. and Behav.* 7:105-138.

Beach, F.A. 1978, Human Sexuality and evolution. IN: *Human Evolution. Biosocial Perspectives* (S.L. Washburn and E.R. McCown, eds.) Menlo Park: Benjamin/Cummings Publ. Co., pp. 123-153.

Beattie, J. 1964, *Other Cultures*. New York: The Free Press.

Beauvoir, S. de 1954, *The Second Sex*. (H.M. Parshley, trans. and ed.) New York: Vintage Books.

326

Becker, W.C. 1964, Consequences of different kinds of parental discipline. IN: *Review of Child Development Research* (M.L. Hoffman and L.W. Hoffman, eds.) New York: Russell Sage Foundation, pp. 169-209.

Benedict, B. 1969, Role analysis in animals and men, *Man* 4:203-214.

Bernstein, I.S. 1963, Social activities related to rhesus monkey consort behavior, *Psych. Reports* 13:375-379.

Bernstein, I.S. 1964a, The integration of rhesus monkeys introduced to a group, *Folia primat.* 2:50-63.

Bernstein, I.S. 1964b, Role of the dominant male rhesus monkey in response to external challenges to the group, *J. Comp. Physiol. Psych.* 57:404-406.

Bernstein, I.S. 1966a, Analysis of a key role in a capuchin *(Cebus albifrons)* group, *Tulane Studies in Zool.* 13:49-54.

Bernstein, I.S. 1966b, Defining the natural habitat, *Primatologen* 12:177-179.

Bernstein, I.S. 1969a, Introductory techniques in the formation of pigtail monkey troops, *Folia primat.* 10:1-19.

Bernstein, I.S. 1969b, Stability of the status hierarchy in a pigtail monkey *(Macaca nemestrina)* group. *Anim. Behav.* 17:452-458.

Bernstein, I.S. 1970, Primate status hierarchies. IN: *Primate Behavior,* vol. 1 (L.A. Rosenblum, ed.) New York: Academic Press, pp. 71-109.

Bernstein, I.S. 1972, Daily activity cycles and weather influences on a pigtail monkey group, *Folia primat.* 18:309-415.

Bernstein, I.S. 1974a, The influence of introductory techniques on the formation of captive mangabey groups, *Primates* 12:33-44.

Bernstein, I.S. 1974b, Activity profiles of primate groups. IN: *Behavior of Nonhuman Primates* vol. 3 (A.M. Schrier and F. Stollnitz, eds.) New York: Academic Press, pp. 69-106.

Bernstein, I.S. 1975, Activity patterns in a gelada monkey group, *Folia primat.* 23:50-71.

Bernstein, I.S. 1976, Dominance, aggression and reproduction in primate societies, *J. Theor. Biol.* 60:459-472.

Bernstein, I.S. 1978, Sex differences in the behavior of nonhuman primates, *Soc. Sci. and Med.* 12:151-154.

Bernstein, I.S. and Gordon, T.P. 1974, The function of aggression in primate societies, *J. Theor. Biol.* 60:459-472.

Bernstein, I.S. and Gordon, T.P. 1980, The social component of dominance relationships in rhesus monkeys *(Macaca mulatta), Anim. Behav.* 28:1033-1039.

Bernstein, I.S., Gordon, T.P. and Peterson, M. 1979, Role behavior of an agonadal alpha-male rhesus monkey in a heterosexual group, *Folia primat.* 32:263-267.

Bernstein, I.S., Gordon, T.P. and Rose, R.M. 1974a, Factors influencing the expression of aggression during introductions to rhesus monkey groups, IN: *Primate Aggression, Territoriality and Xenophobia* (R.L. Holloway, ed.) New York: Academic Press, pp. 211-240.

Bernstein, I.S., Gordon, T.P. and Rose, R.M. 1974b, Behavioral and environmental events influencing primate testosterone levels, *J. Hum. Evol.* 3:517-525.

Bernstein, I.S., Gordon, T.P., Grady, C.L. and Rose, R.M. 1979, Agonistic rank, aggression, social context, and testosterone in male pigtail monkeys, *Aggress. Behav.* 5:329-339.

Bernstein, I.S., Gordon, T.P., Rose, R.M. and Peterson, M.S. 1978, Influences of sexual and social stimuli upon circulating levels of testosterone in male pigtail macaques, *Behav. Biol.* 24:400-404.

Bernstein, I.S. and Mason, W.A. 1963, Group formation by rhesus monkeys, *Anim. Behav.* 11:28-31.

Bernstein, I.S., Rose, R.M. and Gordon, T.P. 1977, Behavioral and hormonal responses of male rhesus monkeys introduced to females in the breeding and nonbreeding season, *Anim. Behav.* 25: 609-614.

Bernstein, I.S. and Sharpe, L.G. 1966, Social roles in a rhesus monkey group, *Behaviour* 26:91-104.

Bethell, T. 1976, Darwin's mistake, *Harper's* 252:70-75.

Bischof, N. 1975, Comparative ethology of incest avoidance. IN: *Biosocial Anthropology* (R. Fox, ed.) London: Malaby Press, pp. 37-67.

Blurton-Jones, N.G. 1967, An ethological study of some aspects of social behavior of children in nursery school. IN: *Primate Ethology* (D. Morris, ed.) Garden City, N.Y.: Anchor Books, pp. 437-463.

Boggess, J. 1979, Troop male membership changes and infant killing in langurs *(Presbytis entullus)*, *Folia primat.* 32:65-107.

Bolin, I. 1980, Parental behavior — a socialization process — in *Alouatta palliata pigra* in natural habitats in Guatemala and Belize, M.A. Thesis, University of Alberta.

Bolwig, N. 1959, A study of the behaviour of the chacma baboon, *Papio ursinus*, *Behaviour* 14:136-163.

Boucher, D.H. 1977, On wasting parental investment, *Am. Nat.* 111:786-788.

Boulding, E. 1976, *The Underside of History: A View of Women Through Time.* Boulder: Westview Press.

Bowlby, J. 1973, *Attachment and Loss, Volume 2, Separation.* London: Hogarth Press.

Box, H.O. 1975a, Quantitative studies of behaviour within captive groups of marmoset monkeys, *Primates* 16:155-174.

Box, H.O. 1975b, A social development study of young monkeys *(Callithrix jacchus)* within a captive family group, *Primates* 16:419-435.

Box, H.O. 1977, Quantitative data on the carrying of young captive monkeys *(Callithrix jacchus)* by other members of their family groups, *Primates* 18:475-484.

Box, H.O. 1978, Social interactions in family groups of captive marmosets *(Callithrix jacchus).* IN: *The Biology and Conservation of the Callitrichidae* (D.G. Kleiman, ed.) Washington, D.C.: Smithsonian Inst. Press, pp. 239-249.

Bramblett, C.A. 1970, Coalitions among gelada baboons, *Primates* 11:327-333.

Bramblett, C.A. 1973, Social organization as an expression of role behavior among Old World monkeys, *Primates* 14:101-112.

Bramblett, C.A. 1976, *Patterns of Primate Behavior.* Palo Alto: Mayfield Press.

Bramblett, C. 1978, Sex differences in the acquisition of play among juvenile vervet monkeys. IN: *Social Play in Primates* (E.O. Smith, ed.) New York: Academic Press, pp. 33-48.

Bramblett, C.A., Pejaver, L.D. and Drickman, D.J. 1975, Reproduction in captive vervet and Syke's monkeys, *J. Mammal.* 56:940-946.

Bremer, J. 1959, *Asexualization, A Follow-up Study of 244 Cases.* New York: MacMillan.

Bressler, M. 1968, Sociology, biology and ideology, IN: *Genetics, Biology and Behavior Series* (D.C. Glass, ed.) New York: Rockefeller University Press, pp. 178-210.

Breuggeman, J.A. 1973, Parental care in a group of free-ranging rhesus monkeys *(Macaca mulatta), Folia primat.* 20:178-210.

Brindley, C.P., Clark, P., Hutt, C., Robinson, I. and Wethli, E. 1973, Sex differences in the activities and social interactions of nursery school children. IN: *Comparative Ecology and Behaviour of Primates* (R.P. Michael and J.H. Crook, eds.) London: Academic Press, pp. 799-828.

Brodie, H.K.H., Gartrell, N., Doering, C., and Rhue, T. 1974, Plasma testosterone levels in heterosexual and homosexual men, *Amer. J. Psychiat.* 131:82-83.

Brown, J.L. 1975, *The Evolution of Behavior.* New York: Norton.

Bruner, J.S., Jolly, A. and Sylva, A. (eds.) 1976, *Play, Its Role in Development and Evolution.* New York: Penguin Books.

Burt, W.H. 1943, Territoriality and home range concepts as applied to mammals. *J. Mammal.* 24:346-352.

Burton, F.D. 1971, Sexual climax in female Macaca mulatta. IN: *Proceedings Third International Congress of Primatology,* vol. 3, Basel: S. Karger, pp. 180-191.

Burton, F.D. 1972, The integration of biology and behavior in the socialization of *Macaca sylvana* of Gibraltar. IN: *Primate Socialization* (F.E. Poirier, ed.) New York: Random House, pp. 29-62.

Burton, F.D. 1977, Ethology and the development of sex and gender identity in nonhuman primates, *Acta Biotheoretica* 26:1-18.

Burton, F.D. and Bick, M.J., 1972, A drift in time can define a deme: the implications of tradition drift in primate societies for hominid evolution, *J. Hum. Evol.* 1:53-59.

Busse, C.D. 1978, Do chimpanzees hunt cooperatively?, *Am. Nat.* 112:767-770 (letter).

Butler, R.J. 1975, Effect of age and sex on dental wearing, *Am. J. Phys. Anthrop.* 40:132.

Bygott, J.D. 1972, Cannibalism among wild chimpanzees, *Nature* (Lond.) 238:410-411.

Bygott, J.D. 1979, Agonistic behavior, dominance and social structure in wild chimpanzees of the Gombe National Park. IN: *The Great Apes: Perspectives on Human Evolution,* vol. 5 (D. Hamburg and E.R. McCown, eds.) Menlo Park: Benjamin/Cummings Publ. Co., pp. 405-427.

Caine, N. and Mitchell, G. 1979, The relationship between maternal rnak and companion choice in immature macaques *(macaca mulatta* and *Macaca radiata), Primates* 20: 583-590.

Callan, H. 1970, *Ethology and Society. Towards an Anthropological View.* Oxford: Clarendon Press.

Candland, D.K., Dresdale, L., Leiphart, J., Bryan, D., Johnson, C. and Nazar, B. 1973, Social structure of the squirrel monkey *(Saimiri sciureus Iquitos), Folia primat.* 20:211-240.

Caplan, A.L. 1977, Tautology, circularity and biological theory (letter), *Am. Nat.* 111: 390-393.

Caplan, A.L. 1978, *The Sociobiology Debate.* New York: Harper and Row.

Carpenter, C.R. 1934, A field study of the behavior and social relations of howling monkeys, *Comp. Psych. Monogr.* 10:1-168.

Carpenter, C.R. 1935, Behavior of red spider monkeys in Panama, *J. Mammal.* 16:171-180.

Carpenter, C.R. 1940, A field study in Siam of the behavior and social relations of the gibbon (*Hylobates lar*), *Comp. Psych. Monogr.* 16:1-212.

Carpenter, C.R. 1942a, Societies of monkeys and apes, *Biological Symposia* 8:177-204.

Carpenter, C.R. 1942b, Sexual behavior of free ranging rhesus monkeys (*Macaca mulatta*), *J. Comp. Psych.* 33:113-162.

Carpenter, C.R. 1964, *Naturalistic Behavior of Nonhuman Primates*. University Park: Penn. State Univ. Press.

Carthy, J.D. and Ebling, F.J. (eds.) 1964, *The Natural History of Aggression*. New York: Academic Press.

Cartmill, M. 1975, *Primate Origins*. Minneapolis: Burgess Publ. Co.

Caspari, E. 1972, Sexual selection in human evolution. IN: *Sexual Selection and the Descent of Man, 1871-1971* (B. Campbell, ed.), Chicago: Aldine, pp. 332-356.

Cassidy, J. 1979, Half a century on the concept of innateness and instinct: survey, synthesis and philosophical implications, *Z. Tierpsychol.* 50:364-386.

Castrodeza, C. 1977, Tautologies, beliefs and empirical knowledge in biology (letter), *Am. Nat.* 111:393-394.

Chagnon, N.A. and Irons, W. 1979, *Evolutionary Biology and Human Social Behavior: An Anthropological Perspective*. North Scituate, Mass: Duxbury Press.

Chalmers, N.R. 1968, The social behavior of free-living mangabeys in Uganda, *Folia primat.* 8:263-281.

Chalmers, N.R. 1972, Comparative aspects of early infant development in some captive cercopithecines. IN: *Primate Socialization* (F.E. Poirier, ed.), New York: Random House, pp. 63-82.

Chalmers, N. 1979, *Social Behaviour in Primates*. London: Edward Arnold.

Chamove, A., Harlow, H.F. and Mitchell, G. 1967, Sex differences in the infant-directed behavior of preadolescent rhesus monkeys, *Child Dev.* 38:329-335.

Chance, M.R.A. 1956, Social structure of a colony of *Macaca mulatta*, *Brit. J. Anim. Behav.* 4:1-13.

Chance, M.R.A. 1963, The social bond of primates, *Primates* 4:1-22.

Chance, M.R.A. 1967, Attention structure as the basis of primate rank orders, *Man* 2:503-518.

Chance, M.R.A. and Jolly, C. 1970, *Social Groups of Monkeys, Apes and Men*. London: Jonathan Cape.

Chance, M.R.A. and Larsen, R.R. (eds.) 1976, *The Social Structure of Attention*. New York: Wiley and Sons.

Chapman, M. and Hausfater, G. 1979, The reproductive consequences of infanticide in langurs, *Behav. Ecol. Sociobiol.* 5:227-240.

Charles-Dominique, P. 1971, Eco-etholgie des prosimiens du Gabon, *Biblogia Gabonica* 7:121-228.

Charles-Dominique, P. 1974, Aggression and territoriality in nocturnal prosimians. IN: *Primate Aggression, Territoriality and Xenophobia* (R.L. Holloway, ed.), New York: Academic Press, pp. 31-48.

Charles-Dominque, P. 1977, *Ecology and Behavior of Nocturnal Prosimians* (Translated by R.D. Martin) New York: Columbia University Press.

Cheney, D.L. 1977, The acquisition of rank and the development of reciprocal alliances among free-ranging immature baboons, *Beh. Eco. Sociobiol.* 2:303-318.

Chepko-Sade, B.D. 1974, Division of group F at Cayo Santiago, *Am. J. Phys. Anthrop.*

41:472 (abstract).

Chepko-Sade, D. and Olivier, T.J. 1979, Coefficient of genetic relationship and the probability of intragenealogical fission in *Macaca mulatta, Behav. Ecol. Sociobiol.* 5:263-278.

Chepko-Sade, D. and Sade, D.S. 1979, Patterns of group splitting within matrilineal kinship groups, *Behav. Ecol. Sociobiol.* 5:67-86.

Chevalier-Skolnikoff, S. 1974, Male-female, female-female, and male-male sexual behavior in the stumptail monkey, with special attention to the female orgasm, *Arch. Sex. Behav.* 3:95-116.

Chevalier-Skolnikoff, S. 1976, Homosexual behavior in a laboratory group of stumptail monkeys *(Macaca arctoides)*: form, contents and possible social functions, *Arch. Sex. Behav.* 5:511-527.

Chevalier-Skolnikoff, C. and Poirier, F.E. (eds.) 1977, *Primate Bio-social Development: Biological, Social and Ecological Determinants.* New York: Garland Publ. Inc.

Chiarelli, A.B. 1972, *Taxonomic Atlas of Living Primates.* New York: Academic Press.

Chikazawa, D., Gordon, T.P., Bean, C.A. and Bernstein, I.S. 1979, Mother-daughter dominance reversals in rhesus monkeys *(Macaca mulatta), Primates,* 20:301-305.

Chism, J. 1978, Relationships between patas infants and group members other than the mother, *Recent Advances in Primatology,* vol. 1 (D. Chivers and J. Herbert, eds.), London: Academic Press, pp. 173-176.

Chism, J., Rowell, T.E. and Richards, S.M. 1978, Daytime births in captive patas monkeys, *Primates* 19:765-767.

Chivers, D.J. 1972, The siamang and the gibbon in the Malay Peninsula. IN: *Gibbon and Siamang, Volume 1* (D.M. Rumbaugh, ed.), Basel: S. Karger, pp. 103-135.

Chivers, D.J. 1973, An introduction to the socio-ecology of Malayan Forest Primates. IN: *Comparative Ecology and Behaviour of Primates* (R.P. Michael and J.H. Crook, eds.), London: Academic Press, pp. 101-146.

Chivers, D.J. 1974, The siamang in Malaya. A field study of a primate in tropical rain forest, *Contrib. Primatol.* vol. 4, Basel: S. Karger.

Chivers, D.J. 1976, Communication within and between family groups of siamang *(Symphalangus syndactylus), Behaviour* 57:116-135.

Chivers, D.J. 1977, The behavior of siamangs in the Krau Game Reserve, *Malay Nat. J.* 29 (1):7-22.

Chivers, D.J. and Chivers, S.T. 1975, Events preceding and following the birth of a wild siamang, *Primates* 16:227-230.

Chivers, D.J., Raemaekers, J.J. and Aldrich-Blake, F.P.G. 1975, Long-term observations of siamang behaviour, *Folia primat.* 23:1-49.

Clark, C.B. 1977, A preliminary report on weaning among chimpanzees of the Gombe National Park, Tanzania. IN: *Primate Bio-social Development: Biological, Social and Ecological Determinants* (S. Chevalier-Skolnikoff and F.E. Poirier, eds.), New York: Garland Publ., Inc., pp. 235-260.

Clark, L.D. and Gay P.E. 1978, Behavioral correlates of social dominance, *Biol. Psychiat.* 13:445-454.

Clarke, M.R. 1978, Social interactions of juvenile bonnet monkeys, *Macaca radiata, Primates* 20:301-305.

Clutton-Brock, T.H. (ed.) 1977, *Primate Ecology: Studies of Feeding and Ranging*

Behaviour in Lemurs, Monkeys and Apes. London: Academic Press.

Clutton-Brock, T.H. and Harvey, P.H. 1977, Primate ecology and social organization, *J. Zool.* (London) 183:1-39.

Clutton-Brock, T.H. and Harvey, P.H. 1978, *Readings in Sociobiology.* San Francisco: W.H. Freeman.

Clutton-Brock, T.H., Harvey, P.H. and Rudder, B. 1977, Sexual dimorphism, socionomic sex ratio and body weight in primates, *Nature* 269:797-800.

Coe, C.L., Mendoza, S.P., Davidson, J.M., Smith, E.L., Dallman, M.F. and Levine, S. 1978, Hormonal response to stress in the squirrel monkey, *Neuroendocrinology* 26:367-377.

Coelho, A.M. Jr. 1974, Socio-bioenergetics and sexual dimorphism in primates, *Primates* 15:263-269.

Coelho, A.M. Jr., Coelho, L., Bramblett, C.R., Bramblett, S., and Quick, L. 1977, Ecology, population characteristics and sympatric association in primates: a socio-bioenergetic analysis of howler and spider monkeys in Tikal, Guatemala, *Yearbook of Physical Anthropology* 20:96-135.

Coimbra-Filho, A.F. and Mittermeier, R.A. 1973, New data on the taxonomy of the Brazilian marmosets of the genus *Callithrix erxleben 1777, Folia Primat.* 20:241-264.

Collingwood, R.G. 1960, *The Idea of Nature.* London: Oxford Press.

Conoway, C.H. and Koford, C.B. 1964, Estrous cycles and mating behavior in a free-ranging band of rhesus monkeys, *J. Mammal.* 45:577-588.

Conoway, C.H. and Sade, D.S. 1965, The seasonal spermatogenic cycle in free-ranging rhesus monkeys, *Folia primat.* 3:1-12.

Count, E.W. 1973, On the idea of protoculture. IN: *Precultural Primate Behavior* (E.W. Menzel, Jr., ed.) Symp. 4th Cong. Int'l. Primat. Soc., vol. l, Basel: S. Karger, pp. 1-26.

Crook, J.H. 1966, Gelada baboon herd structure and movement. A comparative report, *Symp. Zool. Soc. Lond.* 18:237-258.

Crook, J.H. 1968, The nature and function of territorial aggression. IN: *Man and Aggression* (M.F. Ashley-Montagu, ed.), London: Oxford University Press, pp. 141-178.

Crook, J.H. 1970a, Social organization and the environment, aspects of contemporary social ethology, *Anim. Behav.* 18:197-209.

Crook, J.H. 1970b, The socio-ecology of primates. IN: *Social Behaviour in Birds and Mammals* (J.H. Crook, ed.) London: Academic Press, pp. 103-166.

Crook, J.H. 1971, Sources of cooperation in animals and man. IN: *Man and Beast: Comparative Social Behavior* (J.F. Eisenberg and W.S. Dilton, eds.), Washington: Smithsonian Inst. Press, pp. 237-260.

Crook, J.H. 1972, Sexual selection, dimorphism and social organization in the primates. IN: *Sexual Selection and the Descent of Man 1871-1971* (B. Campbell, ed.), Chicago: Aldine, pp. 231-281.

Crook, J.H. 1974, Primate social structure and dynamics, conspectus 1974. IN: *Symp. 5th Congr. Int'l Primat. Soc.*, pp. 3-12.

Crook, J.H. 1975, Problems of inference in the comparison of animal and human social organisations, *Soc. Sci. Inform* 14:89-112.

Crook, J.H. 1977, On the integration of gender strategies in mammalian social systems.

IN: *Reproductive Behavior and Evolution,* vol. 1 (J.S. Rosenblatt and B.R. Komisaruk, eds.), New York: Plenum, pp. 17-39.

Crook, J.H. and Aldrich-Blake, F.P.G. 1968, Ecological and behavioral contrasts between sympatric ground dwelling primates in Ethiopia, *Folia primat.* 8:192-227.

Crook, J.H. and Gartlan, J.C. 1966, Evolution of primate societies, *Nature* 210:1200-1203.

Crook, J.H. and Goss-Custard, J.D. 1972, Social ethology, *Ann. Rev. of Psych.* 23: 277-312.

Cubicciotti, D.D. and Mason, W.A. 1975, Comparative studies of social behavior in Callicebus and Saimiri: male-female emotional attachments, *Behavioral Biology* 16:185-197.

Cubicciotti, D.D. and Mason, W.A. 1978, Comparative studies of social behavior in Callicebus and Saimiri: heterosexual jealousy behavior, *Behav. Eco. Sociobiol.* 3:311-322.

Curtin, R. 1977, Langur social behavior and infant mortality, *Kroeber Anth. Soc. Papers* 50:27-36.

Curtin, R. and Dolhinow, P. 1978, Primate social behavior in a changing world, *Am. Sci.* 66:468-475.

Czaja, J.A. and Bielert, C. 1975, Female rhesus sexual behavior and distance to a male partner: relation to stage of the menstrual cycle, *Arch. Sex. Behav.* 4:583-597.

Daly, M. 1978, The costs of mating, *Am. Nat.* 112:771-774.

Daly, M. and Wilson, M. 1978, *Sex, Evolution and Behavior.* North Scituate, Mass: Duxbury Press.

Darling, F.S. 1937, *A Herd of Red Deer.* London: Oxford Univ. Press.

Darlington, P.J. Jr. 1972, Nonmathematical models for evolution of altruism and for group selection, *Proc. Nat. Acad. Sci.* 69:293-297.

Darwin, C. 1936 (Modern Library Edition), *The Origin of Species, and The Descent of Man.* New York: Random House, Inc.

Darwin, E. Zoonomia 1794, IN: *The Essential Writings of Erasmus Darwin,* 1968 (D. King-Hele, ed.), London: MacGibbon and Kee.

Dawkins, R. 1976, *The Selfish Gene.* Oxford: Oxford University Press.

Dawkins, R. 1979, Twelve misunderstandings of kin selection, *Z. Tierpsychol.* 51:184-200.

Dawkins, R. and Brockmann, H.J. 1979, Do digger wasps commit the Concorde Fallacy? *Anim. Behav.* 28:892-896.

Dawkins, R. and Carlisle, T.R. 1976, Parental investment, mate desertion, and a fallacy, *Nature* (Lond.) 262:131-133.

Dawson, G.A. 1978, Composition and stability of social groups of the tamarin, *Saguinus oedipus geoffroyi,* in Panama: ecological and behavioral implications. IN: *The Biology and Conservation of the Callitrichidae* (D.G. Kleiman, ed.), Washington D.C.: Smithsonian Inst. Press, pp. 23-37.

Dawson, G.A. 1979, The use of time and space by the Panamanian tamarin, *Saguinus oedipus, Folia primat.* 31:253-284.

Deag, J. 1977, Aggression and submission in monkey societies, *Anim. Behav.* 25:465-474.

Deag, J. 1978, The adaptive significance of baboon and macaque social behaviour. IN: *Population Control by Social Behaviour* (F.J. Ebling and D.M. Stoddart, eds.)

London: Institute of Biology, pp. 83-113.

Deag, J. and Crook, J. 1971, Social behavior and 'agonistic buffering' in the wild barbary macque *Macaca sylvana L., Folia primat.* 15:183-200.

Demarest, W.J. 1977, Incest avoidance among human and nonhuman primates. IN: *Primate Bio-social Development: Biological, Social, and Ecological Determinants* (S. Chevalier-Skolnikoff and F.E. Poirier, eds.), New York: Garland Publ., Inc., pp. 323-342.

Demment, M.W. 1978, Nutritional constraints on the evolution of body size in baboons, Paper prepared for participants in "Baboon Field Research: Myths and Models" (Unpublished manuscript).

Demment, M.W. 1979, Reduced weight and restricted diet in female baboons: a hypothesis on sexual dimorphism (Unpublished manuscript).

Denham, W.W. 1971, Energy relations and some basic properties of primate social organization, *Am. Anth.* 73:77-95.

DeVore, I. 1963a, A comparison of the ecology and behavior of monkeys and apes. IN: *Classification and Human Evolution* (S.L. Washburn, ed.), Chicago: Aldine, pp. 301-319.

DeVore, I. 1963b, Mother-infant relations in free-ranging baboons. IN: *Maternal Behavior in Mammals* (H.L. Rheingold, ed.), New York: Wiley, pp. 305-335.

DeVore, I. 1964, Primate behavior. IN: *Horizons of Anthropology* (S. Tax, ed.), Chicago: Aldine, pp. 25-36.

DeVore, I. 1965 (second ed. 1974), Male dominance and mating behavior in baboons. IN: *Sex and Behavior* (F.A. Beach, ed.), New York: Wiley, pp. 305-335.

DeVore, I. 1977, The new science of genetic self interest, *Psych. Today* 10:42-51, 84-88.

DeVore, I. and Washburn, S.L. 1963, Baboon ecology and human evolution. IN: *African Ecology and Human Evolution* (F.C. Howell and F. Bourliere, eds.), Chicago: Aldine, pp. 335-367.

Diamond, M. 1976, Human sexual development: biological foundations for social development. IN: *Human Sexuality in Four Perspectives* (F.A. Beach, ed.), Baltimore: Johns Hopkins Univ. Press, pp. 22-61.

Doering, C.H., Brodie, H.K.H., Kraemer, H., Becker, H., and Hamburg, D.A. 1974, Plasma testosterone levels and psychologic measures in men over a two month period. IN: *Sex Differences in Behavior* (R.C. Friedman, R. Richart and R.L. Vande Wiele, eds.) New York: Wiley and Sons, pp. 413-431.

Doerr, P., Kockott, G., Vogt, H.J., Perke, K.M. and Dittmar, F. 1973, Plasma testosterone, estradiol and semen analysis in male homosexuals, *Arch. Gen. Psychiat.* 29:829-833.

Dolhinow, P. 1977 (letter) Normal monkeys? *Am Sci.* 65:266.

Dolhinow, P. 1980, An experimental study of mother loss in the Indian langur monkey, *Folia primat.* 33:77-128.

Dolhinow, P. and Bishop, N. 1970, The development of motor skills and social relationships among primates through play. IN: *Minnesota Symposia on Child Psychology* (J.P. Hill, ed.), Univ. of Minnesota Press, pp. 141-201.

Downhower, J.F. 1976, Darwin's finches and the evolution of sexual dimorphism in body size, *Nature* 263:558-563.

Downhower, J.F. and Armitage, K.B. 1971, The yellow-bellied marmot and the evolution of polygamy, *Am. Nat.* 105:335-370.

Doyle, G.A. and Martin, R.D. 1979, *The Study of Prosimian Behavior*, New York: Academic Press.

Drew, D.R. 1973, Group formation in captive *Galago crassicandatas*. Notes on the dominance concept. *Z. Tierpsychol.* 32:425-435.

Drickamer, L.C. 1974, Social rank, observability and sexual behavior of rhesus monkeys *(Macaca mulatta)*, *J. Reprod. Fert.* 37:117-120.

Drickamer, L.C. 1975, Quantitative observation of behavior in free-ranging *Macaca mulatta:* methodology and aggression, *Behaviour* 55:202-236.

Dunbar, R.I.M. 1976, Australopithicine diet based on a baboon analogy, *J. Hum. Evol.* 5:161-167.

Dunbar, R.I.M. 1978, Sexual behavior and social relationships among gelada baboons, *Anim. Behav.* 26:167-178.

Dunbar, R.I.M. 1980, Determinants and evolutionary consequences of dominance among female gelada baboons, *Behav. Ecol. Sociobiol.* 7:253-265.

Dunbar, R.I.M. and Dunbar, E.P. 1974a, Ecological relation and niche separation between sympatric terrestrial primates in Ethiopia, *Folia primat.* 21:36-60.

Dunbar, R.I.M. and Dunbar, E.P. 1974b, The reproductive cycle of the gelada baboon, *Anim. Behav.* 22:204-211.

Dunbar, R.I.M. and Dunbar, E.P. 1974c, Behaviour related to birth in wild gelada baboons *(Theropithecus gelada)*, *Behaviour* 50:185-191.

Dunbar, R.I.M. and Dunbar, E.P. 1975, Social dynamics of gelada baboons, *Contrib. Primat.* 6, Basel: S. Karger.

Dunbar, R.I.M. and Dunbar, E.P. 1977, Dominance and reproductive success among female gelada baboons, *Nature* 266:351-352.

Duvall, S.W., Bernstein, I.S. and Gordon, T.P. 1976, Paternity and status in a rhesus monkey group, *J. Reprod. Fert.* 47:25-31.

Eaton, G.G. 1973, Social and endocrine determinants of sexual behaviour in simian and prosimian females. IN: *Primate Reproductive Behaviour*, Vol. 2 (C.H. Phoenix, ed.), Basel: S. Karger, pp. 20-35.

Eaton, G.G. 1974, Male dominance and aggression in Japanese macaque reproduction. IN: *Reproductive Behavior* (W. Montagna, ed.), New York: Plenum Press, pp. 287-297.

Eaton, G.G. 1977, Display behavior in a confined troop of Japanese macaques *(Macaca fuscata)*, *Anim. Behav.* 25:525-535.

Eaton, G.G. 1978, Longitudinal studies of sexual behavior in the Oregon troop of Japanese macaques. IN: *Sex and Behavior. Status and Prospectus.* (T.E. McGill, D.A. Dewsbury and B.D. Sachs, eds.), New York: Plenum Press, pp. 35-59.

Eaton, G.G., Goy, R.W. and Phoenix, C.H. 1973, Effects of testosterone treatment in adulthood on sexual behavior of female pseudohermaphrodite rhesus monkeys, *Nature New Biology* 242:119-120.

Eaton, G.G. and Resko, J.A. 1974a, Ovarian hormones and sexual behavior in *Macaca nemestrina*, *J. Comp. Physiol. Psych.* 86:919-925.

Eaton, G.G. and Resko, J.A. 1974b, Plasma testosterone and male dominance in a Japanese macaque troop compared with repeated measures of testosterone in laboratory males, *Horm. Behav.* 5:251-259.

Ehardt-Seward, C. and Bramblett, C. 1980, The structure of social space among a captive group of vervet monkeys, *Folia primat.* (in press).

Ehrhardt, A.A., 1973, Maternalism in fetal hormone and related syndromes. IN: *Contempory Sexual Behavior: Critical Issues in the 1970's* (J. Zubin and J. Money, eds.), Baltimore: Johns Hopkins Univ. Press, pp. 99-115.

Ehrhardt, A.A. 1974, Androgens in prenatal development: behavioral changes in non-human primates and men, *Advances in the Biosciences* 13:153-162.

Ehrhardt, A.A. and Baker, S.W. 1974, Fetal androgens, human CNS differentiation and behavior sex differences. IN: *Sex Differences in Behavior* (R.C. Friedman, R. Richart and R.L. Vande Wiele, eds.), New York: Wiley and Sons, pp. 53-76.

Eisely, L. 1961, *Darwin's Century.* Garden City: Doubleday Anchor.

Eisenberg, J.F. 1976, Communication mechanisms and social integration in the black spider monkey. *Ateles fusciceps* and related species. *Smithsonian Contr. Zoo.*113.

Eisenberg, J.F. 1978, Comparative ecology and reproduction of New World monkeys. IN: *The Biology and Conservation of the Callitrichidae* (D.G. Kleiman, ed.), Washington D.C.: Smithsonian Inst. Press, pp. 13-22.

Eisenberg, J.F. and Kuehn, R.E. 1966, The behavior of *Ateles geoffroyi* and related species, *Smithsonian Misc. Coll.* 151:1-63.

Eisenberg, J.F., Muckenhirn, N.A. and Rudran, R. 1972, The relation between ecology and social structure in primates, *Science* 176:863-874.

Eisenberg, L. 1972, The "human" nature of human nature, *Science* 176:123-128.

Ellefson, J.O. 1974, A natural history of white-handed gibbons in the Malayan peninsula. IN: *Gibbon and Siamang*, volume 3 (D.M. Rumbaugh, ed.), Basel: S Karger, pp. 2-136.

Embler, W. 1951, Metaphor and social belief, *ETC.* 8:83-93.

Enomoto, T. 1974, The sexual behavior of Japanese monkeys, *J. Hum. Evol.* 3:351-372.

Enomoto, T. 1978, On social preference in sexual behavior of Japanese monkeys *(Macaca fuscata), J. Hum. Evol.* 7:283-293.

Epple, G. 1970a, Quantitative studies on scent marking in the marmoset *Callithrix jacchus), Folia primat.* 13:48-62.

Epple, G. 1970b, Maintenance, breeding and development of marmoset monkeys (Callitrichidae) in captivity, *Folia primat.* 12:56-76.

Epple, G. 1973, The role of pheromones in the social communication of marmoset monkeys (Callitrichidae), *J. Reprod. Fert. Suppl.* 19:447-454.

Epple, G. 1975a, The behavior of marmoset monkeys (Callitrichidae). IN: *Primate Behavior. Developments in Field and Laboratory Research*, volume 4 (L.A. Rosenblum, ed.), New York: Academic Press, pp. 195-239.

Epple, G. 1975b, Parental behavior in *Saguinus fusciocollis* ssp. (Callitrichidae), *Folia primat.* 24:221-238.

Epple, G. 1978, Notes on the establishment and maintenance of the pair bond in *Saguinus fusciocollis.* IN: *The Biology and Conservation of the Callitrichidae* (D.G. Kleiman, ed.) Washington D.C.: Smithsonian Inst. Press, pp. 231-237.

Erwin, J. and Maple, T. 1976, Ambisexual behavior with male-male anal penetration in male rhesus monkeys, *Arch. Sex Behav.* 5:9.

Erwin, J., Maple, T. and Welles, J.F. 1975, Responses of rhesus monkeys to reunion, *Proc. V. Inter'l Congr. Prim.*:254-262.

Fady, J.C. 1969, Le jeux sociaux: le compagnon de jeux chez les jeunes. Observation chez *Macaca irus, Folia primat.* 11:134-143.

Fairbanks, L.A. 1980, Relationships among adult females in captive vervet monkeys: testing a model of rank-related attractiveness, *Anim. Behav.* 28:853-859.

Fairbanks, L.A., McGuire, M.T. and Page, N. 1978, Social roles in captive vervet monkeys (*Cerocopithecus aethiops sabaeus*), *Behavioral Processes* 3:335-352.

Fedigan, L.M. 1972a, Roles and activities of male geladas *(Theropithecus gelada), Behaviour* 41:82-90.

Fedigan, L.M. 1972b, Social and solitary play in a colony of vervet monkeys *(Cercopithecus aethiops), Primates* 13:347-364.

Fedigan, L.M. 1976, A study of roles in the Arashiyama West Troop of Japanese macaques *(Macaca fuscata), Contrib. Primat.* 9, Basel: S. Karger.

Fedigan, L.M. and Fedigan, L. 1977, The social development of a handicapped infant in a free-living troop of Japanese monkeys. IN: *Primate Bio-social Development: Biological, Social and Ecological Determinants* (S. Chevalier-Skolnikoff and F.E. Poirier, eds.), New York: Garland Publ., Inc., pp. 205-224.

Fedigan, L.M. and Gouzoules, H. 1978, The consort relationship in a troop of Japanese monkeys. IN: *Recent Advances in Primatology Volume 1: Behaviour* (D. Chivers, ed.), London: Academic Press, pp. 493-495.

Fischer, R. B. and Nadler, R.D. 1978, Affiliative, playful and homosexual interactions of adult female lowland gorillas, *Primates* 19:657-664.

Fisher, R.A. 1930, *The Genetical Theory of Natural Selection.* Oxford: Clarendon Press. (Second edition, 1958).

Fleagle, J.G. 1977, Amazing grace, *Animal Kingdom* 80:24-29.

Ford. S.M. 1980, Callitrichids as phyletic dwarfs, and the place of the Callitrichidae in Platyrrhini, *Primates* 21:31-43.

Fossey, D. 1979, Development of the mountain gorilla *(Gorilla gorilla beringei):* the first thirty-six months. IN: *The Great Apes* (D.A. Hamburg and E.R. McCown, ed.), Menlo Park: Benjamin/Cummings, pp. 139-184.

Fouts, R.S. 1975, Capacities for language in great apes. IN: *Socioecology and Psychology of Primates* (R.H.Tuttle, ed.) The Hague, Mouton Publishers, pp. 371-390.

Fox, G.J. 1972, Some comparisons between siamang and gibbon behaviour, *Folia primat.* 18:122-139.

Fox, R. 1972, Alliance and constraint: sexual selection and the evolution of human kinship systems. IN: *Sexual Selection and the Descent of Man, 1871-1971* (B. Campbell, ed.), Chicago: Aldine, pp. 282-331.

Fox, R. 1975, Primate kin and human kinship. IN: *Biosocial Anthropology* (R. Fox, ed.), London: Malaby Press, pp. 9-35.

Fragazy, D.M. and Mitchell, G.D. 1974, Infant socialization in primates, *J. Hum. Evol.* 3:563-574.

Freud, S. 1935, *A General Introduction to Psycho-analysis.* New York: Liveright Publ. Corp.

Friedman, R.C., Richart, R.M., Vande-Wiele, R.L. 1974, *Sex Differences in Behavior.* New York: Wiley and Sons.

Frisch, J. 1963, Japan's contribution to modern anthropology. IN: *Studies in Japanese Culture* (J. Roggendorf, ed.), Tokyo: Sophia University Press, pp. 225-244.

Furuya, Y. 1969, On the fission of troops of Japanese monkeys. II. General view of troop fission of Japanese monkeys, *Primates* 10:47-69.

Galat, G. and Galat-Luong, A. 1976, La colonisation de la mangrove par *Cercopithecus aethiops sabaeus* au Senegal, *Terre et la vie* 30:3-30.

Galat-Luong, A. and Galat, G. 1979, Conséquences comportmentales de perturbations sociales répetées sur une troup de Mones de Lowe, *Cercopithecus campbelli lowei* de Côte d'Ivoire, *Terre et Vie* 33:49-58.

Galdikas, B. 1979a, Orangutan adaptation at Tanjung Puting Reserve, Ph.D. Dissertation, University of California.

Galdikas, B. 1970b, Orangutan adaptation at Tanjung Puting Reserve: mating and ecology. IN: *The Great Apes* (D.A. Hamburg and E.R. McCown, eds.) Menlo Park: Benjamin/Cummings, pp. 195-233.

Gallup, G.G. Jr. 1975, Towards an operational definition of self-awareness. IN: *Socioecology and psychology of primates* (R.H. Tuttle, ed.), The Hague: Mouton Publ., pp. 309-341.

Gardner, B.T. and Gardner, R.A. 1971, Two-way communication with an infant chimpanzee. IN: *Behavior of Non-Human Primates, volume 4* (A.M. Schrier and T. Stollnitz, eds.), New York: Academic Press, pp. 117-184.

Gartlan, J.S. 1964, Dominance in east African monkeys, *Proc. E.A. Afr. Acad.* 2:75-79.

Gartlan, J.S. 1968, Structure and function in primate society, *Folia primat.* 8:89-120.

Gartlan, J.S. 1974, Adaptive aspects of social structure in *Erythrocebus patas, Symp. 5th Cong. Int'l. Primat. Soc.*, pp. 161-171.

Gartlan, J.S. 1975, Ecology and behavior of the patas monkey, *Erythrocebus patas,* Film distr. by Rockefeller University.

Gartlan, J.S. and Brain, C.K. 1968, Ecology and social variability in *Cercopithecus aethiops* and *C. mitis.* IN: *Primates. Studies in Adaptation and Variability* (P.C. Jay, ed.), NewYork: Holt, Rinehart & Winston, pp. 253-292.

Gaulin, S.J.C. 1979, A Jarman/Bell model of primate feeding niches, *Hum. Ecol.* 7:1-20.

Gautier-Hion, A. 1975, Dimorphisme sexuel et organisation social chez les Cercopithecines forestières africains, *Mammalia* 39:365-374.

Gautier-Hion, A. 1980, Seasonal variations of diet related to species and sex in a community of *Cercopithecus* monkeys, *J. Anim. Ecol.* 49:237-269.

Goldberg, S. and Lewis, M. 1969, Play behavior in the year-old infant: early sex differences, *Child Dev.* 40:21-31.

Goldberg, S. 1973, *The Inevitability of Patriarchy.* New York: Wm. Morrow & Co.

Goldfoot, D.A. 1977, Rearing conditions which support or inhibit later sexual potential of laboratory-born rhesus monkeys: hypotheses and diagnostic behaviors, *Lab. Anim. Sci.* 27:548-556.

Goldfoot, D.A., Westerborn-Van-Loon, H., Groenveld, W. and Slob, A.K. 1980, Behavioral and physiological evidence of sexual climax in the female stumptailed macaque, *Science* 208:1477-1478.

Goodall, J. 1965, Chimpanzees of the Gombe Stream Reserve. IN: *Primate Behavior. Field Studies of Monkeys and Apes* (I. DeVore, ed.), New York: Holt, Rinehart & Winston, pp. 425-473.

Goodall, J. 1967, Mother-offspring relationships in free-ranging chimpanzees. IN: *Primate Ethology* (D. Morris, ed.), Chicago: Aldine, pp. 365-436.

Goodall, J. 1968a, A preliminary report on expressive movements and communication in the Gombe Stream chimpanzees. IN: *Primates. Studies in Adaptation and Variability* (P.C. Jay, ed.), New York: Holt, Rinehart & Winston, pp. 313-374.

Goodall, J. 1968b, The behavior of free-living chimpanzees of the Gombe Stream Reserve, *Anim. Behav. Monogr.* 1:161-311.

Goodall, J. 1971a, *In the Shadow of Man.* Boston: Houghton Mifflin.

Goodall, J. 1971b, Some aspects of aggressive behavior in a group of free-living chimpanzees, *Int. Soc. Sci. J.* 23:89-97.

Goodall, J. 1973a, Cultural elements in a chimpanzee community. IN: *Precultural Primate Behavior* (E.W. Menzel, ed.), Basel: S. Karger, pp. 144-184.

Goodall, J. 1973b, The behaviour of chimpanzees in their natural habitat, *Am. J. Psychiat.* 130:1-12.

Goodall, J. 1975, The behaviour of the chimpanzee. IN: *Hominisation und Verhalten* (G.Kurth and I. Eibl-Eibesfeldt, eds.), Stuttgart: Gustav Fischer, pp. 74-136.

Goodall, J. 1977, Infant killing and cannibalism in free-living chimpanzees, *Folia primat.* 28:259-282.

Goodall, J. 1979, Life and death at Gombe, *Nat. Geo.* 155:593-621.

Goodall, J., Bandora, A., Bergmann, E., Busse, C., Matama, H., Mpongo, E., Pierce, A. and Riss, D. 1979, Intercommunity interactions in the chimpanzee population of the Gombe National Park. IN: *The Great Apes. Perspectives on Human Evolution,* vol. 5 (D.A. Hamburg and E.R. McCown, eds.), Menlo Park: Benajmin/Cummings Publ. Co., pp. 11-53.

Gordon, J.E. and Smith, E. 1965, Children's aggression, parental attitudes, and the effects of an affiliation-arousing study, *J. Personality Soc. Psychol.* 1:654-659.

Gordon, T.P., Bernstein, I.S. and Rose, R.M. 1978, Social and seasonal influences on testosterone secretion in the adult rhesus monkey, *Physiol. Behav.* 21:623-627.

Gordon, T.P., Rose, R.M., Bernstein, I.S. 1976, Seasonal rhythms in plasma testosterone levels in the rhesus monkey: a three year study, *Horm. Beh.* 7:229-243.

Gordon, T.P., Rose, R.M., Grady, C.L. and Bernstein, I.S. 1979, Effects of increased testosterone secretion on the behavior of adult male rhesus living in social groups, *Folia primat.* 32:149-160.

Goss-Custard, J.D., Dunbar, R.I.M., Aldrich-Blake, F.P.G. 1972, Survival, mating and rearing strategies in the evolution of primate social structure, *Folia primat.* 17:1-19.

Gough, K. 1979, The origin of the family. IN: *Women: A Feminist Perspective,* 2nd Edition (J. Freeman, ed.), Palo Alto: Mayfield Press, pp, 83-105.

Gould, S.J. 1966, Allometry and size in ontogeny and phylogeny, *Biol. Rev.* 41:587-640.

Gould, S.J. 1975, Allometry in primates, with emphasis on scaling and the evolution of the brain. IN: *Approaches to Primate Paleobiology* (F.S. Szalay, ed.) *Contrib. Primatol.* 5:244-292.

Gould, S.J. 1977, *Ever Since Darwin.* New York: W.W. Norton & Co.

Gould, S.J. 1980, Wallace's fatal flaw, *Nat. Hist.* 89:26-40.

Gouzoules, H. 1975, Maternal rank and early social interactions of infant stumptail macaques, *Macaca arctoides, Primates* 16:405-418.

Gouzoules, H. 1980a, A description of genealogical rank changes in a troop of Japanese monkeys *(Macaca fuscata), Primates* 21:262-267.

Gouzoules, H. 1980b, Biosocial determinants of behavioral variability in infant Japanese monkeys, Ph.D. dissertation. The University of Wisconsin, Madison.

Gouzoules, H., Fedigan, L.M. and Fedigan, L. 1975, Responses of a transplanted troop of Japanese macaques (*Macaca fuscata*) to bobcat (*Lynx rufus*) predation, *Primates* 16:335-349.

Gouzoules, S. (in preparation, a), Social relationships of adult female Japanese monkeys, *Macaca fuscata*, Ph.D. dissertation. The University of Chicago.

Gouzoules, S. (in preparation, b), Demographic, seasonal and activity pattern influences on the spatial associations of female Japanese monkeys.

Goy, R.W. 1966, Role of androgens in the establishment and regulation of behavioral sex differences in mammals, *J. Anim. Sci.* 25:Suppl. 21-35.

Goy, R.W. 1968, Organizing effects of androgen on the behavior of rhesus macaques. IN: *Endocrinology and Human Behavior* (R.P. Michael, ed.), London: Oxford Univ. Press, pp. 12-31.

Goy, R.W. 1970, Experimental control of psychosexuality, *Phil. Trans. Roy. Soc. Lond.* 259:149-162.

Goy, R.W. and Goldfoot, D.A. 1973, Experiential and hormonal factors influencing development of sexual behavior in the rhesus monkey. IN: *The Neurosciences, Third Study Program,* volume 3 (F.O. Schmitt and F.G. Worden, eds.), Cambridge, Mass.: M.I.T. Press, pp. 571-582.

Goy, R.W. and Goldfoot, R.A. 1975, Neuroendocrinology: animal models and problems of human sexuality, *Arch. Sex. Behav.* 4:405-420.

Goy, R.W. and Resko, J.A. 1972, Gonadal hormones and behavior of normal and pseudohermaphroditic nonhuman primate females, *Recent Progress in Hormone Research* 28:707-733.

Grafen, A. 1980, Opportunity, cost, benefit, and degree of relatedness, *Anim. Behav.* 28:967-968.

Gray, J.A. 1971, Sex differences in emotional behavior in mammals including man: endocrine bases, *Acta Psycholog.* 35:29-46.

Gregory, M., Silvers, A. and Sutch, D. (eds.) 1978, *Sociobiology and Human Nature.* San Francisco: Jossey-Bass.

Grewal, B.S. 1980, Social relationships between adult central males and kinship groups of Japanese monkeys at Arashiyama with some aspects of troop organization, *Primates* 21:161-180.

Griffin, D.R. 1976, *The Question of Animal Awareness. Evolutionary Continuity of Mental Experience.* New York: Rockefeller University Press.

Groves, C.P. 1972, Systematics and phylogeny of gibbons. IN: *Gibbon and Siamang,* vol. 1 (D.M. Rimbaugh, ed.) Basel: S. Karger, pp. 2-89.

Haldane, J.B.S. 1932, *The Causes of Evolution.* London: Longman, Green and Co.

Hall, K.R.L. 1967, Social interactions of the adult male and adult females of a patas monkey group. IN: *Social Communication Among Primates* (S.A. Altmann, ed.), Chicago: University of Chicago Press, pp. 261-280.

Hall, K.R.L. 1968a, Behaviour and ecology of the wild patas monkey, *Erythrocebus patas,* in Uganda. IN: *Primates. Studies in Adaptation and Variability* (P.C. Jay, ed.), New York: Holt, Rinehart & Winston, pp. 32-119.

Hall, K.R.L. 1968b, Social learning in monkeys. IN: *Primates. Studies in Adaptation and Variability* (P.C. Jay, ed.), New York: Holt, Rinehart & Winston, pp. 383-392.

Hall, K.R.L. 1968b, Aggression in monkey and ape societies. IN: *Primates. Studies in Adaptation and Variability* (P.C. Jay, ed.), New York: Holt, Rinehart & Winston, pp. 149-161.

Hall, K.R.L. and DeVore, I. 1965, Baboon social behavior. IN: *Primate Behavior. Field Studies of Monkeys and Apes* (I. DeVore, ed.), New York: Holt, Rinehart & Winston, pp. 53-110.

Hall, K.R.L., Boelkins, R.C. and Goswell, M.J. 1965, Behaviour of patas monkeys, *Erythrocebus patas,* in captivity, with notes on the natural habitat, *Folia primat.* 3:22-49.

Hall, K.R.L. and Mayer, B. 1967, Social interactions in a group of captive patas monkeys (*Erythrocebus patas*), *Folia primat.* 5:213-236.

Halperin, S.D. 1979, Temporary association patterns in free-ranging chimpanzees: an assessment of individual grouping preferences. IN: *The Great Apes. Perspectives on Human Evolution,* volume 5 (D.A. Hamburg and E.R. McCown, eds.), Menlo Park: Benjamin/Cummings Publ. Co., pp. 491-499.

Hamburg, D.A. 1972, Evolution of emotional responses: evidence from recent researches. IN: *Primates on Primates* (D.D. Quiatt, ed.) Minneapolis: Burgess, pp. 61-71.

Hamburg, D.A. 1973, An evolutionary and developmental approach to human aggressiveness, *The Psychoanalytic Quart.* XLII:185-196.

Hamburg, D.A. and McCown, E.R. 1979, Introduction. IN: *The Great Apes. Perspectives on Human Evolution,* volume 5 (D.A. Hamburg and E.R. McCown, eds.) Menlo Park: Benjamin/Cummings Publ. Co., pp. 1-10.

Hamilton, W.D. 1963, The evolution of altruistic behaviour, *Am. Nat.* 97:354-356.

Hamilton, W.D. 1964, The genetical evolution of social behaviour I and II, *J. Theoret. Biol.* 7:1-52.

Hammer, K.H. and Missakian, E. 1978, A longitudinal study of social play in Synanon peer-reared children. IN: *Social Play in Primates* (E.O. Smith, ed.) New York: Academic Press, pp. 297-319.

Hampson, J.L. and Hampson, J. 1961, The ontogenesis of sexual behavior in man. IN: *Sex and Internal Secretions,* volume 2 (N.C. Young, ed.) Baltimore: Williams & Wilkins, pp. 1401-1432.

Hanby, J.P. 1974, Male-male mounting in Japanese monkeys (*Macaca fuscata*), *Anim. Behav.* 22:836-849.

Hanby, J.P., Robertson, L.T. and Phoenix, C.H. 1971, The sexual behavior of a confined troop of Japanese macaques, *Folia primat.* 16:123-143.

Hansen, E.W. 1962, The development of infant and maternal behavior in rhesus monkeys. Unpublished doctoral dissertation, Univ. of Wisconsin, Madison, Wisconsin.

Hansen, E.W. 1966, The development of maternal and infant behaviour in the rhesus monkey, *Behaviour* 27:107-149.

Haraway, D. 1978, Animal sociology and a natural economy of the body politic, Part II: The past is the contested zone: Human nature and theories of production and reproduction in primate behavior studies, *Signs* 4:37-60.

Harcourt, A.H. 1979, Social relationships among adult female mountain gorillas, *Anim. Behav.* 28:967-968.

Harcourt, A.H. and Stewart, K.J. 1977, Apes, sex and society, *New Sci.* (Oct. 20): 160-162.

Harcourt, A.H., Stewart, K.J. and Fossey, D. 1976, Male emigration and female transfer in wild mountain gorilla, *Nature* 263:226-227.

Harding, R.S.O. 1975, Meat-eating and hunting in baboons. IN: *Socioecology and Psychology of Primates* (R.H. Tuttle, ed.) The Hague: Mouton, pp. 245-257.

Harding, R.S.O. 1980, Agonism, ranking and the social behavior of adult male baboons, *Am. J. Phys. Anthrop.* 53:203-216.

Harlow, H.F. 1959, Love in infant monkeys, *Sci. Amer.* 200:68-74.

Harlow, H.F. 1962, The heterosexual affectional system in monkeys, *Amer. Psych.* 17:1-9.

Harlow, H.F. 1965, Sexual behavior in rhesus monkeys. IN: *Sex and Behavior* (F.A. Beach, ed.) New York: Wiley & Sons, pp. 234-265 (second edition 1974).

Harlow, H.F. 1969, Age-mate or peer affectional system. IN: *Advances in the Study of Behavior*, volume 2 (D.S. Lehrman, R.A. Hinde and E. Shaw, eds.) New York: Academic Press, pp. 334-383.

Harlow, H.F. 1975, Lust, latency and love: simian secrets of successful sex, *J. Sex Res.* 11:79-90.

Harlow, H.F. and Harlow, M.K. 1962, Social deprivation in monkeys, *Sci. Amer.* 207:136-146.

Harlow, H.F. and Harlow, M.K. 1965, The affectional systems. IN: *Behavior of Nonhuman Primates*, volume 2 (A.M. Schrier, H.F. Harlow and F. Stollnitz, eds.) New York: Academic Press, pp. 287-334.

Harlow, H.F. and Harlow, M.K. 1966, Learning to love, *Amer. Sci.* 54:244-272.

Harlow, H.F., Harlow, M.K. and Hansen, E.W. 1963, The maternal affectional system of rhesus monkeys. IN: *Maternal Behavior in Mammals* (H. Rheingold, ed.) New York: Wiley & Sons, pp. 254-281.

Harlow, H.F., Harlow, M.K., Hansen, E.W. and Suomi, S.J. 1972, Infantile sexuality in monkeys, *Arch. Sex. Beh.* 2:1-7.

Harlow, H.F., Harlow, M.K. and Suomi, S.J. 1971, From thought to therapy: lessons from a primate laboratory, *Amer. Sci.* 59:538-549.

Harlow, H.F. and Lauersdorf, H.E. 1974, Sex differences in passion and play, *Perspect. Biol. Med.* 17:348-360.

Harlow, H.F. and Mears, C. 1979, *The Human Model: Primate Perspectives.* Washington D.C.: V.H. Winston and Sons.

Harrison, L. 1975, Cro-magnon woman — in eclipse, *The Science Teacher* 42:8-11.

Hartman, C.G. 1932, Studies in the reproduction of the monkey, *M. rhesus*, with special reference to menstruation and pregnancy, *Contrib. Embroy.* Carnegie Inst. Wash. 23:1-161.

Harvey, P.H., Kavanaugh, M. and Clutton-Brock, T.H. 1978a, Sexual dimorphism in primate teeth, *J. Zool.* (Lond.) 186:475-485.

Harvey, P.H., Kavanaugh, M. and Clutton-Brock, T.H. 1978b, Canine tooth size in female primates, *Nature* 276:817-818.

Hausfater, G. 1975, Dominance and reproduction in baboons, *Contrib. Primatol.* 7 Basel: S. Karger.

Hausfater, G. 1977, Primate infanticide (letter to the editor), *Am. Sci.* 65:404.

Hayes, K.J. and Nissen, C.H. 1971, Higher mental functions of a home-raised chimpanzee. IN: *Behavior of Nonhuman Primates*, volume 4 (A.M. Schrier and F. Stollnitz, eds.) New York: Academic Press, pp. 60-115.

Hazama, H. 1964, Weighing wild Japanese monkeys in Arashiyama, *Primates* 5:81-104.

Hearn, J.P. 1978, The endocrinology of reproduction in the common marmoset *(Callithrix jacchus)*. IN: *The Biology and Conservation of the Callitrichidae* (D.G. Kleiman, ed.) Washington D.C.: Smithsonian Inst. Press, pp. 163-171.

Hebb, D.O. 1953, Heredity and environment in mammalian behaviour, *Brit. J. Anim. Behav.* 1:43-47.

Heim, A.H. 1970, *Intelligence and Personality*. Penguin Books.

Hendricks, D.E., Seay, D.M. and Barnes, B. 1975, The effects of the removal of dominant animals in a small group of *Macaca fascicularis, J. Gen. Psych.* 92:157-168.

Herbert, J. 1967, The social modification of sexual and other behavior in the rhesus monkey. IN: *Neue Ergebnisse der Primatologie* (D. Starck, R. Schneider and H.J. Kuhn, eds.) Stuttgart: Fischer Verlag, pp. 232-246.

Herbert J. 1968, Sexual preference in the rhesus monkey *(Macaca mulatta)* in the laboratory, *Anim. Behav.* 16:120-128.

Herbert, J. 1970, Hormones and reproductive behavior in rhesus and talapoin monkeys, *J. Reprod. Fert.*, Suppl. 11:119-140.

Herbert, J. and Trimble, M.R. 1967, Effects of oestradiol and testosterone on the sexual receptivity and attractiveness of the female rhesus monkey, *Nature* (Lond.) 216: 165.

Hershkovitz, P. 1979, The species of sakis, genus *Pithecia* (Cebidae, Primates) with notes on sexual dichromotism, *Folia primat.* 31:1-22.

Hinde, R.A. 1974, *Biological Bases of Human Social Behavior*. New York: McGraw-Hill Book Co.

Hinde, R.A. 1978, Dominance and role — two concepts with dual meanings, *J. Social Biol. Struct.* 1:27-38.

Hinde, R.A. 1979, The nature of social structure. IN: *The Great Apes* (D.A. Hamburg and E.R. McCown, eds.) Menlo Park: Benjamin/Cummings, pp. 295-315.

Hinde, R.A. and Atkinson, S. 1970, Assessing the roles of social partners in maintaining mutual proximity, as exemplified by mother-infant relations in rhesus monkeys, *Anim. Behav.* 18:169-176.

Hinde, R.A. and Spencer-Booth, Y. 1967a, The behavior of socially-living rhesus monkeys in their first two and a half years, *Anim. Behav.* 15:169-196.

Hinde, R.A. and Spencer-Booth, Y. 1967b, The effect of social companions on mother-infant relations in rhesus monkeys. IN: *Primate Ethology* (D. Morris, ed.) Chicago: Aldine, pp. 343-364.

Hinde, R.A. and Spencer-Booth, Y. 1970, Individual differences in the responses of rhesus monkeys to a period of separation from their mothers, *J. Child Psychiatr.* 11:159-176.

Hinde, R.A. and Spencer-Booth, Y. 1971, Effects of separation from mother on rhesus monkeys, *Science* 173:111-118.

Hinde, R.A. and Stevenson-Hinde, J. 1976, Towards understanding relationships: dynamic stability. IN: *Growing Points in Ethology* (P.P.G. Bateson and R.A. Hinde, eds.) Cambridge: Cambridge University Press, pp..451-479.

Hinde, R.A. and White, L. 1974, The dynamics of a relationship — rhesus monkey ventro-ventral contact, *J. Comp. Physiol. Psych.* 86:8-23.

343

Hofstadter, R. 1959, *Social Darwinism in American Thought*. New York: Braziller.

Holloway, R.L. (ed.) 1974, *Primate Aggression, Territoriality and Xenophobia*. New York: Academic Press.

Hopkins, B. 1965, *Forest and Savannah*. London: Heinemann.

Horwich, R.H. 1972, Home range and food habits of the Nilgiri langur *Presbytis johnii*, *J. Bom. Nat. Hist. Soc.* 69:255-267.

Horwich, R.H. 1976, The whooping display in Nilgiri langurs: an example of daily fluctuations superimposed on a general trend, *Primates* 17:419-431.

Horwich, R.H. 1980, Behavioral rhythms in the Nilgiri langur, *Presbytis johnii, Primates* 21:220-229.

Hrdy, S.B. 1974, Male-male competition and infanticide among the langurs (*Presbytis entellus*) of Abu, Pajasthan, *Folia primat.* 22:19-58.

Hrdy, S.B. 1976, Care and exploitation of nonhuman primate infants by conspecifics other than the mother. IN: *Advances in the Study of Behavior*, volume 6 (D.S. Lehrman, R.A. Hinde and E. Shaw, eds.) New York: Academic Press, pp. 101-158.

Hrdy, S.B. 1977a, Infanticide as a primate reproductive strategy, *Am. Sci.* 65:40-49.

Hrdy, S.B. 1977b, *The Langurs of Abu. Male and Female Strategies of Reproduction.* Cambridge: Harvard Univ. Press.

Hrdy, S.B. 1979a, Infanticide among animals: a review, classification and examination of the implications for the reproductive strategies of females, *Ethol. and Sociobiol.* 1:13-40.

Hrdy, S.B. 1979b, The evolution of human sexuality: the latest word and the last, *Quart. Rev. Biol.* 54:309-314.

Hull, D.L. 1978, Scientific bandwagon or traveling medicine show? *Transaction Society* 15:50-59.

Hutchinson, R.R. and Renfrew, J.W. 1966, Stalking, attack and eating behaviors elicited from the same sites in the hypothalumus, *J. Comp, Physiol. Psychol.* 61:360-367.

Hutt, C. 1972, *Males and Females*. Penguin Books.

Huxley, J.S. 1938, Darwin's theory of sexual selection and the data subsumed by it, in the light of recent research, *Am. Nat.* 72:416-433.

Imanishi, K. 1957, Social behaviour in Japanese monkeys, *Macaca fuscata, Psychologia* 1:47-54.

Imanishi, K. 1960, Social organization of subhuman primates in their natural habitat, *Curr. Anthrop.* 1:390-405.

Imanishi, K. 1961, The origin of the human family — a primatological approach, *Japanese Journal of Ethnology* 25:119-130.

Imanishi, K. 1965, Identification: a process of socialization in the subhuman society of *Macaca fuscata*. IN: *Japanese Monkeys* (K. Imanishi and S. Altmann, eds.) Atlanta: Emory Univ. Press, pp. 30-51.

Ingram, J.C. 1977, Interactions between parents and infants, and the development of independence in the common marmoset, *Anim. Behav.* 25:811-827.

Ingram, J.C. 1978, Parent-infant interactions in the common marmoset (*Callithrix jacchus*). IN: *The Biology and Conservation of the Callitrichidae* (D.G. Kleiman, ed.) Washington D.C.: Smithsonian Inst. Press, pp. 281-291.

Itani, J. 1956, Food habits of the wild Japanese monkey. Publication of Primates Research Group, pp. 1-14.

Itani, J. 1958, On the acquisition and propagation of a new food habit in the natural group of the Japanese monkey at Takasakiyama, *Primates* 1:84-100.

Itani, J. Paternal care in the wild Japanese monkey (*Macaca fuscata*), *Primates* 2:61-93.

Itani, J. 1963, Vocal communication in the wild Japanese monkey, *Primates* 4:11-66.

Itani, J. 1965, On the acquisition and propagation of a new food habit in the troop of Japanese monkeys at Takasakiyama. IN: *Japanese Monkeys* (K. Imanishi and S.A. Altmann, eds.) Atlanta: Emory Univ. Press, pp. 52-65.

Itani, J. 1977, Evolution of primate social structure, *J. Hum. Evol.* 6:235-243.

Itani, J. and Nishimura, A. 1973, The study of infrahuman culture in Japan. IN: *Precultural Primate Behavior* (E.W. Menzel Jr., ed.) *Symp. of 4th Cong. Int'l Soc. Prim.*, Volume 1, Basel: S. Karger, pp. 26-50.

Itani, J. and Suzuki, A. 1967, The social unit of chimpanzees, *Primates* 8:355-382.

Itani, J., Tokuda, K., Furuya, Y., Kano, K. and Shin, Y. 1963, The social construction of natural troops of Japanese monkeys at Takasakiyama, *Primates* 4:2-43.

Itoigawa, N. 1974, Variables in male leaving a group of Japanese monkeys. IN: *Proc. Symp. 5th Cong. Int'l Primat. Soc.* (S. Kondo, M. Kawai, A. Ehara and S. Kawamura, eds.) Tokyo, Japan: Japan Science Press, pp. 233-245.

Iwamoto, T. 1979, Feeding ecology. IN: *Ecological and Sociological Studies of Gelada Baboons, Contrib. Primat.* 16 (M. Kawai, ed.) Basel: S. Karger, pp. 280-330.

Izawa, K. 1970, Unit groups of chimpanzees and their nomadism in the savannah woodland, *Primates* 11:1-45.

Izawa, K. 1978, Ecological study of Japanese snow monkey in Hakusan National Park, Report # 4, Special Printing, Ishikawa Prefecture, Hakusan Nature Protection Centre.

Izawa, K. and Nishida, T. 1963, Monkeys living in the northern limits of their distribution, *Primates* 4:67-87.

Jacklin, C.N., Maccoby, E.E. and Dick, A. 1973, Barrier behavior and toy preference: sex differences and their absence in the year-old child, *Child Dev.* 44:196-200.

Janetos, A.C. 1980, Strategies of female mate choice: a theoretical analysis, *Behav. Ecol. Sociobiol.* 7:102-112.

Jay, P. 1963, Mother-infant relations in langurs. IN: *Maternal Behavior in Mammals* (H.L. Rheingold, ed.) New York: John Wiley, pp. 282-304.

Jay, P.C. 1965, The common langur in northern India. IN: *Primate Behavior. Field Studies of Monkeys and Apes* (I. DeVore, ed.) New York: Holt, Rinehart & Winston, pp. 197-249.

Jensen, G.D., Bobbitt, R.A. and Gordon, B.N. 1967, Sex differences in social interaction between infant monkeys and their mothers. IN: *Recent Advances in Biological Psychiatry*, Volume IX (J. Wortis, ed.) New York: Plenum Press, pp. 283-293.

Jensen, G.D., Bobbitt, R.A. and Gordon, B.N. 1968, Sex differences in the development of independence of infant monkeys, *Behaviour* 30:1-14.

Jensen, G.D., Bobbitt, R.A. and Gordon, B.N. 1973, Mothers' and infants' roles in the development of independence of *Macaca nemestrina*, *Primates* 14:79-88.

Johnson, R.N. 1972, *Aggression in Man and Animals*. Philadelphia: Saunders.

Jolly, A. 1966, *Lemur Behavior*. Chicago: Univ. of Chicago Press.

Jolly, A. 1972, *The Evolution of Primate Behavior*. New York: MacMillan.

Jolly, C. 1970, The seed-eaters: a new model of hominid differentiation based on a baboon analogy, *Man* 5:5-26.

Jolly, C. 1972, The classification and natural history of *Theropithecus (Simopithecus)* (Andrews, 1916) baboons of the African Plio-Pleistocene, *Bull. Brit. Mus. Nat. Hist. Geol.* 22:1-23.

Jorde, L.B. and Spuhler, J.N. 1974, A Statistical analysis of selected aspects of primate demography, ecology and social behavior, *J. Anth. Res.* 30:199-224.

Jouventin, P. 1975, Observations sur la socio-ecologie du mandrill, *Terre et Vie* 29:493-532.

Kagan, J. and Moss, H.A. 1960, The stability of passive and dependent behaviour from childhood through adulthood, *Child Dev.* 31:577-591.

Kano, T. 1971, The chimpanzee of Filabanga, western Tanzania, *Primates* 12:229-246.

Kano, T. 1972, Distribution and adaptation of the chimpanzee on the eastern shore of Lake Tanganyika, *Kyoto Univ. African Studies* 7:37-129.

Kaplan, J.R. 1977, Patterns of fight interference in free-ranging rhesus monkeys, *Amer. J. Phys. Anthrop.* 47:278-288.

Kaplan, J.R. 1978, Fight interference and altruism in rhesus macaques, *Am. J. Phys. Anthrop.* 49:241-250.

Kaufman, I.C. 1967, Depression in infant monkeys separated from their mothers, *Science* 155:1030-1031.

Kaufman, I.C. 1975, Learning what comes naturally: the role of life experience in the establishment of species typical behavior, *Ethos* 3:129-142.

Kaufman, I.C. and Rosenblum, L.A. 1966, A behavioral taxonomy for *Macaca nemestrina* and *Macaca radiata* based on longitudinal observation of family groups in the laboratory, *Primates* 7:205-258.

Kaufman, I.C. and Rosenblum, L.A. 1969a, The waning of the mother-infant bond in two species of macaque. IN: *Determinants of Infant Behavior*, vol. IV (B.M. Foss, ed.) London: Metheun, pp. 41-59.

Kaufman, I.C. and Rosenblum, L.A. 1969b, Effects of separation from mother on the emotional behavior of infant monkeys, *Ann. N.Y. Acad. Sci.* 159:681-695.

Kaufman, J.H. 1965, A three year study of mating behavior in a free-ranging band of rhesus monkeys, *Ecology* 46:500-512.

Kaufman, J.H. 1966, Behavior of infant rhesus monkeys and their mothers in a free-ranging band, *Zoologica* 51:17-28.

Kawabe, M. 1966, One observed case of hunting behavior among wild chimpanzees living in the savannah woodland of western Tanzania, *Primates* 7:393-396.

Kawai, M. 1960, A field experiment on the process of group formation in the Japanese monkey (*Macaca fuscata*) and the releasing of the group at O'Hirayama, *Primates* 2:181-253.

Kawai, M. 1965a, On the system of social ranks in a natural troop of Japanese monkeys. I. Basic rank and dependent rank. IN: *Japanese Monkeys* (K. Imanishi and S.A. Altmann, eds.) Atlanta: Emory Univ. Press, pp. 66-86.

Kawai, M. 1965b, On the system of social ranks in a natural troop of Japanese monkeys. II. Ranking order as observed among the monkeys on and near the test box. IN: *Japanese Monkeys* (K. Imanishi and S.A. Altmann, eds.) Atlanta: Emory Univ. Press, pp. 87-104.

Kawai, M. 1965c, Newly acquired precultural behaviour of the natural troop of Japanese monkeys on Koshima Islet, *Primates* 6:1-30.

Kawai, M. 1967, Ecological studies of reproduction in Japanese monkeys (*Macaca fuscata*) I. Problems of the birth season, *Primates* 8:35-74.

Kawai, M. (ed.) 1979, Ecological and sociological studies of gelada baboons, *Contrib. Primat.* 16, Basel: S. Karger.

Kawamura, S. 1958, A study on the rank system of Japanese monkeys, *Primates* 1:149-156.

Kawamura, S. 1959, The process of sub-culture propagation among Japanese macaques, *Primates* 2:43-54.

Kawamura, S. 1965, Matriarchal social ranks in the Minoo-B troop. A study of the rank system of Japanese monkeys, IN: *Japanese Monkeys* (K. Imanishi and S.A. Altmann, eds.) Atlanta: Emory Univ. Press, pp. 105-112.

Kedenburg, D. 1979, Testosterone and human aggressiveness: an analysis, *J. Hum. Evol.* 8:407-410.

Kessler, S.J. and McKenna W. 1978, *Gender: An Ethno-Methodological Approach.* New York: Wiley & Sons.

Kingdon, J.S. 1980, The role of visual signals and face patterns in African monkeys (guenons) of the genus *Cercopithecus, Trans. Zool. Soc. Lond.* 35:431-475.

Kinzey, W.G. 1972, Canine teeth of the monkey, *Callicebus moloch:* lack of sexual dimorphism, *Primates* 13:365-369.

Kinzey, W.G. 1977a, Diet and feeding behaviour of *Callicebus torquatus.* IN: *Primate Ecology* (T.H. Clutton-Brock, ed.) London: Academic Press, pp. 127-152.

Kinzey, W.G. 1977b, Positional behavior and ecology in Callicebus, *Yearbook of Physical Anthropology* 20:468-480.

Kinzey, W.G. 1978, Feeding behavior and the molar features in two species of titi monkeys. IN: *Recent Advances in Primatology,* vol. 1: Behaviour (D.J. Chivers and J. Herbert, eds.) London: Academic Press, pp. 373-384.

Kinzey, W.G. 1979, Marmosets and tamarins, *Science* 203:879.

Kinzey, W.G. and Gentry, A.H. 1979, Habitat utilization in two species of Callicebus. IN: *Primate Ecology: Problem-Oriented Field Studies* (R.W. Sussman, ed.) New York: John Wiley & Sons, pp. 89-100.

Kinzey, W.G. Rosenberger, A.L., Heisler, P.S., Prowse, D.L. and Trilling, J.S. 1977, A preliminary field investigation of the yellow handed titi monkey, *Callicebus torquatus torquatus* in Northern Peru, *Primates* 18:159-181.

Kleiman, D.G. 1977, Monogamy in mammals, *Quart. Rev. Biol.* 52:39-69.

Kleiman, D.G. 1978, The development of pair preferences in the lion tamarin: male competition or female choice? IN: *Biology and Behavior of Marmosets* (H. Rothe, H.J. Wolters, and J.P. Hearn, eds.) Göttingen: Eigenverlag, pp. 203-207.

Kleiman, D.G. 1979, Parent-offspring and sibling competition in a monogamous primate, *Am. Nat.* 114:753-760.

Klein, L.L. 1974, Agonistic behavior in neotropical primates. IN: *Primate Aggression, Territoriality and Xenophobia* (R.L. Holloway, ed.) New York: Academic Press, pp. 77-122.

Kling, A. 1975, Testosterone and aggressive behavior in man and nonhuman primates. IN: *Hormonal Correlates and Behavior,* vol. 1 (B.E. Eleftheriout and R.L. Sprott, eds.) New York: Plenum Press, pp. 305-321.

Kling, A. and Dunne, K. 1976, Social-environmental factors affecting behavior and plasma testosterone in normal and amygdala lesioned *Macaca speciosa, Primates* 17:23-42.

Klopfer, P, 1973, Does behavior evolve? *Ann. N.Y. Acad. Sci.* 223:113-119.

Koford, C.B. 1965, Population dynamics of rhesus monkesy on Cayo Santiago. IN: *Primate Behavior* (I. DeVore, ed.) New York: Holt, Rinehart & Winston, pp. 160-174.

Kolata, G.B. 1976, Sex and the dominant male, *Science* 191:55-56.

Kolodny, R.C., Masters, W.H., Hendryx, J. and Toro, G. 1971, Plasma testosterone and semen analysis in male homosexuals, *New Eng. J. Med.* 285:1170-1174.

Korner, A.F. 1974, Methodological considerations in studying sex differences in the behavioral functioning of newborns. IN: *Sex Differences in Behavior* (R.C. Friedman, R.M. Richart and R.L. Vande Wiele, eds.) New York: John Wiley & Sons, pp. 197-208.

Kortlandt, A. 1962, Chimpanzees in the wild, *Sci Am.* 206:128-138.

Kortmulder, K. 1968, An ethological theory of the incest taboo and exogamy, *Curr. Anthr.* 9:437-449.

Koyama, N. 1967, On dominance rank and kinship of a wild Japanese monkey troop in Arashiyama, *Primates* 8:189-216.

Koyama, N. 1970, Changes in dominance rank and division of a wild Japanese monkey troop in Arashiyama, *Primates* 11:335-390.

Krige, P.D. and Lucas, J.W. 1974, Aunting behaviour in an urban troop of *Cercopithecus aethiops, J.Behav. Sci.* 2:55-61.

Kuhn, T.S. 1970, *The Nature of Scientific Revolution.* Chicago: The Univ. of Chicago Press.

Kummer, H. 1967, Tripartite relations in hamadryas baboons, IN: *Social Communication Among Primates* (S.A. Altmann, ed.) Chicago: Univ. of Chicago Press, pp. 63-71.

Kummer, H. 1968, *Social Organization of Hamadryas Baboons.* Chicago: Univ. of Chicago Press.

Kummer, H. 1971a, *Primate Societies. Group Techniques of Ecological Adaptation.* Chicago: Aldine-Atherton.

Kummer, H. 1971b, Immediate causes of primate social structures, *Proc. Third Int'l Cong. Primat.*, vol. 3, Basel: S. Karger, pp. 1-11.

Kummer, H. 1973, Dominance versus possession. An experiment on hamadryas baboons. IN: *Precultural Primate Behavior. Symp. of the 4th Int'l Congr. of Primat.*, vol. 1 (E.W. Menzel, ed.) Basel: S. Karger, pp. 226-231.

Kummer, H. 1974, Distribution of interindividual distances in patas monkeys and gelada baboons, *Folia primat.* 21:153-160.

Kummer, H., Goetz, W. and Angst, W. 1970, Cross-species modifications of social behavior in baboons. IN: *Old World Monkeys. Evolution, Systematics and Behavior* (J.R. & P.R. Napier, eds.) London: Academic Press, pp. 351-363.

Kummer, H., Goetz, W. and Angst, W. 1974, Triadic differentiation: an inhibitory process protecting pair bonds in baboons, *Behaviour* 49:62-87.

Kurland, J.A. 1977, Kin selection in the Japanese monkey, *Contrib. Primat.* 12, Basel: S. Karger.

Lack, D. 1969, *Ecological Adaptations for Breeding in Birds.* London: Methuen.

Lancaster, J.B. 1972, Play-mothering: the relations between juvenile females and young infants among free-ranging vervet monkeys. IN: *Primate Socialization* (F.E. Poirier, ed.) New York: Random House, pp. 83-104.

Lancaster, J.B. 1973, In praise of the achieving female monkey, *Psych. Today*, pp. 30-36, 99.

Lancaster, J.B. 1975, *Primate Behavior and the Emergence of Human Culture.* New York: Holt, Rinehart & Winston.

Lancaster, J.B. 1976, Sex roles in primate societies. IN: *Sex Differences. Social and Biological Perspectives* (M.S. Teitelbaum, ed.) Garden City, New York: Anchor Press/Doubleday, pp. 22-61.

Lancaster, J.B. 1978a, Sex and gender in evolutionary perspective. IN: *Human Sexuality: A Comparative and Developmental Perspective* (H. Katchadourian, ed.) Berkeley, California: Univ. of Calif. Press, pp. 51-80.

Lancaster, J.B. 1978b, Carrying and sharing in human evolution, *Human Nature* 1:82-89.

Lancaster, J.B. and Lee, R.B. 1965, The annual reproductive cycle in monkeys and apes. IN: *Primate Behavior. Field Studies of Monkeys and Apes* (I. DeVore, ed.) New York: Holt, Rinehart & Winston, pp. 486-513.

Langford, J.B. 1963, Breeding behavior of *Hapale jacchus* (common marmoset), *South Afr. J. Sci.* 59:299-300.

Laughlin, W.S. 1968, Hunting: an integrating biobehavioral system and its evolutionary importance. IN: *Man the Hunter* (R.B. Lee and I. DeVore, eds.) Chicago: Aldine, pp. 304-320.

Lee, R.B. and DeVore, I. (eds.) 1968, *Man the Hunter*. Chicago: Aldine.

LeGros Clark, W.E. 1962, *The Antecedents of Man* (Second Edition) Edinburgh Press.

Leibowitz, L. 1975, Perspectives on the evolution of sex differences. IN: *Toward an Anthropology of Women* (R.R. Reiter, ed.) New York: Monthly Review Press, pp. 22-35.

Leibowitz, L. 1978, *Females, Males, Families. A Biosocial Approach.* North Scituate, Mass.: Duxbury Press.

Lehrman, D.S. 1970, Semantic and conceptual issues in the nature-nurture problem. IN: *Development and Evolution of Behavior* (L.R. Aronson, E. Tobach, D.S. Lehrman and J.S. Rosenblatt, eds.) San Francisco: W.H. Freeman, pp. 17-52.

Leonard, J.W. 1979, A strategy approach to the study of primate dominance behaviour, *Behav. Proc.* 4:155-172.

Leutenegger, W. 1973, Maternal-fetal weight relationships in primates, *Folia primat.* 20:280-293.

Leutenegger, W. 1977, Sociobiological correlates of sexual dimorphism in body weight in South African australopiths. *S. Afr. J. Sci.* 73:143-144.

Leutenegger, W. 1978, Scaling of sexual dimorphism in body size and breeding system in primates, *Nature* 272:610-611.

Leutenegger, W. 1979, Evolution of litter size in primates, *Am. Nat.* 114:525-531.

Leutenegger, W. and Kelley, J.T. 1977, Relationship of sexual dimorphism in canine size and body size to social, behavioral, and ecological correlates in anthropoid primates, *Primates* 18:117-136.

Levi-Strauss, C. 1968, The concept of primitiveness. IN: *Man the Hunter* (R.B. Lee and I. DeVore, eds.) Chicago: Aldine, pp. 349-352.

Levi-Strauss, C. 1970, *The Elementary Structures of Kinship*. London: Social Science Paperbacks.

Lewontin, R.C. 1970, The units of selection, *Ann. Rev. Ecol. Sys.* 1:1-18.

Lewontin, R.C. 1974, *The Genetic Basis of Evolutionary Change*. New York: Columbia Univ. Press.

Lewontin, R.C. 1977, Caricature of Darwinism, *Nature* 266:283-284.

Lewontin, R.C. 1978, Adaptation, *Sci. Am.* 239:212-230.

Lewontin, R.C. 1979, Sociobiology as an adaptionist program, *Behav. Sci.* 24:5-14.

Lindburgh, D.G. 1971, The rhesus monkey in North India: an ecological and behavioral study. IN: *Primate Behavior*, vol. 2 (L.A. Rosenblum, ed.) New York: Academic Press, pp. 2-106.

Lindburg, D.G. 1975, Mate selection in the rhesus monkey, *Macaca mulatta*, Paper read at American Association of Physical Anthripologists meetings, Denver, Colorado.

Lindburg, D.G. 1980, Mating behavior and estrus in the Indian rhesus monkey. IN: *Perspectives in Primate Biology* (P.K. Seth, ed.) IN PRESS.

Linton, R. 1945, *The Cultural Background of Personality*. New York: Appleton-Century Crofts.

Loizos, C. 1967, Play behaviour in higher primates: a review. IN: *Primate Ethology* (D. Morris, ed.) New York: Anchor Books, pp. 226-282.

Lorenz, K. 1950, The comparative method in studying innate behaviour patterns, *Symp. Soc. Exper. Biol.* 4:221-268.

Lorenz, K. 1966, *On Aggression*. London: Methuen.

Lorenz, K. 1970, *Studies in Animal and Human Behavior*, vol. 1 (trans. R.D. Martin) London: Methuen.

Loy, J.D. 1970a, Behavioral responses of free-ranging rhesus monkeys to food shortage, *Am. J. Phys. Anthrop.* 33:263-271.

Loy, J.D. 1970b, Peri-menstrual sexual behavior among rhesus monkeys, *Folia primat.* 13:286-297.

Loy, J.D. 1975a, The descent of dominance in *Macaca:* insights into the structure of human societies. IN: *Socioecology and Psychology of Primates* (R.H. Tuttle, ed.) The Hague: Mouton.

Loy, J. 1975b, The copulatory behavior of adult male patas monkeys, *Erythrocebus patas, J. Reprod. Fert.* 45:193-195.

Loy, J. and Loy, K. 1974, Behavior of an all-juvenile group of rhesus monkeys, *Amer. J. of Phys. Anthrop.* 40:83-96.

Loy,J.D., Loy, K., Patterson, D., Keifer, G. and Conoway, C. 1978, The behavior of gonadectomized rhesus monkeys. I. Play. IN: *Social Play in Primates* (E.O. Smith, ed.) New York: Academic Press, pp. 49-78.

MacKinnon, J. 1979, Reproductive behavior in wild orangutan populations. IN: *The Great Apes* (D.A. Hamburg & E.R. McCown, eds.) Menlo Park: Benjamin/Cummings, pp. 257-273.

MacRoberts, M.H. 1970, The social organization of Barbary apes (*Macaca sylvana*) on Gibraltar, *Am. J. of Phys. Anthrop.* 33:83-100.

McGinnis, L.M. 1979, Maternal separation versus removal from group companions in rhesus monkeys, *J. Child Psychol.* 20:15-27.

McGinnis, P.R. 1979, Sexual behavior in free-living chimpanzees: consort relationships.

350

IN: *The Great Apes. Perspectives on Human Evolution,* vol. 5 (D.A. Hamburg and E.R. McCown, eds.) Menlo Park: Benjamin/Cummings Publ. Co., pp. 429-439.

McGrew, W.C. 1972, *An Ethological Study of Children's Behavior.* New York: Academic Press.

McGrew, W.C. 1977, Socialization and object manipulation of wild chimpanzees. IN: *Primate Biosocial Development: Biological, Social and Ecological Determinants* (S. Chevalier-Skolnikoff and F.E. Poirier, eds.) New York: Garland Pub. Inc., pp. 261-288.

McGrew, W.C. 1979, Evolutionary implications of sex differences in chimpanzee predation and tool use. IN: *The Great Apes. Perspectives on Human Evolution,* vol. 5 (D.A. Hamburg and E.R. McCown, eds.) Menlo Park: Benjamin Cummings Pub. Co., pp. 441-463.

McGuire, M.T. 1974, The St. Kitts vervet, *Contrib. Primat.* 1, Basel: S. Karger.

McKenna, J.J. 1978, Biosocial functions of grooming behavior among the common Indian langur monkey *(Presbytis entellus), Am. J. Phys. Anthrop.* 48:503-510.

McKenna, J.J. 1979a, Aspects of infant socialization, attachment, and maternal caregiving patterns among primates: a cross-disciplinary review, *Yearbook of Physical Anthropology* 22:250-286.

McKenna, J.J. 1979b, The evolution of allomothering among colobine monkeys: function and opportunism in evolution, *Am. Anth.* 81:818-840.

McKinney, W.T., Jr. 1974, Primate social isolation, *Arch. Gen. Psychiat.* 31:422-426.

Maccoby, E.E. and Jacklin, C.M. 1973, Stress, activity and proximity seeking: sex differences in the year-old child, *Child. Dev.* 44:34-42.

Maccoby, E.E. and Jacklin, C.N. 1978, *The Psychology of Sex Differences,* vol. 1. Stanford: Stanford Univ. Press.

Makurius, R. 1968, Comments on Incest Taboo and Exogamy, *Curr. Anth.* 9:443.

Makwana, S.C. 1979, Infanticide and social change in two groups of the Hanuman langur, *Presbytis entellus* at Jodhpur, *Primates* 20:293-300.

Malina, R. M. 1975, *Growth and Development. The First Twenty Years in Man.* Minneapolis: Burgess Publ. Co.

Marler, P. 1976, *On Aggression.* New York: Harcourt, Brace & World.

Marler, P. and Hamilton, W.J. 1966, *Mechanisms of Animal Behavior.* New York: John Wiley and Sons.

Marsden, H.M. 1968, Agonistic behavior of young rhesus monkeys after changes induced in social rank of their mothers, *Anim. Behav.* 16:38-44.

Marsh, C.W. 1979, Female transference and mate choice among Tana River red colobus, *Nature* 281:568-569.

Martin, R.D. 1974, The biological basis of human behavior. IN: *The Biology of Brains* (W.B. Broughton, ed.) London: Institute of Biology, pp. 215-250.

Martin, M.K. and Voorhies, B. 1975, *Female of the Species.* New York: Columbia Univ. Press.

Maslow, A.H. 1936a, The role of dominance in the social and sexual behavior of infrahuman primates: II. An experimental determination of the behavior syndrome of dominance, *J. Genetic Psych.* 48:310-338.

Maslow, A.H. 1936b, The role of dominance in the social and sexual behavior of infrahuman primates: IV. The determination of hierarchy in pairs and in a group, *J.*

Genetic Psych. 49:161-198.

Mason, W.A. 1960, The effects of social restriction on the behavior of rhesus monkeys: I. Free social behavior, *J. Comp. Physiol. Psychol.* 53:582-589.

Mason, W.A. 1968a, Early social deprivation in the nonhuman primates: implications for human behavior. IN: *Environmental Influences* (D.C. Glass, ed.) New York: Rockefeller Univ. Press, pp. 70-100.

Mason, W.A. 1968b, Use of space by Callicebus groups. IN: *Primates. Studies in Adaptation and Variability* (P.C. Jay, ed.) New York: Holt, Rinehart & Winston, pp. 200-216.

Mason, W.A. 1971, Field and laboratory studies of social organization in Saimiri and Callicebus. IN: *Primate Behavior Developments in Field and Laboratory Research,* vol. 2 (L.A. Rosenblum, ed.) New York: Academic Press, pp. 107-137.

Mason, W.A. 1974, Comparative studies of social behavior in Callicebus and Saimiri: behavior of male-female pairs, *Folia primat.* 22:1-8.

Mason, W.A. 1975, Comparative studies of social behavior in Callicebus and Saimiri: strength and specificity of attraction between male-female cagemates, *Folia primat.* 23:113-123.

Mason, W.A. 1976, Primate social behavior: pattern and process. IN: *Evolution of Brain and Behavior in Vertebrates* (R.B. Masterton, ed.) Hillsdale, New Jersey: Lawrence Erlbaum Associates, pp. 425-455.

Mason, W.A. 1978, Ontogeny of social systems. IN: *Recent Advances in Primatology,* vol. 1 (D. Chivers and J. Herbert, eds.) London: Academic Press, pp. 5-14.

Mason, W.A. and Epple, G. 1969, Social organization in experimental groups of Saimiri and Callicebus. IN: *Proc. Second Int'l. Cong. Primat. Vol. 1, Behavior* (C.R. Carpenter, ed.) Basel: S. Karger, pp. 59-65.

Mason, W.A., Green, P.C. and Posepanko, C.J. 1960, Sex differences in affective social responses of rhesus monkeys, *Behaviour* 16:74-83.

Mason, W.A. and Kenney, M.D.1974, Redirection of filial attachments in rhesus monkeys: dogs as mother surrogates, *Science* 183:1209-1211.

Massey, A. 1977, Agonistic aids and kinship in a group of pigtail macaques, *Behav. Ecol. Sociobiol.* 2:31-40.

Matson, F.W. 1976, *The Idea of Man.* New York: Delacorte Press.

Maynard-Smith, J. 1977, Parental investment: a prospective analysis, *Anim. Behav.* 25: 1-9.

Maynard-Smith, J. 1978a, Group selection. IN: *Readings in Sociobiology* (T.H. Clutton-Brock and P. Harvey, eds.) San Francisco: W.H. Freeman, pp. 20-30.

Maynard-Smith, J. 1978b, *The Evolution of Sex.* Cambridge Mass.: Cambridge Univ. Press.

Mayr, E. 1972, Sexual selection and natural selection. IN: *Sexual Selection and the Descent of Man, 1871-1971* (B. Campbell, ed.) Chicago: Aldine, pp. 87-104.

Mead, M. 1949, *Male and Female. A Study of the Sexes in a Changing World.* New York: William Morrow.

Mead, M. 1963, *Sex and Temperament in Three Primitive Societies.* New York: William Morrow.

Mead, M. 1971, Innate behavior and building new cultures: a commentary. IN: *Man and Beast: Comparative Social Behavior* (J.F. Eisenberg and W.S. Dillion, eds.)

Washington D.C.: Smithsonian Inst. Press, pp. 371-381.

Mech, L.D. 1970, *The Wolf: Ecology and Social Behavior of an Endangered Species.* New York: Natural History Press.

Megargee, E.I. and Hokanson, J.E. 1970, *The Dynamics of Aggression.* New York: Harper & Row.

Menzel, E.W. Jr. 1966, Responsiveness to objects in free-ranging Japanese monkeys, *Behaviour* 25:130-149.

Menzel, E.W. Jr. 1968, Primate naturalistic research and problems of early experience, *Dev. Psych.* 1:175-184.

Menzel, E.W. Jr., Davenport, R.K. and Rogers, C.M. 1970, The development of tool-using in wild-born and restriction-reared chimpanzees, *Folia primat.* 12:273-283.

Meyer-Bahlburg, H.F.L. 1974, Aggression, androgens and the XYY syndrome. IN: *Sex Differences in Behavior* (R.C. Friedman, R.M. Richart and R.L. Vande Wiele, eds.) New York: Wiley & Sons, pp. 433-453.

Michael, R.P. 1975 (Dec.), Hormones and sexual behavior in the female, *Hospital Practice*, pp. 69-76.

Michael, R.P. and Bonsall, R.W. 1979, Hormones and the sexual behavior of rhesus monkeys. IN: *Endocrine Control of Sexual Behavior* (C. Beyer, ed.) New York: Raven Press, pp. 279-302.

Michael, R.P., Bonsall, R.W. and Warner, P. 1974, Human vaginal secretions: volatile fatty acid content, *Science* 186:1217.

Michael, R.P., Bonsall, R.W. and Zumpe, D. 1978, Consort bonding and operant behavior by female rhesus monkeys, *J. Comp. Physiol. Psych.* 92:837-845.

Michael, R.P. and Herbert, J. 1963, Menstrual cycle influences grooming behavior and sexual activity in the rhesus monkey, *Science* 140:500-501.

Michael, R.P. and Keverne, E.M. 1968, Pheromones in the communication of sexual status in primates, *Nature* (Lond.) 218:746-749.

Michael, R.P. and Saayman, G.S. 1967, Individual differences in the sexual behavior of male rhesus monkeys (*Macaca mulatta*) under laboratory conditions, *Anim. Behav.* 15:460-466.

Michael, R.P., Saayman, G. and Zumpe, D. 1968, The suppression of mounting behavior and ejaculation in male rhesus monkeys (*Macaca mulatta*) by administration of progesterones to their female partners, *J. Endocr.* 41:421.

Michael, R.P. and Welegalla, J. 1967, Ovarian hormones and sexual behavior of the male rhesus monkey (*Macaca mulatta*) under laboratory conditions, *J. Endocr.* 39:81.

Michael, R.P. and Welegalla, J. 1968, Ovarian hormones and the sexual behavior of the female rhesus monkey, *J. Endocr.* 41:407-420.

Michael, R.P. and Wilson, M. 1974, Effects of castration and hormone replacement in fully adult male rhesus monkeys (*Macaca mulatta*), *Endocrinology* 5:150-159.

Michael, R.P., Wilson, M.I. and Zumpe, D. 1974, The bisexual behavior of female rhesus monkeys. IN: *Sex Differences in Behavior* (R.C. Friedman, R.M. Richart and R.L. Vande Wiele, eds.) New York: John Wiley & Sons, pp. 399-412.

Michael, R.P. and Zumpe, D. 1970, Sexual initiating behavior by female rhesus monkeys (*Macaca mulatta*) under laboratory conditions, *Behaviour* 36:168-186.

Michael, R.P. and Zumpe, D. 1971, Patterns of reproductive behavior. IN: *Comparative Reproduction of Nonhuman Primates* (E.S.E. Hafez, ed.) Springfield: C.C. Thomas, pp. 205-242.

Miller, C. and Swift, K. 1976, *Words and Women*. Garden City: Doubleday, Anchor Press.

Miller, M.H., Kling, A. and Dicks, D. 1973, Familial interactions of male rhesus monkeys in a semi-free-ranging troop, *Am. J. Phys. Anthrop.* 38:605-612.

Miller, R.E., Caul, W.F. and Mirsky, I.A. 1967, Communication of affect between feral and socially isolated monkeys, *J. Personality and Soc. Psychol.* 7:231-239.

Millet, K. 1970, *Sexual Politics*. Garden City, New York: Doubleday & Co.

Mineka, S. and Suomi, S.J. 1978, Social separation in monkeys, *Psych. Bulletin* 85:1376-1400.

Missakian, E.A. 1972, Genealogical and cross-genealogical dominance relations in a group of free-ranging rhesus monkeys (*Macaca mulatta*) on Cayo Santiago, *Primates* 13:169-180.

Missakian, E.A. 1973, Genealogical mating activity in free-ranging groups of rhesus monkeys (*Macaca mulatta*) on Cayo Santiago, *Behaviour* 45:225-241.

Mitchell, G.D. 1968a, Attachment differences in male and female infant monkeys, *Child Dev.* 39:611-620.

Mitchell, G.D. 1968b, Persistent behavior pathology in rhesus monkeys following early social isolation, *Folia primat.* 8:132-147.

Mitchell, G.D. 1969, Abnormal behavior in primates. IN: *Primate Behavior. Developments in Field and Laboratory Research*, vol. 1 (L.A. Rosenblum, ed.) New York: Academic Press, pp. 195-249.

Mitchell, G.D. 1977, Parental behavior in nonhuman primates. IN: *Handbook of Sexology* (J. Money and H. Musaph, eds.) Elsevier: North Holland Biomedical Press, pp. 749-759.

Mitchell, G.D. 1979, *Behavioral Sex Differences in Nonhuman Primates*. New York: Van Nostrand Reinholt Co.

Mitchell, G.D., Arling, G.L. and Moller, G.W. 1967, Long-term effects of maternal punishment on the behavior of monkeys, *Psychodynamic Science* 8:209-210.

Mitchell, G.D. and Brandt, E. 1972, Paternal behavior in primates. IN: *Primate Socialization* (F.E. Poirier, ed.) New York: Random House, pp. 173-206.

Mitchell, G.D., Redican, W.K. and Gombe, J. 1974, Lessons from a primate: males can raise babies, *Psych. Today* 7:63-70.

Mitchell, G.D., Ruppenthal, G.C., Raymond, E.J. and Harlow, H.F. 1966, Long-term effects of multiparous and primiparous monkey mother rearing, *Child Dev.* 37:781-791.

Mitchell, G.D. and Stevens, C. 1969, Primiparous and multiparous monkey mothers in a mildly stressful social situation: first three months, *Dev. Psychobiol.* 1:280-286.

Mitchell, G.D. and Tokunaga, D.H. 1976, Sex differences in nonhuman primate grooming, *Behav. Processes* 1:335-345.

Mohnot, S.M. 1971, Some aspects of social changes and infant-killing in the Hanuman langur, *Presbytis entellus* (Primates: Cercopithecidae) in Western India, *Mammalia* 35:175-198.

Money, J. 1963, Cytogenetic and psychosexual incongruities with a note on space-form blindness, *Am. J. Psychiat.* 119:820-827.

Money, J. 1968, *Sex Errors of the Body*. Baltimore: Johns Hopkins Univ. Press.

Money, J. 1980, *Love and Love-Sickness. The Science of Sex, Gender Difference and Pair-Bonding*. Baltimore: Johns Hopkins Univ. Press.

Money, J. and Ehrhardt, A.A. 1968, Prenatal hormone exposure. IN: *Endocrinology and Human Behaviour* (R.P. Michael, ed.) London: Oxford Univ. Press, pp. 32-47.

Money, J. and Ehrhardt, A.A. 1972, *Man and Woman. Boy and Girl*. Baltimore: Johns Hopkins Univ. Press.

Montagu, A. (ed.) 1968, *Man and Aggression*. London: Oxford Univ. Press.

Montague, A. (ed.) 1980, *Sociobiology Examined*. Oxford: Oxford University Press.

Mori, U. 1979a, Individual relationships within a unit. IN: *Ecological and Sociological Studies of Gelada Baboons, Contrib. Primat.* 16 (M. Kawai, ed.) Basel: S. Karger, pp. 93-124.

Mori, U. 1979b, Development of sociability and social status. IN: *Ecological and Sociological Studies of Gelada Baboons, Contrib. Primat.* 16 (M. Kawai, ed.) Basel: S. Karger, pp. 125-154.

Mori, U. and Kawai, M. 1974, Social relations and behavior of gelada baboons. IN: *Contemporary Primatology* (S. Kondo, M. Kawai and A. Ehara, eds.) Basel: S. Karger, pp. 470-474.

Morris, D. 1967, *The Naked Ape*. Corgi Books.

Moss, H.A. 1967, Sex, age and state as determinants of mother-infant interaction, *Merrill-Palmer Quart.* 13:19-36.

Moss, H.A. 1974, Early sex differences and mother-infant interaction. IN: *Sex Differences in Behavior* (R.C. Friedman, R.M. Richart and R.L. Vande Wiele, eds.) New York: John Wiley & Sons, pp. 149-163.

Moyer, K.E. 1974, Sex differences in aggression. IN: *Sex Differences in Behavior* (R.C. Friedman, R.M. Richart and R.L. Vande Wiele, eds.) New York: Wiley & Sons, pp. 335-372.

Moynihan, M. 1966, Communication in the titi monkey, Callicebus, *J. Zool.* 150:77-127.

Moynihan, M. 1976, *The New World Primates*. Princeton: Princeton Univ. Press.

Murray, R.D. 1979, The evolution and functional significance of incest avoidance, *J. Hum. Evol.* 9:173-178.

Myers, R.E. 1978, Comparative neurology of vocalization and speech: proof of a dichotomy. IN: *Human Evolution. Biosocial Perspectives* (S.L. Washburn and E.R. McCown, eds.) Menlo Park: Benjamin/Cummings Publ. Co., pp. 59-75.

Nadel, S.F. 1957, *The Theory of Social Structure*. Glencoe: Free Press.

Nadler, R.D. and Braggio, J.T. 1974, Sex and species differences in captive-reared juvenile chimpanzees and orang-utans, *J. Hum. Evol.* 3:541-550.

Nagy, K.A. and Milton, K. 1979, Energy metabolism and food consumption by wild howler monkeys *(Alouatta palliata)*, *Ecology* 60:475-480.

Napier, J.R. and Napier, P.H. 1967, *A Handbook of Living Primates*. London: Academic Press.

Nash, L. 1976, Troop fission in free-ranging baboons in the Gombe Stream National Park, Tanzania, *Am. J. Phys. Anthrop.* 44:63-77.

Nash, L. 1978a, Kin preference in the behavior of young baboons. IN: *Recent Advances in Primatology*, vol. 1 (D.J. Chivers and J. Herbert, eds.) London: Academic Press, pp. 71-73.

Nash, L. 1978b, The development of the mother-infant relationship in wild baboons

(Papio anubis), Anim. Behav. 26:746-759.

Nelson, T.W. 1975, Quantitative observations on feeding behavior in *Saguinus geoffroyi* (Callitrichidae, Primates), *Primates* 16:223-226.

Neville, M.K. 1968, A free-ranging rhesus monkey troop lacking adult males, *J. Mammal.* 49:771-773.

Neville, M.K. 1972, Social relations within troops of red howler monkeys *(Alouatta seniculus), Folia primat.* 18:47-77.

Neyman, P.F. 1978, Aspects of the ecology and social organization of free-ranging cotton-top tamarins *(Saguinus oedipus)* and the conservation status of the species. IN: *The Biology Conservation of the Callitrichidae* (D.G. Kleiman, ed.) Washington, D.C.: Smithsonian Inst. Press, pp. 39-71.

Nicholson, N.A. 1977, A comparison of early behavioral development in wild and captive chimpanzees. IN: *Primate Bio-social Development. Biological, Social and Ecological Determinants* (S. Chevalier-Skolnikoff and F.E. Poirier, eds.) New York: Garland Publ., Inc., pp. 529-563.

Nilsen, A.P. 1973, Grammatical gender and its relationship to the equal treatment of males and females in children's books, Doctoral Dissertation: Univ. of Iowa.

Nishida, T. 1966, A sociological study of solitary male monkeys, *Primates* 7:141-204.

Nishida, T. 1968, The social group of wild chimpanzees in the Mahali Mountains, *Primates* 9:167-224.

Nishida, T. 1970, Social behavior and relationship among wild chimpanzees of the Mahali Mountains, *Primates* 11:47-87.

Nishida, T. 1973, The ant-gathering behavior by the use of tools among wild chimpanzees of the Mahali Mountains, *J. Hum. Evol.* 2:357-370.

Nishida, T. 1979, The social structure of chimpanzees of the Mahali Mountains. IN: *The Great Apes. Perspectives on Human Evolution,* vol. 5 (D. Hamburg and E.R. McCown, eds) Menlo Park: Benjamin/Cummings Publ. Co., pp. 73-121.

Nishida, T. and Kawanaka, K. 1972, Inter-unit-group relationship among wild chimpanzees of the Mahali Mountains, *Kyoto Univ. African Studies* 7:131-169.

Nishida, T., Vehara, S. and Ramadhani, N. 1979, Predatory behavior among chimpanzees of the Mahali Mountains, *Primates* 20:1-20.

Norikoshi, K. 1971, Tests to determine the responsiveness of free-ranging Japanese monkeys in food-getting situations, *Primates* 12:113-124.

Norikoshi, K. and Koyama, N. 1975, Group shifting and social organization among Japanese monkeys, *Proc. Symp. 5th Int'l. Prim. Soc.* (S. Kondo, M. Kawai, A. Ehara and S. Kawamura, eds.) Tokyo: Japan Science Press, pp. 43-61.

Nottebohm, M. and Arnold, A. 1976, Sexual dimorphism in vocal control areas of the songbird brain, *Science* 194:211-213.

O'Connor, J.F., Shelley, E.M. and Stern, L.O. 1974, Behavioral rhythms related to the menstrual cycle. IN: *Biorhythms and Human Reproduction* (M. Ferin, F. Halberg, R.M. Richart and R.L. Vande Wiele, eds.) New York: Wiley & Sons, pp. 309-324.

O'Donald, P. 1962, Theory of sexual selection, *Heredity* 17:541-552.

Oates, J.F. 1977, The guereza and man. IN: *Primate Conservation* (Prince Rainier III and G.H. Bourne, eds.) New York: Academic Press, pp. 419-467.

Oates, J.F. 1978, The social life of a black-and-white colobus monkey, *Colobus guereza, Z. Tierpsychol.* 45:1-60.

Oates, J.F. 1979, Comments on the geographical distribution and status of the South Indian black leaf-monkey, *Mammalia* 43:485-493.

Ohsawa, H. 1979, Herd dynamics. IN: *Ecological and Sociological Studies of Gelada Baboons, Contrib. Primat.* 16 (M. Kawai, ed.) Basel: S. Karger, pp. 47-80.

Orians, G.H. 1969, On the evolution of mating systems in birds and mammals, *Am. Nat.* 103:589-603.

Owens, N.W. 1975, Social play behavior in free-living baboons, *Papio anubis, Anim. Behav.* 23:387-408.

Oxnard, C.E. 1979, Some methodological factors in studying the morphological-behavioral interface. IN: *Environment, Behavior and Morphology* (M.E. Morbeck, H. Preuschoft and J. Gomberg, eds.) New York: Fischer, pp. 183-207.

Packer, C. 1977, Reciprocal altruism in *Papio anubis, Nature* (Lond.) 265:441-443.

Packer, C. 1978, The relative importance of group selection in the evolution of primate societies, *Symp. Soc. Stud. Hum. Biol.* 18:103-112.

Packer, C. 1979a, Inter-troop transfer and inbreeding avoidance in *Papio anubis, Anim. Behav.* 27:1-26.

Packer, C. 1979b, Male dominance and reproductive activity in *Papio anubis, Anim. Behav.* 27:37-45.

Packer, C. and Pusey, A. 1979, Female aggression and male membership in troops of Japanese macaques and olive baboons, *Folia. primat.* 31:212-218.

Parker, G.A., Baker, R.R. and Smith, V.G.F. 1972, The origin and evolution of gamete dimorphism and the male-female phenomenon, *J. Theor. Biol.* 36:529-533.

Parker, S. 1976, The precultural basis of the incest taboo: toward a biosocial theory, *Am. Anth.* 78:285-305.

Parsons, T. 1951, *The Social System.* New York: The Free Press.

Paterson, J.D. 1973, Ecologically differentiated patterns of aggressive and sexual behavior in two troops of Ugandan baboons, *Papio anubis, Am. J. of Phys. Anthrop.* 38: 641-647.

Paterson, J.D. 1978, Variations in ecology and adaptation of Ugandan baboons, *Papio cynocephalus anubis:* with special reference to forest environments and analog models for early hominids, Ph.D. dissertation, University of Toronto.

Peters, R.H. 1976, Tautology in evolution and ecology, *Am. Nat.* 110:1-12.

Peterson, R.S. 1968, Social behavior in pinnipeds with particular reference to the northern fur seal. IN: *The Behavior and Physiology of Pinnipeds* (R.S. Harrison, ed.) New York: Appleton-Century Crofts, pp. 3-53.

Petter, J.J. 1965, The lemurs of Madagascar. IN: *Primate Behavior. Field Studies of Monkeys and Apes* (I. DeVore, ed.) New York: Holt, Rinehart & Winston, pp. 292-319.

Pfeiffer, J.E. 1972, *The Emergence of Man.* New York: Harper & Row.

Phoenix, C.H. 1974a, The role of androgens in the sexual behavior of adult male rhesus monkeys. IN: *Reproductive Behavior* (W. Montagna and W.A. Sadler, eds.) New York: Plenum, pp. 249-258.

Phoenix, C.H. 1974b, Prenatal testosterone in the nonhuman primate and its consequences for behavior. IN: *Sex Differences in Behavior* (R.C. Friedman, R.M. Richart and R.L. Vande Wiele, eds.) New York: Wiley & Sons, pp. 19-32.

Phoenix, C.H. 1978a, Steroids and sexual behavior in castrated male rhesus, *Horm. and*

Behav. 10:1-9.

Phoenix, C.H. 1978b, Sexual behavior of laboratory and wild-born male rhesus monkeys, *Horm. and Behav.* 10:178-192.

Phoenix, C.H., Goy, R.W., Resko, J.A. and Koering, M. 1968, Probability of mating during various stages of the ovarian cycle in *Macaca mulatta, Anat. Rec.* 160:490.

Phoenix, C.H., Goy, R.W. and Young, W.C. 1967, Sexual behavior: general aspects. IN: *Neuroendocrinology*, vol. II, New York: Academic Press, pp. 163-196.

Phoenix, C.H., Resko, J.A. and Goy, R.W. 1968, Psychosexual differentiation as a function of androgenic stimulation. IN: *Perspectives in Reproductive and Sexual Behavior* (M. Diamond, ed.) Bloomington: Indiana Univ. Press, pp. 33-49.

Piaget, J. 1971, *Biology & Knowledge.* Edinburgh: Edinburgh Univ. Press.

Piaget, J. 1979, *Behaviour and Evolution.* London: Routledge & Kegan Paul.

Pilbeam, D. 1979, Toward a concept of man, *Nat. Hist.* 88:100-104.

Poirier, F.E. 1968a, Analysis of a Nilgiri langur (*Presbytis johnii*) home range change, *Primates* 9:29-43.

Poirier, F.E. 1968b, The Nilgiri langur (*Presbytis johnii*) mother-infant dyad. *Primates* 9:45-68.

Poirier, F.E. 1968c, Nilgiri langur territorial behavior, *Primates* 9:351-361.

Poirier, F.E. 1969a, The Nilgiri langur (*Presbytis johnii*) troop: its composition, structure, function and change, *Folia primat.* 10:20-47.

Poirier, F.E. 1969b, Behavioral flexibility and intertroop variation among Nilgiri langurs (*Presbytis johnii*) of South India, *Folia primat.* 11:119-133.

Poirier, F.E. 1970a, The Nilgiri langur (*Presbytis johnii*) of South India. IN: *Primate Behavior. Developments in Field and Laboratory Research*, vol. 1 (L.A. Rosenblum, ed.) New York: Academic Press, pp. 254-383.

Poirier, F.E. 1970b, Dominance structure of the Nilgiri langur (*Presbytis johnii*) of South India, *Folia primat.* 12:161-186.

Poirier, F.E. 1972a, Introduction. IN: *Primate Socialization* (F.E. Poirier, ed.) New York: Random House pp. 3-28.

Poirier, F.E. 1972b, The St. Kitts green monkey (*Cercopithecus aethiops sabaeus*): ecology, population dynamics, and selected behavioral traits, *Folia primat.* 17:20-55.

Poirier, F.E. 1977, Introduction. IN: *Primate Biosocial Development: Biological, Social, and Ecological Determinants* (S. Chevalier-Skolnikoff and F.E. Poirier, eds.) New York: Garland Publ. Inc., pp. 1-39.

Poirier, F.E. and Smith, E.O. 1974, Socializing function of primate play, *Amer. Zool.* 14:275-287.

Pollock, J.I. 1979, Female dominance in *Indri indri, Folia primat.* 31:143-164.

Popp, J.L. and DeVore, I. 1979, Aggressive competition and social dominance theory: synopsis. IN: *The Great Apes* (D.A. Hamburg and E.R. McCown, eds.) Menlo Park: Benjamin/Cummings, pp. 317-338.

Post, D. 1977, Baboon feeding behavior and the evolution of sexual dimorphism. Paper presented at American Association of Physical Anthropologists Meeting, Seattle, Washington.

Post, D. 1980, Sexual dimorphism in the anthropoid primates: some thoughts on causes, correlates, and the relationship to body size. IN: *Sexual Dimorphism in Primates* (C. Eastman and P. Heisler, eds.) New York: Garland STPM Press (in press).

Pratt, C.L. 1969, The developmental consequences of variation in early social stimulation. Ph.D. dissertation, University of Wisconsin.

Premack, D. 1971, On the assessment of language competence in the chimpanzee. IN: *Behavior of Non-Human Primates*, vol. 4 (A.M. Schrier and F. Stollinitz, eds.) New York: Academic Press, pp. 186-228.

Pusey, A. 1979, Intercommunity transfer of chimpanzees in Gombe National Park. IN: *The Great Apes. Perspectives on Human Evolution*, vol. 5, Menlo Park: Benjamin/Cummings Publ. Co., pp. 465-479.

Quadango, D.M., Briscoe, R. and Quadango, J.S. 1977, Effect of perinatal gonadal hormones on selected nonsexual behavior patterns: a critical assessment of the nonhuman and human literature, *Psych. Bull.* 84(1):62-80.

Quiatt, D. 1979, Aunts and mothers: adaptive implications of allomaternal behavior of nonhuman primates, *Am. Anth.* 81:310-319.

Raemakers, J. 1979, Ecology of sympatric gibbons, *Folia primat.* 31:227-245.

Raisnam, G. and Field, P.M. 1971, Sexual dimorphism in the preoptic arch of the rat, *Science* 173:731-733.

Raisnam, G. and Field, P.M. 1973, Sexual dimorphism in the neuropil of the preoptic area of the rat and its dependence on neonatal androgen, *Brain Research* 54:1-29.

Raleigh, M.J. Flannery, J.W. and Ervin, F.R. 1979, Sex differences in behavior among juvenile vervet monkeys (*Cercopithecus aethiops sabaeus*) *Beh. Neur. Biol.* 26: 455-465.

Ralls, K. 1976, Mammals in which females are larger than males, *Quart. Rev. Biol.* 51: 245-276.

Ralls, K. 1977, Sexual dimorphism in mammals: avian models and unanswered questions, *Am. Nat.* 111:917-938.

Ransom, T.W. and Ransom, B.S. 1971, Adult male-infant relations among baboons (*Papio anubis*), *Folia primat.* 16:179-195.

Ransom, T.W. and Rowell, T.E. 1972, Early social development of feral baboons. IN: *Primate Socialization* (F.E. Poirier, ed.) New York: Random House, pp. 105-144.

Rattray, J.M. 1960, *The Grass Cover of Africa.* Rome: Food & Agri. Org. of the U.N.

Rayor, L.S. and Chizar, D. 1978, Comparability of dominance indices in captive pigtail macaques (*Macaca nemestrina*), *Bull. Psychon. Soc.* 12:468-479.

Redican, W.K. 1976, Adult male-infant interactions in non-human primates. IN: *The Role of the Father in Child Development* (M.E. Lamb, ed.) New York: John Wiley & Sons, pp. 345-385.

Resko, J.A. 1967, Plasma androgen levels of the rhesus monkey: the effects of age and season, *Endocrinology* 81:1203-1212.

Resko, J.A. 1974, The relationship between fetal hormones and the differentiation of the CNS in primates. IN: *Reproductive Behavior* (W. Montagna and W. Sadler, eds.) New York: Plenum Press, pp. 211-222.

Resko, J.A. and Phoenix, C.H. 1972, Sex behavior and testosterone concentrations in the plasma of the rhesus monkey before and after castration, *Endocrinology* 91:499-503.

Reynolds, P.C. 1976, The emergence of early hominid social organization: 1. The attachment system, *Yearbook of Physical Anthropology* 20:73-95.

Reynolds, V. 1966, Open groups in hominid evolution, *Man* 1(n.s.):441-452.

Reynolds, V. 1968, Kinship and the family in monkeys, apes and man, *Man* 3:209-223.

Reynolds, V. 1970, Roles and role change in monkey society: the consort relationship of rhesus monkeys, *Man* (n.s.)5:449-465.

Reynolds, V. 1975, How wild are the Gombe chimpanzees? *Man* 10:123-125.

Reynolds, V. 1976, *The Biology of Human Action*. Reading: W.H. Freeman & Co.

Reynolds, V. and Luscombe, G. 1976, Greeting behaviour, displays and rank order in a group of free-ranging chimpanzees. IN: *The Social Structure of Attention* (M.R.H. Chance and R.R. Larsen, eds.) London: John Wiley & Sons, pp. 105-115.

Reynolds, V. and Reynolds, F. 1965, Chimpanzees in the Budongo Forest. IN: *Primate Behavior: Field Studies of Monkeys and Apes* (I. DeVore, ed.) New York: Holt, Rinehart & Winston, pp. 368-424.

Richard, A.F. 1978, *Behavioral Variation. Case Study of a Malagasy Lemur.* Lewisburg: Bucknell University Press.

Richards, M.P.M., Bernal, J.F. and Brackbill, Y. 1976, Early behavioral differences: gender or circumcision? *Develop. Psychobiol.* 9:89-95.

Richards, S.M. 1974, The concept of dominance and methods of assessment, *Anim. Behav.* 22:914-930.

Rijksen, H.D. 1978, *A Field Study on Sumatran Orangutans (Pongo pygmaeus belli Lesson 1827)* Wageningen: H. Veenman & Zonen.

Ripley, S. 1967, Intertroop encounters among Ceylon gray langurs (*Presbytis entellus*). IN: *Social Communication Among Primates* (S. Altmann, ed.) Chicago: Univ. Chicago Press, pp. 237-253.

Ripley, S. 1980, Infanticide in langurs and man: adaptive advantage or social pathology? IN: *Biosocial Mechanisms of Population Regulation* (M.N. Cohen, R.S. Malpass, H. G. Klein, eds.) New Haven: Yale Univ. Press, pp. 349-390.

Riss, D. and Goodall, J. 1977, The recent rise to the alpha rank in a population of free-living chimpanzees, *Folia primat.* 27:134-151.

Robertson, R.J. and Biermann, G.C. 1979, Parental investment determined by expected benefits, *Z. Tierpsychol.* 50:124-128.

Robinson, J.C. 1979, Vocal regulation of use of space by groups of titi monkeys, *Callicebus moloch, Behav. Ecol. Sociobiol.* 5:1-15.

Rodman, P.S. 1973, Population composition and adaptive organization among orangutans of the Kutai Reserve. IN: *Comparative Ecology and Behavior of Primates* (R.P. Michael and J.H. Crook, eds.) New York: Academic Press, pp. 171-209.

Rondinelli, R. and Klein, L.L. 1976, An analysis of adult social spacing tendencies and related social mechanisms in a colony of spider monkeys (*Ateles geoffroyi*) at the San Francisco Zoo, *Folia primat.* 25:122-142.

Roonwal, M.L. and Mohnot, S.M. 1977, *Primates of South Asia. Ecology, Sociobiology and Behavior.* Cambridge: Harvard Univ. Press.

Rose, R.M., Bernstein, I.S. and Gordon, T.P. 1975, Consequences of social conflict on plasma testosterone levels in rhesus monkeys, *Psychosom. Med.* 37:50-61.

Rose, R.M., Bernstein, I.S. and Gordon, T.P. 1978, Plasma testosterone and cortisol in rhesus monkeys living in a social group, *J. Endocrinol.* 76:67-74.

Rose, R.M., Bernstein, I.S., Gordon, T.P. and Lindsley, J.G. 1978, Changes in testosterone and behavior during adolescence in the male rhesus monkey, *Psychosom. Med.* 40:60-70.

Rose, R.M., Gordon, T.P. and Bernstein, I.S. 1972, Plasma testosterone levels in the male rhesus: influences of sexual and social stimuli, *Science* 178:643-645.

Rose, R.M., Gordon, T.P., Bernstein, I.S. and Catlin, S.F. 1974, Androgens and aggression: a review and recent findings in primates. IN: *Primate Aggression, Territoriality and Xenophobia* (R.L. Holloway, ed.) New York: Academic Press, pp. 275-304.

Rose, R.M., Holaday, J.W. and Bernstein, I.S. 1971, Plasma testosterone, dominance rank and aggressive behavior in male rhesus monkeys, *Nature* (Lond.) 231:366-368.

Rosen, S.I. 1974, *Introduction to the Primates, Living and Fossil.* Englewood Cliffs: Prentice-Hall.

Rosenblum, L.A. 1971a, Kinship interaction patterns in pigtail and bonnet macaques, *Proc. 3rd Int. Congr. Primatol. Zurich* 3:79-84.

Rosenblum, L.A. 1971b, The ontogeny of mother-infant relations in macaques. IN: *Ontogeny of Vertebrate Behavior* (H. Maltz, ed.) New York: Academic Press, pp. 315-367.

Rosenblum, L.A. 1974, Sex differences in mother-infant attachment in monkeys. IN: *Sex Differences in Behavior* (R.C. Friedman, R.M. Richart and R.L. Vande Wiele, eds.) John Wiley & Sons, pp. 123-141.

Rosenblum, L.A. and Alpert, S. 1977, Responses to mother and stranger. IN: *Primate Biosocial Development: Biological, Social, and Ecological Determinants* (S. Chevalier-Skolnikoff and F.E. Poirier, eds.) New York: Garland Publ. Inc., pp. 463-478.

Rosenblum, L.A. and Coe, C.L. 1977, The influence of social structure on squirrel monkey socialization. IN: *Primate Biosocial Development: Biological, Social and Ecological Determinants* (S. Chevalier-Skolnikoff and F.E. Poirier, eds.) New York: Garland Publ. Inc., pp. 479-500.

Rosenblum, L.A. and Kaufman, I.C. 1967, Laboratory observations of early mother-infant relations in pigtail and bonnet macaques. IN: *Social Communication Among Primates* (S.A. Altmann, ed.) Chicago: Univ. of Chicago Press, pp. 33-43.

Rossi, A.S. 1973, Maternalism, sexuality and the new feminism. IN: *Contemporary Sexual Behavior: Critical Issues in the 1970's,* Baltimore: Johns Hopkins Univ. Press, pp. 145-173.

Rothe, H. 1973, Beobachtungen zur Geburt beim Weissbuschelaffschen (*Callithrix jacchus* Erxleben 1777), *Folia primat.* 19:257-285.

Rothe, H. 1974, Further observations on the delivery behaviour of the common marmoset (*Callithrix jacchus*), *Z. Saugetierkunde* 39:135-142.

Rothe, H. 1975a, Some aspects of sexuality and reproduction in groups of captive marmosets (*Callithrix jacchus*), *Z. Tierpsychol.* 37:255-273.

Rothe, H. 1975b, Influence of newborn marmosets (*Callithrix jacchus*) behaviour on the expression and efficiency of maternal and paternal care. IN: *Contemporary Primatology* (S. Kondo, M. Kawai, and A. Ehara, eds.) Basel: S. Karger, pp. 315-320.

Rothe, H. 1978, Parturition and related behavior in *Callithrix jacchus* (Ceboidea, Callitrichidae). IN: *The Biology and Conservation of the Callitrichidae* (D.G. Kleiman, ed.) Washington, D.C.: Smithsonian Inst. Press, pp. 193-206.

Rowell, T.E. 1963, Behavior and reproductive cycles of rhesus macaques, *J. Reprod. Fert.* 6:193-203.

Rowell, T.E. 1966a, Hierarchy in the organization of a captive baboon group, *Anim. Behav.* 14:430-443.

Rowell, T.E. 1966b, Forest-living baboons in Uganda, *J. Zool. Lond.* 149:344-364.

Rowell, T.E. 1967, Female reproductive cycles and the behavior of baboons and rhesus macacques. IN: *Social Communication Among Primates* (S. Altmann, ed.) Chicago: Univ. of Chicago Press, pp. 15-32.

Rowell, T.E. 1968, The effect of temporary separation from their group on the mother-infant relationship of baboons, *Folia primat.* 9:114-122.

Rowell, T. 1969a, Long-term changes in a population of Ugandan baboons, *Folia primat.* 11:241-254.

Rowell, T.E. 1969b, Intrasexual behavior and female reproductive cycles of baboons *(Papio anubis), Anim. Behav.* 17:159-167.

Rowell, T.E. 1969c, Variability in the social organization of primates. IN: *Primate Ethology* (D. Morris, ed.) Garden City, New York: Anchor Books, pp. 283-305.

Rowell, T.E. 1970, Reproductive cycles of Cercopithecus monkeys, *J. Reprod. Fert.* 22:321-338.

Rowell, T.E. 1971, Organization of caged groups of Cercopithecus monkeys, *Anim. Behav.* 19:625-645.

Rowell, T. 1972a, *The Social Behaviour of Monkeys.* Baltimore: Penguin.

Rowell, T. 1972b, Female reproduction cycles and social behavior in primates. IN: *Advances in the Study of Behavior,* vol. 4 (D.S. Lehrman, R.A. Hinde and E. Shaw, eds.) New York: Academic Press, pp. 69-103.

Rowell, T.E. 1974a, The concept of dominance, *Behav. Biol.* 11:131-154.

Rowell, T.E. 1974b, Contrasting adult male roles in different species of nonhuman primates, *Arch. Sex. Beh.* 3:143-149.

Rowell, T.E. 1977, Variation in age at puberty in monkeys, *Folia primat.* 27:284-296.

Rowell, T.E., Din, N.A. and Omar, A. 1968, The social development of baboons in their first three months, *J. Zool.* 155:461-483.

Rowell, T.E. and Hartwell, K.M. 1978, The interaction of behavior and reproductive cycles in patas monkeys, *Behav. Biol.* 24:141-167.

Rowell, T.E., Hinde, R.A. and Spencer-Booth, Y. 1964, "Aunt"-infant interaction in captive rhesus monkeys, *Anim. Behav.* 12:219-226.

Rowell, T.E. and Richards, S.M. 1979, Reproductive strategies of some African monkeys, *J. Mammal.* 60:58-69.

Rudran, R. 1973, Adult male replacement in one-male troops of purple-faced langurs *(Presbytis senex senex)* and its effect on population structure, *Folia primat.* 19: 166-192.

Rumbaugh, D.M., von Glaserfield, E.C., Gill, T.V., Warner, H., Pisani, P., Brown, J.V., and Bill, C.L. 1975, The language skills of a young chimpanzee in a computer-controlled training situation. IN: *Socioecology and Psychology of Primates* (R.H. Tuttle, ed.) Paris: Mouton Publ., pp. 391-401.

Ruppenthal, G.C., Arling, G.L., Harlow, H.F., Sackett, G.P., and Suomi, S.J. 1976, A ten-year perspective of motherless mother monkey behavior, *J. Abnorm. Psychol.* 85:341-349.

Ruse, M. 1979, *Sociobiology: Sense or Nonsense.* Boston: D. Reidel.

Saayman, G.S. 1970, The menstrual cycles and sexual behavior in a troop of free-ranging chacma baboons *(Papio ursinus), Folia primat.* 12:81-110.

Saayman, G.S. 1971, Behaviour of the adult males in a troop of free-ranging chacma baboons *(Papio ursinus), Folia primat.* 15:36-57.

Saayman, G.S. 1972, Effects of ovarian hormones upon the sexual skin and mounting behavior in the free-ranging chacma baboon *(Papio ursinus), Folia primat.* 17:297-303.

Saayman, G.S. 1973, Effects of ovarian hormones upon the sexual skin and behavior of ovariectomized baboons *(Papio ursinus)* under free-ranging conditions. IN: *Primate Reproductive Behavior* (C.H. Phoenix, ed.) Basel: S. Karger, pp. 64-98.

Saayman, G.S. 1975, The influence of hormonal and ecological factors upon sexual behavior and social organization in Old World primates. IN: *Socioecology and Psychology of Primates* (R.H. Tuttle, ed.) Paris: Mouton Publ., pp. 181-204.

Sabater-Pi 1979, Feeding behaviour and diet of chimpanzees *(Pan troglodytes)* in the Okorobiko Mountains of Rio Muni (West Africa), *Z. Tierpsychol.* 50:265-281.

Sackett, G.P. 1972, Exploratory behavior of rhesus monkeys as a function of rearing experiences and sex, *Dev. Psychol.* 6:260-270.

Sackett, G.P. 1974, Sex differences in rhesus monkeys following varied rearing experiences. IN: *Sex Differences in Behavior* (R.C. Friedman, R.M. Richart and R.L. Vande Wiele, eds.) New York: John Wiley & Sons, pp. 99-122.

Sackett, G.P., Holm, R.A. and Ruppenthal, G.C. 1976, Social isolation rearing: species differences in behavior of macaque monkeys, *Dev. Psychol.* 12:283-288.

Sade, D.S. 1965, Some aspects of parent-offspring and sibling relations in a group of rhesus monkeys, with a discussion of grooming, *Am. J. Phys. Anthrop.* 23:1-18.

Sade, D.S. 1968, Inhibition of mother-son mating among free-ranging rhesus monkeys, *Sci. and Psychoanal.* XII:18-38.

Sade, D.S. 1972, A longitudinal study of social relations of rhesus monkeys. IN: *Functional and Evolutionary Biology of Primates* (R. Tuttle, ed.) Chicago: Aldine Atherton, pp. 378-398.

Sahlins, M. 1976, *The Use and Abuse of Biology. An Anthropological Critique of Sociobiology.* Ann Arbor: Univ. of Michigan Press.

Sarbin, T.R. and Allen, V.L. 1968, Role theory. IN: *The Handbook of Social Psychology,* vol. 1 (G. Lindzey and E. Aronson, eds.) Reading: Addison-Wesley Publ. Co., pp. 488-567.

Schaller, G.B. 1964, *The Year of the Gorilla.* Chicago: Univ. of Chicago Press.

Schaller, G.B. 1965, The behavior of the mountain gorilla. IN: *Primate Behavior. Field Studies of Monkeys and Apes* (I. DeVore, ed.) New York: Holt, Rinehart & Winston, pp. 324-367.

Scheffler, I. 1967, *Science and Subjectivity.* Indianapolis: Bobbs-Merrill Co., Inc.

Schjelderup-Ebbe, T. 1922, Beitrage zur sozialpsychologie das haushuhns, *Z. Psychol.* 88:225-252.

Schultz, A.H. 1956, Postembryonic age changes. IN: *Primatologia,* vol. I (H. Hofer, A.H. Schultz and D. Starck, eds.) Basel: S. Karger, pp. 886-964.

Schultz, A.H. 1969, *The Life of Primates.* New York: Universe Books.

Scott, J.P. 1974, Agonistic behavior of primates: a comparative perspective. IN: *Primate Aggression, Territoriality and Xenophobia* (R.L. Holloway, ed.) New York: Academic Press, pp. 417-434.

Scott, J.P. and Frederickson, E. 1951, The causes of fighting in mice and rats, *Physiol. Zool.* 24:273-309.

Searcy, W.A. 1979, Female choice of mates: a general model for birds and its application to red-winged blackbirds *(Agelaius phoeniceus) Am. Nat.* 114:77-100.

Selander, R.K. 1966, Sexual dimorphism and differential niche utilization in birds, *Condor* 68:113-151.

Selander, R.K. 1972, Sexual selection and dimorphism in birds. IN: *Sexual Selection and the Descent of Man, 1871-1971* (B. Campbell, ed.) Chicago: Aldine, pp. 180-230.

Seyfarth, R.M. 1977, A model of social grooming among adult female monkeys, *J. Theor. Biol.* 65:671-698.

Seyfarth, R.M. 1978a, Social relationships among adult male and female baboons. I. Behaviour during sexual consortship, *Behaviour* 64:204-226.

Seyfarth, R.M. 1978b, Social relationships among adult male and female baboons. II. Behaviour throughout the female reproductive cycle, *Behaviour* 64:227-247.

Seyfarth, R.M. and Cheney, D.L. 1979, The ontogeny of vervet monkey alarm calling behaviour: a preliminary report, *Z. Tierpsychol.* 54:37-56.

Seyfarth, R.M., Cheney, D.L. and Marler, P. 1980a, Vervet monkey alarm calls: semantic communication in a free-ranging primate, *Anim. Behav.* 28:1070-1094.

Seyfarth, R.M., Cheney, D.L. and Marler, P. 1980b, Monkeys' responses to three different alarm calls: evidence of predator classification, *Science* 210:801-803.

Shafton, A. 1976, *Conditions of Awareness: Subjective Factors in the Social Adaptations of Man and Other Primates.* Portland: Riverstone Press.

Simonds, P.E. 1965, The bonnet macaque in south India. IN: *Primate Behavior, Field Studies of Monkeys and Apes* (I. DeVore, ed.) New York: Holt, Rinehart & Winston, pp. 175-196.

Simonds, P.E. 1974, Sex differences in bonnet macaque networks and social structure, *Arch. Sex. Beh.* 3:151-166.

Simpson, M.J.A. 1973, The social grooming of male chimpanzees. IN: *Comparative Ecology and Behavior of Primates* (R.P. Michael and J.H. Crook, eds.) London: Academic Press, pp. 411-505.

Slater, J.B. 1974, A reassessment of ethology, *The Biology of Brains* (W.B. Broughton, ed.) London: Institute of Biology, pp. 89-113.

Slob, A.K., Baum, M.J., DeJong, F.H. and Westbrock, D.L. 1978, Persistence of sexual behavior in ovariectomized stumptail macaques following dexamethasone treatment or adrenalectomy, *Exp. Brain Res.* 32:R21-R22.

Slob, A.K., Wiegand, S.J., Goy, R.W. and Robinson, J.A. 1978, Heterosexual interactions in laboratory housed stumptail macques *(Macaca arctoides)*: observations during the menstrual cycle and after ovariectomy, *Horm. and Behav.* 10:193-211.

Slocum, S. 1975, Woman the gatherer: male bias in anthropology. IN: *Toward an Anthropology of Women* (R.R. Reiter, ed.) New York: Monthly Review Press, pp. 36-50.

Smith, E.O. (ed.) 1978, *Social Play in Primates.* New York: Academic Press.

Smith, R.J. 1980, Interspecific scaling of maxillary canine size and shape in female primates: relationships to social structure and diet, *J. Hum. Evol.* 10:165-173.

Southwick, C.H. 1972, *Aggression Among Nonhuman Primates.* Reading, Mass.: Addison Wesley Modular Publ.

Southwick, C.H., Beg, M.A. and Siddiqi, M.R. 1965, Rhesus monkeys in north India. IN: *Primate Behavior* (I. DeVore, ed.) New York: Holt, Rinehart & Winston, pp. 111-159

Southwick, C.H. and Siddiqi, M.R. 1967, The roles of social tradition in the maintenance of dominance in a wild rhesus group, *Primates* 8:341-354.

Southwick, C.H. and Siddiqi, M.R. 1974, Contrasts in primate social behavior, *Bio. Sci.* 24:398-406.

Spencer-Booth, Y. 1968, The behavior of group companions towards rhesus monkey infants, *Anim. Behav.* 16:541-557.

Spencer-Booth, Y. 1977, The relationships between mammalian young and conspecifics other than mothers and peers. IN: *Advances in the Study of Behavior*, vol. 3 (D.S. Lehrman, R.A. Hinde and E. Shaw, eds.) New York: Academic Press, pp. 120-194.

Spencer-Booth, Y. and Hinde, R.A. 1971, Effects of 6 days separation from mother on 18-to-32 week old rhesus monkeys, *Anim. Behav.* 19:174-191.

Spivak, H. 1971, Ausdruckstormen und soziale Beziehungen in einer Dschelada-Gruppe *(Theropithecus gelada)*, *Z. Tierpsychol.* 28:279-296.

Stammbach, E. 1978, On social differentiation in groups of captive female hamadryas baboons, *Behaviour* 67:322-338.

Stanley, S.M. 1975, A theory of evolution above the species level, *Proc. Nat. Acad. Sci. U.S.A.* 72:646-650.

Starin, E.D. 1978, Food transfer by wild titi monkeys *(Callicebus torquatus torquatus)*, *Folia primat.* 30:145-151.

Stebbins, G.L. 1977, In defense of evolution: tautology or theory? (letter) *Am. Nat.* 111:386-390.

Stebbins, C.A., Kelly, B.R., Tolor, A. and Power, M.E. 1977, Sex differences in assertiveness in college students, *J. Psych.* 95:309-315.

Stent, G.S. 1977, You can take the ethics out of altruism but you can't take the altruism out of ethics, *Hastings Ctr. Rep.* 7:33-36.

Stephenson, G.R. 1973, Testing for group specific communication patterns in Japanese macaques. IN: *Symp. IV Congr. Primatol. Volume I: Precultural Primate Behavior*, Basel: S. Karger, pp. 51-75.

Stephenson, G.R. 1974, Social structure of mating activity in Japanese macaques, *Symp. 5th Congr. Int'l Prim. Soc.* (S. Kondo, M. Kawai, A. Ehara and S. Kawamura, eds.) Tokyo, Japan: Japan Science Press, pp. 65-115.

Stevenson, M.F. and Poole, T.B. 1976, An ethogram of the common marmoset *(Callithrix jacchus)*: general behavioural repertoire, *Anim. Behav.* 24:428-451.

Stoller, R.J. 1968, *Sex and Gender.* New York: Science House.

Struhsaker, T.T. 1967a, Social structure among vervet monkeys *(Cercopithecus aethiops)*, *Behaviour* 29:6-121.

Struhsaker, T.T. 1967b, Auditory communication among vervet monkeys *(Cercopithecus aethiops)*. IN: *Social Communication Among Primates* (S.A. Altmann, ed.), Chicago: Univ. of Chicago Press, pp. 281-324.

Struhsaker, T.T. 1967c, Ecology of vervet monkeys *(Cercopithecus aethiops)* in the Masai-Amboseli Game Reserve, Kenya, *Ecology* 48:891-904.

Struhsaker, T.T. 1967d, Behavior of vervet monkeys *(Cercopithecus aethiops)*, *Univ. Calif. Publ. Zool.* 82:1-74.

Struhsaker, T.T. 1969, Correlates of ecology and social organization among African cercopithecines, *Folia primat.* 9:123-134.

Struhsaker, T.T. 1971, Social behaviour of mother and infant vervet monkeys (*Cercopithecus aethiops*), *Anim. Behav.* 19:233-250.

Struhsaker, T. 1977, Infanticide and social organization in the redtail monkey (*Cercopithecus ascanius schmidti*) in the Kibale Forest, Uganda, *Z. Tierpsychol.* 45: 75-84.

Struhsaker, T.T. and Gartlan, J.S. 1970, Observations on the behavior and ecology of the patas monkey, *J. Zool. Lond.* 161:49-63.

Strum, S.C. 1975, Primate predation: interim report on the development of a tradition in a troop of olive baboons, *Science* 187:755-757.

Sturup, G.K. 1972, Castration: the total treatment. IN: *Sexual Behaviors. Social, Clinical and Legal Aspects* (H.L.P. Resnick and M.E. Wolfgang, eds.) Boston: Little Brown & Co., pp. 361-382.

Sugiyama, Y. 1965, On the social change of Hanuman langurs (*Presbytis entellus*) in their natural condition, *Primates* 6:381-418.

Sugiyama, Y. 1967, Social organization of hunuman langurs. IN: *Social Communication Among Primates* (S.A. Altmann, ed.) Chicago: Univ. of Chicago Press, pp. 207-219.

Sugiyama, Y. 1968, Social organization of chimpanzees in the Budongo forest, *Primates* 9:225-258.

Sugiyama, Y. 1969, Social behavior of chimpanzees in the Budongo forest, *Primates* 10:197-225.

Sugiyama, Y. 1972, Social characteristics and socialization of wild chimpanzees. IN: *Primate Socialization* (F.E. Poirier, ed.) New York: Random House, pp. 145-163.

Sugiyama, Y. 1973, The social structure of wild chimpanzees: a review of field studies. IN: *Comparative Ecology and Behavior of Primates* (R.P. Michael and J.H. Crook, eds,) London: Academic Press, pp. 375-410.

Sugiyama, Y. 1976, Life history of male Japanese monkeys. IN: *Advances in the Study of Behavior*, vol. 7 (J.S. Rosenblatt, R.A. Hinde, E. Shaw and C. Beer, eds.) New York: Academic Press, pp. 255-285.

Sugiyama, Y. and Parthasarathy, M.D. 1979, Population change of the Hanuman langur (*Presbytis entellus*) 1961-1976, in Dharwar area, India, *J. Bomb. Nat. Hist. Soc.* 75:860-867.

Suomi, S.J. 1976, Mechanisms underlying social development. A re-examination of mother-infant interactions in monkeys. IN: *Minnesota Symposium on Child Development*, vol. 10 (A. Pick, ed.) Minneapolis: Univ. of Minnesota Press, pp. 201-228.

Suomi, S.J. 1977, Development of attachment and other social behaviors in rhesus monkeys. IN: *Attachment Behavior* (T. Alloway, P. Pilner and L. Krames, eds.) New York: Plenum Press, pp. 197-224.

Suomi, S.J. and Harlow H. 1976, Monkeys without play. IN: *Play. Its Role in Development and Evolution* (J.S. Bruner, A. Jolly, and K. Sylva, eds.) Penguin Books, pp. 490-495.

Suomi, S.J. and Harlow, H.F. 1977, Early separation and behavioral maturation. IN: *Genetics, Environment and Intelligence* (A. Oliverio, ed.) Elsevier: North Holland Biomedical Press, pp. 197-214.

Sussman, R.W. 1977, Socialization, social structure and ecology of two sympatric species of lemur. IN: *Primate Bio-Social Development: Biological, Social and Ecological*

Determinants (S. Chevalier-Skolnikoff and F.E. Poirier, eds.) New York: Garland Publ., Inc., pp. 515-528.

Suzuki, A. 1975, The origin of hominid hunting: a primatological perspective. IN: *Socio-ecology and Psychology of Primates* (R.H. Tuttle, ed.) The Hague: Mouton Press, pp. 259-278.

Swyer, G.I.M. 1968, Clinical effects of agents affecting fertility. IN: *Endocrinology and Human Behavior* (R.P. Michael, ed.) New York: Oxford Univ. Press, pp. 161-172.

Syme, G.J. 1974, Competitive orders as measures of social dominance, *Anim. Behav.* 22: 931-940.

Symons, D. 1978, The question of function: dominance and play. IN: *Social Play in Primates* (E.O. Smith, ed.) New York: Academic Press, pp. 193-230.

Szalay, F.S. 1975, Hunting-scavenging protohominids: a model for hominid origins, *Man* 10:420-429.

Talmage-Riggs, G. and Anschel, S. 1973, Homosexual behavior and dominance in a group of captive squirrel monkeys *(Saimiri sciureus), Folia primat.* 19:61-72.

Tanaka, J. 1965, Social structure of Nilgiri langurs, *Primates* 6:107-122.

Tanner, J.M. 1977 (2nd ed.), Human growth and constitution. IN: *Human Biology* (G.A. Harrison, J.S. Weiner, J.M. Tanner and N.A. Barnicot, eds.) Oxford: Oxford Univ. Press, pp. 301-385.

Tanner, N. and Zihlman, A. 1976, Women in evolution. Part I: Innovation and selection in human origins, *Signs* 1:585-608.

Taub, D.M. 1980a, Female choice and mating strategies among wild Barbary macaques *(Macaca sylvanus L.).* IN: *The Macaques: Studies in Ecology, Behavior and Evolution* (D.G. Lindburg, ed.) New York: Van Nostrand Reinhold Co.

Taub, D.M. 1980, Testing the agonistic buffering hypothesis. I. The dynamics of participation in the triadic interaction, *Behav. Ecol. Sociobiol.* 6:187-197.

Taylor, H., Teas, J., Richie, T. and Southwick, C. 1978, Social interactions between adult male and infant rhesus monkeys in Nepal, *Primates* 19:343-351.

Teleki, G. 1973a, *The Predatory Behavior of Wild Chimpanzees.* Lewisburg: Bucknell Press.

Teleki, G. 1973b, The onmivorous chimpanzee, *Sci. Am.* 228:32-42.

Teleki, G., Hunt, E.E. Jr., and Pfifferling, J.H. 1976, Demographic observations (1963-1973) on the chimpanzees of Gombe National Park, Tanzania, *J. Hum. Evol.* 5:559-598.

Thomas, J. 1977, Adam and Eve revisited — the making of a myth or the reflection of reality? *Hum. Dev.* 20:326-351.

Tiger, L. 1969, *Men in Groups.* New York: Random House.

Tiger, L. and Fox, R. 1971, *The Imperial Animal.* New York: Dell Publ. Co.

Tinbergen, N. 1953, *The Herring Gull's World. A Study of the Social Behaviour of Birds.* London: Collins.

Tokuda, K. 1961-2, A study on the sexual behavior in a Japanese monkey troop, *Primates* 3:1-40.

Tokunaga, D.H. and Mitchell, G. 1977, Factors affecting sex differences in nonhuman primate grooming, *J. Behav. Sci.* 2:259-269.

Trivers, R.L. 1971, The evolution of reciprocal altruism, *Quart. Rev. Biol.* 46:35-57.

Trivers, R.L. 1972, Parental investment and sexual selection. IN: *Sexual Selection and the*

Descent of man (B. Campbell, ed.) Chicago: Aldine, pp. 136-179.

Tsumori, A. 1967, Newly acquired behavior and social interactions of Japanese monkeys, IN: *Social Communication Among Primates* (S.A. Altmann, ed.) Chicago: Univ. of Chicago Press, pp. 207-219.

Tsumori, A., Kawai, M. and Motoyishi, R. 1965, Delayed response of wild Japanese monkeys by the sand-digging test (I) — Case of the Koshima troop, *Primates* 6: 195-212.

Tutin, C.E.G. 1975, Exceptions to promiscuity in a feral chimpanzee community. IN: *Contemporary Primatology* (S. Kondo, M. Kawai and A. Ehara, eds.) Basel: S. Karger, pp. 445-449.

Tutin, C.E.G. 1979a, Mating patterns and reproductive strategies in a community of wild chimpanzees *(Pan troglodytes schweinfurthii)*, *Behav. Ecol. Sociobiol.* 6:29-38.

Tutin, C.E.G. 1979b, Responses of chimpanzees to copulation, with special reference to interference by immature individuals, *Anim. Behav.* 27:845-854.

Udry, J.R. and Morris, N.M. 1968, Distribution of coitus in the menstrual cycle, *Nature* 220:593-596.

Udry, J.R. and Morris, N.M. 1970, Effect of contraceptive pills on the distribution of sexual activity in the menstrual cycle, *Nature* 227:502-503.

Udry, J.R. and Morris, N.M. 1977, The distribution of events in the human menstrual cycle, *J. Reprod. Fert.* 51:419-425.

Udry, J.R., Morris, N.M. and Waller, C. 1973, Effect of contraceptive pills on sexual activity in the phase of the human menstrual cycle, *Arch. Sex. Behav.* 2:205-214.

Van den Berghe, P.L. 1973, *Age and Sex in Human Societies: A Biosocial Perspective.* Belmont: Wadsworth Publ. Co.

Vandenbergh, J.G. and Drickamer, L.C. 1974, Reproductive coordination among free-ranging rhesus monekys, *Physiol and Behav.* 13:373-376.

Vandenbergh, J. G. and Post, W. 1976, Endocrine coordination in rhesus monkeys: female responses to the male, *Physiol. and Behav.* 17:979-984.

Vessey, S.H. 1971, Free-ranging rhesus monkeys: behavioral effects of removal, separation and reintroduction of group members. *Behaviour* 40:216-227.

Voland, E. 1977, Social play behavior of the common marmoset (*Callithrix jacchus* ERXC. 1777) in captivity, *Primates* 18:883-902.

deWaal, F.B.M., van Hooff, J.A.R.A.M. and Netto, W.J. 1976, An ethological analysis of types of agonistic interaction in a captive group of Java monkeys *(Macaca fascicularis)*, *Primates* 17:257-290.

deWaal, F.B.M. and Hoekstra, J.A. 1980, Contexts and predictability of aggression in chimpanzees, *Anim. Behav.* 28:929-937.

Wade, M.J. and Breden, F. 1980, The evolution of selfish and cheating behavior, *Behav. Ecol. Sociobiol.* 7:167-172.

Wade, T.D. 1977, Status and hierarchy in nonhuman primate societies. IN: *Perspectives in Ethology,* vol. 3 (P.P.G. Bateson and P.H. Klopfer, eds.) New York: Plenum Press, pp. 109-134.

Wallen, K., Bielert, C. and Slimp, J. 1977, Foot clasp mounting in the prepubertal rhesus monkey: social and hormonal influences. IN: *Primate Bio-Social Development: Biological, Social and Ecological Determinants* (S. Chevalier-Skolnikoff and F.E. Poirier, eds.) New York: Garland Publ., Inc., pp. 439-462.

Walters, J. 1980, Interventions and the development of dominance relationships in female baboons, *Folia primat.* 34:61-89.

Waser, P.M. and Homewood, K. 1979, Cost-benefit approaches to territoriality: a test with forest primates, *Behav. Ecol. Sociobiol.* 6:115-119.

Washburn, S.L. and Ciochon, R.L. 1974, Canine teeth: notes on controversies in the study of human evolution, *Am. Anth.* 76:765-784.

Washburn, S.L. and DeVore, I. 1961, Social behavior of baboons and early man. IN: *Social Life of Early Man* (S.L. Washburn, ed.) Chicago: Aldine, pp. 91-103.

Washburn, S.L. and Dolhinow, P.C. (eds.) 1972, *Perspectives on Human Evolution,* vol. II. New York: Holt, Rinehart & Winston.

Washburn, S.L. and Lancaster, C.S. 1968, The evolution of hunting. IN: *Man the Hunter* (R.B. Lee and I. DeVore, eds.) Chicago: Aldine, pp. 293-303.

Washburn, S.L. and Strum, S.C. 1972, Concluding comments. IN: *Perspectives on Human Evolution,* vol. II (S.L. Washburn and P. Dolhinow, eds.) New York: Holt, Rinehart & Winston, pp. 469-491.

Watanabe, K. 1979, Alliance formation in a free-ranging troop of Japanese macaques, *Primates* 20:459-474.

Watanabe, M. 1974, The conception of nature in Japanese culture, *Science* 183:279-282.

Webster, P. 1975, Matriarchy: a vision of power. IN: *Toward an Anthropology of Women* (R.R. Reiter, ed.) New York: Monthly Review Press, pp. 141-156.

West Eberhard, M.J. 1975, The evolution of social behavior by kin selection, *Quart. Rev. Biol.* 50:1-33.

Westermarck, E. 1922, *The History of Human Marriage.* (3 volumes) London: MacMillan & Co.

Western, D. 1979, Size, life history and ecology in mammals, *Afr. J. Ecol.* 17:185-204.

White, L. 1959, *The Evolution of Culture.* New York: McGraw Hill.

Wickler, W. 1967, Socio-sexual signals and their intra-specific imitation among primates. IN: *Primate Ethology* (D. Morris, ed.) Garden City, New York: Doubleday & Co., pp. 89-189.

Wiens, J.A. 1977, On competition and variable environments, *Am. Sci.* 65:590-597.

Williams, G.C. 1966, *Adaptation and Natural Selection.* Princeton: Princeton Univ. Press.

Williams, G.C. 1975, *Sex and Evolution.* Princeton: Princeton Univ. Press.

Williams, J. 1977, *Psychology of Women. Behavior in a Biosocial Context.* New York: Norton.

Wilson, A.P. and Vessey, S.H. 1968, Behavior of free-ranging castrated rhesus monkeys, *J. Endocrinol.* 52:1-14.

Wilson, E.O. 1975a, *Sociobiology, The New Synthesis.* Cambridge: The Belknap Press of Harvard Univ. Press.

Wilson, E.O. 1975b, Human decency is animal, *New York Times Magazine* (October 12), pp. 36-7; 47-50.

Wilson, M., Plant, T.M. and Michael, R.P. 1972, Androgens and the sexual behavior of male rhesus monkeys, *J. Endocrinol.* 52:11.

Winterbottom, J.M. 1929, Studies in sexual phenomena. VII. The transference of male secondary sexual display characters to the female, *J. Genet.* 21:367-387.

Wolf, K.E. and Fleagle, J.G. 1977, Adult male replacement in a group of silvered leaf-monkeys (*Presbytis cristata*) at Kuala Selangor, Malaysia, *Primates* 18:949-955.

Wolfe, L. 1979, Behavioral patterns of estrous females of the Arashiyama West troop of Japanese macaques, *Primates* 20:525-534.

Wolfheim, J.H. 1978, Sex differences in behaviour in a group of captive juvenile talapoin monkeys *(Miopithecus talapoin), Behaviour* 63:110-128.

Wolfheim, J.H., Jensen, G.D. and Bobbitt, R.A. 1970, Effects of group environment on the mother-infant relationship in pigtailed monkeys *(Macaca nemestrina), Primates* 11:119-124.

Wolfheim, J.H. and Rowell, T.E. 1972, Communication among captive talapoin monkeys *(Miopithecus talapoin), Folia primat.* 18:224-255.

Wrangham, R.W. 1974, Artificial feeding of chimpanzees and baboons in their natural habitat, *Anim. Behav.* 22:83-93.

Wrangham, R.W. 1979a, Sex differences in chimpanzee dispersion. IN: *The Great Apes. Perspectives on Human Evolution,* vol. 5 (D.A. Hamburg and E.R. McCown, eds.) Menlo Park: Benjamin Cummings Publ. Co., pp. 481-489.

Wrangham, R.W. 1979b, On the evolution of ape social systems, *Soc. Sci. Inform.* 18: 335-368.

Wynne-Edwards, V.C. 1962, *Animal Dispersion in Relation to Social Behaviour.* Edinburgh: Oliver & Boyd.

Yamada, M. 1957, A case of acculturation in the subhuman society of Japanese monkeys, *Primates* 1:30-46.

Yamada, M. 1963, A study of blood-relationship in the natural society of the Japanese macaque, *Primates* 4:43-65.

Yamada, M. 1966, Five natural troops of Japanese monkeys in Shodoshima Island, *Primates* 7:315-361.

Yamada, M. 1971, Five natural troops of Japanese monkeys in Shodoshima Island. II. A comparison of social structure, *Primates* 12:81-104.

Yerkes, R.M. and Yerkes, A. 1929, *The Great Apes.* New Haven: Yale Univeristy Press.

Yoshiba, K. 1968, Local and intertroop variability in ecology and social behavior of common Indian langurs. IN: *Primates. Studies in Adaptation and Variability* (P.C. Jay, ed.) New York: Holt, Rinehart & Winston, pp. 217-242.

Young, G.H. and Bramblett, C.A. 1977, Gender and environment as determinants of behavior in infant common baboons *(Papio cynocephalus), Arch. Sex. Behav.* 6:365-385.

Young, G.H. and Hankins, R.J. 1979, Infant behaviors in mother-reared and harem-reared baboons *(Papio cynocephalus), Primates* 20:87-93.

Young, W.C., Goy, R.W. and Phoenix, C.H. 1964, Hormones and sexual behavior, *Science* 143:212-218.

Zihlman, A. 1974, Review of sexual selection and the descent of man, 1871-1971, *Am. Anth.* 76:475-478.

Zihlman, A. 1976, Sexual dimorphism and its behavioral implications in early hominids, *IX Congr. Union Int'l. Sci. Prehist. Protohist.* Colloque VI:1-36.

Zihlman, A. 1978a, Women in evolution. Part II: Subsistence and social organization among early hominids, *Signs* 4:1-24.

Zihlman, A. 1978b, Gathering and the hominid adaptation. IN: *Female Hierarchies* (L. Tiger and H.T. Fowler, eds.) Chicago: Beresford Books, pp. 163-194.

Zuckerman, S. 1932, *The Social Life of Monkeys and Apes.* London: Routledge.

Zumpe, D. and Michael, R.P. 1968, The clutching reaction and orgasm in the female rhesus monkey *(Macaca mulatta), J. Endocrinol.* 40:117-123.

Zumpe, D. and Michael, R.P. 1978, Interrelationships between aggression and sexuality in rhesus monkeys: relevance of the primate model. IN: *Society, Stress and Disease,* vol. 3 *The Productive and Reproductive Ages* (L. Levi, ed.) New York: Oxford Univ. Press, pp. 30-41.

INDEX

McGuire, M.T., 226, 234
McKenna, J.J., 99, 100, 176, 177, 188
McKenna, W., 174
McKinney, W.T., Jr., 201
Maccoby, E.E., 87, 88, 171, 196
Makurius, R., 5
Makwana, S,C., 85
Malina, R.M., 63
Maple, T., 142
Marler, P., 31, 73, 81
Marsden, H.M. 127
Marsh, C., 86, 128
Martin, R.D., 6, 18, 20, 37, 84
Martin, M.K., 87, 134, 170, 309, 311, 314, 318
Maslow, A.H. 97, 105, 202
Mason, W.A. 40, 49, 88, 97, 102, 105, 119, 123, 161, 165, 193, 206, 210, 260, 261, 265
Massey, A., 124, 126, 182, 292, 293
Matson, F.W., 20, 21
Mayer, B., 244, 252
Maynard-Smith, J., 244, 252, 275, 289, 303, 304, 306
Mayr, E., 271, 272, 273, 275, 276
Mead, M., 28, 38, 170
Mears, C., 49
Mech, L.D., 256
Megargee, E.I., 72
Menzel, E.W., Jr., 210, 211, 220, 240
Meyer-Bahlburg, H.F.L., 165
Michael, R.P., 139, 140, 141, 142, 145, 147, 156, 158, 159, 160, 162, 163, 283
Miller, C., 22
Miller, M.H., 124
Miller, R.E., 33
Millet, K., 38
Milton, K., 58
Mineka, S., 199
Missakian, E.A., 124, 127, 129, 131, 196
Mitchell, G.D., 1, 64, 87, 123, 177, 178, 179,

180, 182, 184, 191, 193, 203, 205, 208, 209, 210
Mittermeier, R.A., 256
Mohnot, S.M., 85, 252
Money, J., 23, 24, 165, 168, 169, 170, 171, 172, 173, 189
Montagu, A., 72
Mori, U., 124, 176, 190, 191, 193, 249, 251, 252
Morris, D., 309, 312, 313
Morris, N.M., 162
Moss, H.A., 172, 184, 185
Moyer, K.E., 87
Moynihan, M., 260, 261, 265
Muckenhirn, N.A., 114
Murray, R.D., 130
Myers, R.E., 35

Nadel, S.F., 111
Nadler, R.D., 145, 179, 181
Nagy, K.A., 58
Napier, J.R., 8, 12, 55, 255
Napier, P.H., 8, 12, 55, 255
Nash, L., 123, 124, 129, 180
Nelson, T.W., 265
Neville, M.K., 102, 114, 120, 123
Neyman, P.F., 256, 257, 265
Nicholson, N.A., 181
Nilsen, A.P., 23
Nishida, T., 84, 128, 129, 218, 232, 233, 234, 235
Nishimura, A., 132, 133, 142
Nissen, C.H., 35
Norikoshi, K., 102, 128, 234
Nottebohm, M., 171

O'Connor, J.F., 162
O'Donald, P., 274
Oates, J.F., 55, 86, 249, 252
Ohsawa, H., 249, 252
Olivier, T.J., 129
Orians, G.H., 240
Owens, N.W., 195

Oxnard, C.E., 65

Packer, C., 107, 128, 129, 280, 284, 289, 295, 296
Parker, G.A., 304
Parker, S., 130
Parsons, T., 111, 113, 115
Parthasarathy, M.D., 86
Paterson, J.D., 45
Peters, R.H., 271, 272
Peterson, R.S., 238
Petter, J.J., 145
Pfeiffer, J.E., 309, 310
Pfifferling, J.H., 83
Phoenix, C.H., 157, 160, 163, 166, 167, 168, 205, 206
Piaget, J., 29
Pilbeam, D., 300
Poirier, F.E., 81, 134, 176, 177, 179, 188, 190, 194, 197, 234, 246, 247, 248, 252
Pollock, J.I., 88
Poole, T.E., 265
Popp, J.L., 83
Post, D., 59, 60, 62, 63, 140
Pratt, C.L., 208
Premack, D., 35
Pusey, A., 73, 74, 107, 209, 215, 282

Quadango, D.M., 168, 172, 189
Quiatt, D., 189

Raemakers, J.J., 59, 265
Raisman, G., 172
Raleigh, M.J., 180, 195
Ralls, K., 53, 56, 59, 65, 66, 238, 302
Ransom, B.S., 284
Ransom, T.W., 177, 180, 188, 193, 284
Rattray, J.M., 312
Rayor, L.S., 95
Redican, W.K., 190, 191, 192, 193, 226
Renfrew, J.W., 75
Resko, J.A., 95, 161, 163, 165, 166, 173
Reynolds, F., 233, 235
Reynolds, P.C., 318, 319

378

381